T0133733

EMPIRICAL RESEARCH IN SOFTWARE ENGINEERING

CONCEPTS, ANALYSIS, AND APPLICATIONS

EMPIRICAL RESEARCH IN SOFTWARE ENGINEERING

CONCEPTS, ANALYSIS, AND APPLICATIONS

Ruchika Malhotra

CRC Press
Taylor & Francis Group
Boca Raton London New York

CRC Press is an imprint of the
Taylor & Francis Group, an **informa** business

CRC Press
Taylor & Francis Group
6000 Broken Sound Parkway NW, Suite 300
Boca Raton, FL 33487-2742

First issued in hardback 2019

© 2016 by Taylor & Francis Group, LLC
CRC Press is an imprint of Taylor & Francis Group, an Informa business

No claim to original U.S. Government works

ISBN-13: 978-1-4987-1972-8 (hbk)

Visit the Taylor & Francis Web site at
http://www.taylorandfrancis.com

and the CRC Press Web site at
http://www.crcpress.com

I dedicate this book to my grandmother, the late Shrimati Shakuntala Rani Malhotra,

for her infinite love, understanding, and support. Without her none of my success would

have been possible and I would not be who I am today. I miss her very much.

Contents

Foreword xix
Preface xxi
Acknowledgments xxiii
Author xxv

1. Introduction 1
  1.1   What Is Empirical Software Engineering? 1
  1.2   Overview of Empirical Studies 2
  1.3   Types of Empirical Studies 3
      1.3.1   Experiment 4
      1.3.2   Case Study 5
      1.3.3   Survey Research 6
      1.3.4   Systematic Reviews 7
      1.3.5   Postmortem Analysis 8
  1.4   Empirical Study Process 8
      1.4.1   Study Definition 9
      1.4.2   Experiment Design 10
      1.4.3   Research Conduct and Analysis 11
      1.4.4   Results Interpretation 12
      1.4.5   Reporting 12
      1.4.6   Characteristics of a Good Empirical Study 12
  1.5   Ethics of Empirical Research 13
      1.5.1   Informed Content 14
      1.5.2   Scientific Value 15
      1.5.3   Confidentiality 15
      1.5.4   Beneficence 15
      1.5.5   Ethics and Open Source Software 15
      1.5.6   Concluding Remarks 15
  1.6   Importance of Empirical Research 16
      1.6.1   Software Industry 16
      1.6.2   Academicians 16
      1.6.3   Researchers 17
  1.7   Basic Elements of Empirical Research 17
  1.8   Some Terminologies 18
      1.8.1   Software Quality and Software Evolution 18
      1.8.2   Software Quality Attributes 20
      1.8.3   Measures, Measurements, and Metrics 20
      1.8.4   Descriptive, Correlational, and Cause–Effect Research 22
      1.8.5   Classification and Prediction 22
      1.8.6   Quantitative and Qualitative Data 22
      1.8.7   Independent, Dependent, and Confounding Variables 23
      1.8.8   Proprietary, Open Source, and University Software 24
      1.8.9   Within-Company and Cross-Company Analysis 25

	1.8.10	Parametric and Nonparametric Tests	26
	1.8.11	Goal/Question/Metric Method	26
	1.8.12	Software Archive or Repositories	28
1.9	Concluding Remarks		28
Exercises			28
Further Readings			29

2. Systematic Literature Reviews 33
2.1	Basic Concepts		33
	2.1.1	Survey versus SRs	33
	2.1.2	Characteristics of SRs	34
	2.1.3	Importance of SRs	35
	2.1.4	Stages of SRs	35
2.2	Case Study		36
2.3	Planning the Review		37
	2.3.1	Identify the Need for SR	37
	2.3.2	Formation of Research Questions	38
	2.3.3	Develop Review Protocol	39
	2.3.4	Evaluate Review Protocol	45
2.4	Methods for Presenting Results		46
	2.4.1	Tools and Techniques	46
	2.4.2	Forest Plots	49
	2.4.3	Publication Bias	51
2.5	Conducting the Review		51
	2.5.1	Search Strategy Execution	51
	2.5.2	Selection of Primary Studies	53
	2.5.3	Study Quality Assessment	53
	2.5.4	Data Extraction	53
	2.5.5	Data Synthesis	53
2.6	Reporting the Review		56
2.7	SRs in Software Engineering		58
Exercises			60
Further Readings			61

3. Software Metrics 65
3.1	Introduction		65
	3.1.1	What Are Software Metrics?	66
	3.1.2	Application Areas of Metrics	66
	3.1.3	Characteristics of Software Metrics	67
3.2	Measurement Basics		67
	3.2.1	Product and Process Metrics	68
	3.2.2	Measurement Scale	69
3.3	Measuring Size		71
3.4	Measuring Software Quality		72
	3.4.1	Software Quality Metrics Based on Defects	72
		3.4.1.1	Defect Density	72
		3.4.1.2	Phase-Based Defect Density	73
		3.4.1.3	Defect Removal Effectiveness	73

		3.4.2	Usability Metrics	74
		3.4.3	Testing Metrics	75
	3.5	OO Metrics		76
		3.5.1	Popular OO Metric Suites	77
		3.5.2	Coupling Metrics	79
		3.5.3	Cohesion Metrics	83
		3.5.4	Inheritance Metrics	85
		3.5.5	Reuse Metrics	86
		3.5.6	Size Metrics	88
	3.6	Dynamic Software Metrics		89
		3.6.1	Dynamic Coupling Metrics	89
		3.6.2	Dynamic Cohesion Metrics	89
		3.6.3	Dynamic Complexity Metrics	90
	3.7	System Evolution and Evolutionary Metrics		90
		3.7.1	Revisions, Refactorings, and Bug-Fixes	91
		3.7.2	LOC Based	91
		3.7.3	Code Churn Based	91
		3.7.4	Miscellaneous	92
	3.8	Validation of Metrics		92
	3.9	Practical Relevance		93
		3.9.1	Designing a Good Quality System	93
		3.9.2	Which Software Metrics to Select?	94
		3.9.3	Computing Thresholds	95
			3.9.3.1 Statistical Model to Compute Threshold	96
			3.9.3.2 Usage of ROC Curve to Calculate the Threshold Values	98
		3.9.4	Practical Relevance and Use of Software Metrics in Research	99
		3.9.5	Industrial Relevance of Software Metrics	100
	Exercises			100
	Further Readings			101
4.	**Experimental Design**			103
	4.1	Overview of Experimental Design		103
	4.2	Case Study: Fault Prediction Systems		103
		4.2.1	Objective of the Study	104
		4.2.2	Motivation	104
		4.2.3	Study Context	105
		4.2.4	Results	105
	4.3	Research Questions		106
		4.3.1	How to Form RQ?	106
		4.3.2	Characteristics of an RQ	107
		4.3.3	Example: RQs Related to FPS	108
	4.4	Reviewing the Literature		109
		4.4.1	What Is a Literature Review?	109
		4.4.2	Steps in a Literature Review	110
		4.4.3	Guidelines for Writing a Literature Review	111
		4.4.4	Example: Literature Review in FPS	112
	4.5	Research Variables		117
		4.5.1	Independent and Dependent Variables	117

4.5.2   Selection of Variables.. 118
4.5.3   Variables Used in Software Engineering 118
4.5.4   Example: Variables Used in the FPS.. 118
4.6   Terminology Used in Study Types 118
4.7   Hypothesis Formulation 120
      4.7.1   Experiment Design Types.. 120
      4.7.2   What Is Hypothesis?.. 121
      4.7.3   Purpose and Importance of Hypotheses in an Empirical Research 121
      4.7.4   How to Form a Hypothesis?.. 122
      4.7.5   Steps in Hypothesis Testing 124
          4.7.5.1   Step 1: State the Null and Alternative Hypothesis................. 124
          4.7.5.2   Step 2: Choose the Test of Significance 126
          4.7.5.3   Step 3: Compute the Test Statistic and Associated p-Value.... 126
          4.7.5.4   Step 4: Define Significance Level 127
          4.7.5.5   Step 5: Derive Conclusions.............................. 127
      4.7.6   Example: Hypothesis Formulation in FPS 129
          4.7.6.1   Hypothesis Set A.. 130
          4.7.6.2   Hypothesis Set B... 130
4.8   Data Collection 131
      4.8.1   Data-Collection Strategies 131
      4.8.2   Data Collection from Repositories 132
      4.8.3   Example: Data Collection in FPS 134
4.9   Selection of Data Analysis Methods 136
      4.9.1   Type of Dependent Variable... 137
      4.9.2   Nature of the Data Set... 137
      4.9.3   Aspects of Data Analysis Methods 138
Exercises 139
Further Readings.. 140

5. Mining Data from Software Repositories... 143
5.1   Configuration Management Systems... 143
      5.1.1   Configuration Identification... 144
      5.1.2   Configuration Control... 144
      5.1.3   Configuration Accounting... 146
5.2   Importance of Mining Software Repositories 147
5.3   Common Types of Software Repositories 147
      5.3.1   Historical Repositories.. 148
      5.3.2   Run-Time Repositories or Deployment Logs.............................. 149
      5.3.3   Source Code Repositories 150
5.4   Understanding Systems 150
      5.4.1   System Characteristics 150
      5.4.2   System Evolution.. 151
5.5   Version Control Systems 151
      5.5.1   Introduction.. 151
      5.5.2   Classification of VCS... 153
          5.5.2.1   Local VCS 153
          5.5.2.2   Centralized VCS 153
          5.5.2.3   Distributed VCS... 154
5.6   Bug Tracking Systems.. 155

5.7 Extracting Data from Software Repositories 156
    5.7.1 CVS 157
    5.7.2 SVN 159
    5.7.3 Git 162
    5.7.4 Bugzilla 166
    5.7.5 Integrating Bugzilla with Other VCS 169
5.8 Static Source Code Analysis 170
    5.8.1 Level of Detail 171
        5.8.1.1 Method Level 171
        5.8.1.2 Class Level 171
        5.8.1.3 File Level 171
        5.8.1.4 System Level 172
    5.8.2 Metrics 172
    5.8.3 Software Metrics Calculation Tools 173
5.9 Software Historical Analysis 175
    5.9.1 Understanding Dependencies in a System 176
    5.9.2 Change Impact Analysis 177
    5.9.3 Change Propagation 177
    5.9.4 Defect Proneness 178
    5.9.5 User and Team Dynamics Understanding 178
    5.9.6 Change Prediction 178
    5.9.7 Mining Textual Descriptions 179
    5.9.8 Social Network Analysis 179
    5.9.9 Change Smells and Refactoring 180
    5.9.10 Effort Estimation 180
5.10 Software Engineering Repositories and Open Research Data Sets 180
    5.10.1 FLOSSmole 180
    5.10.2 FLOSSMetrics 181
    5.10.3 PRedictOr Models In Software Engineering 182
    5.10.4 Qualitas Corpus 182
    5.10.5 Sourcerer Project 182
    5.10.6 Ultimate Debian Database 183
    5.10.7 Bug Prediction Data Set 183
    5.10.8 International Software Benchmarking Standards Group 184
    5.10.9 Eclipse Bug Data 184
    5.10.10 Software-Artifact Infrastructure Repository 184
    5.10.11 Ohloh 185
    5.10.12 SourceForge Research Data Archive 185
    5.10.13 Helix Data Set 186
    5.10.14 Tukutuku 186
    5.10.15 Source Code ECO System Linked Data 186
5.11 Case Study: Defect Collection and Reporting System for Git Repository 187
    5.11.1 Introduction 187
    5.11.2 Motivation 188
    5.11.3 Working Mechanism 188
    5.11.4 Data Source and Dependencies 190
    5.11.5 Defect Reports 191
        5.11.5.1 Defect Details Report 191
        5.11.5.2 Defect Count and Metrics Report 193

		5.11.5.3	LOC Changes Report	194
		5.11.5.4	Newly Added Source Files	195
		5.11.5.5	Deleted Source Files	196
		5.11.5.6	Consolidated Defect and Change Report	197
		5.11.5.7	Descriptive Statistics Report for the Incorporated Metrics	198
	5.11.6	Additional Features		199
		5.11.6.1	Cloning of Git-Based Software Repositories	199
		5.11.6.2	Self-Logging	201
	5.11.7	Potential Applications of DCRS		202
		5.11.7.1	Defect Prediction Studies	202
		5.11.7.2	Change-Proneness Studies	203
		5.11.7.3	Statistical Comparison	203
	5.11.8	Concluding Remarks		203

Exercises 203
Further Readings 204

6. Data Analysis and Statistical Testing 207
6.1 Analyzing the Metric Data 207
     6.1.1 Measures of Central Tendency 207
            6.1.1.1 Mean 207
            6.1.1.2 Median 208
            6.1.1.3 Mode 209
            6.1.1.4 Choice of Measures of Central Tendency 209
     6.1.2 Measures of Dispersion 211
     6.1.3 Data Distributions 212
     6.1.4 Histogram Analysis 213
     6.1.5 Outlier Analysis 213
            6.1.5.1 Box Plots 214
            6.1.5.2 Z-Score 216
     6.1.6 Correlation Analysis 218
     6.1.7 Example—Descriptive Statistics of Fault Prediction System 218
6.2 Attribute Reduction Methods 219
     6.2.1 Attribute Selection 221
            6.2.1.1 Univariate Analysis 222
            6.2.1.2 Correlation-Based Feature Selection 222
     6.2.2 Attribute Extraction 222
            6.2.2.1 Principal Component Method 222
     6.2.3 Discussion 223
6.3 Hypothesis Testing 223
     6.3.1 Introduction 224
     6.3.2 Steps in Hypothesis Testing 224
6.4 Statistical Testing 225
     6.4.1 Overview of Statistical Tests 225
     6.4.2 Categories of Statistical Tests 225
     6.4.3 One-Tailed and Two-Tailed Tests 226
     6.4.4 Type I and Type II Errors 228
     6.4.5 Interpreting Significance Results 229

	6.4.6	*t*-Test	229
		6.4.6.1 One Sample *t*-Test	229
		6.4.6.2 Two Sample *t*-Test	232
		6.4.6.3 Paired *t*-Test	233
	6.4.7	Chi-Squared Test	235
	6.4.8	*F*-Test	242
	6.4.9	Analysis of Variance Test	244
		6.4.9.1 One-Way ANOVA	244
	6.4.10	Wilcoxon Signed Test	247
	6.4.11	Wilcoxon–Mann–Whitney Test (*U*-Test)	250
	6.4.12	Kruskal–Wallis Test	254
	6.4.13	Friedman Test	257
	6.4.14	Nemenyi Test	259
	6.4.15	Bonferroni–Dunn Test	261
	6.4.16	Univariate Analysis	263
6.5	Example—Univariate Analysis Results for Fault Prediction System		263
Exercises			265
Further Readings			271

7. Model Development and Interpretation ... 275

7.1	Model Development		275
	7.1.1	Data Partition	276
	7.1.2	Attribute Reduction	277
	7.1.3	Model Construction using Learning Algorithms/Techniques	277
	7.1.4	Validating the Model Predicted	277
	7.1.5	Hypothesis Testing	278
	7.1.6	Interpretation of Results	278
	7.1.7	Example—Software Quality Prediction System	278
7.2	Statistical Multiple Regression Techniques		280
	7.2.1	Multivariate Analysis	280
	7.2.2	Coefficients and Selection of Variables	280
7.3	Machine Learning Techniques		281
	7.3.1	Categories of ML Techniques	281
	7.3.2	Decision Trees	282
	7.3.3	Bayesian Learners	282
	7.3.4	Ensemble Learners	282
	7.3.5	Neural Networks	283
	7.3.6	Support Vector Machines	285
	7.3.7	Rule-Based Learning	286
	7.3.8	Search-Based Techniques	287
7.4	Concerns in Model Prediction		290
	7.4.1	Problems with Model Prediction	290
	7.4.2	Multicollinearity Analysis	290
	7.4.3	Guidelines for Selecting Learning Techniques	291
	7.4.4	Dealing with Imbalanced Data	291
	7.4.5	Parameter Tuning	292
7.5	Performance Measures for Categorical Dependent Variable		292
	7.5.1	Confusion Matrix	292
	7.5.2	Sensitivity and Specificity	294

	7.5.3	Accuracy and Precision	295
	7.5.4	Kappa Coefficient	295
	7.5.5	*F*-measure, *G*-measure, and *G*-mean	295
	7.5.6	Receiver Operating Characteristics Analysis	298
		7.5.6.1 ROC Curve	298
		7.5.6.2 Area Under the ROC Curve	300
		7.5.6.3 Cutoff Point and Co-Ordinates of the ROC Curve	300
	7.5.7	Guidelines for Using Performance Measures	302
7.6	Performance Measures for Continuous Dependent Variable		304
	7.6.1	Mean Relative Error	304
	7.6.2	Mean Absolute Relative Error	304
	7.6.3	PRED (A)	305
7.7	Cross-Validation		306
	7.7.1	Hold-Out Validation	307
	7.7.2	*K*-Fold Cross-Validation	307
	7.7.3	Leave-One-Out Validation	307
7.8	Model Comparison Tests		309
7.9	Interpreting the Results		310
	7.9.1	Analyzing Performance Measures	310
	7.9.2	Presenting Qualitative and Quantitative Results	312
	7.9.3	Drawing Conclusions from Hypothesis Testing	312
	7.9.4	Example—Discussion of Results in Hypothesis Testing Using Univariate Analysis for Fault Prediction System	312
7.10	Example—Comparing ML Techniques for Fault Prediction		315
Exercises			323
Further Readings			324

8. Validity Threats			331
8.1	Categories of Threats to Validity		331
	8.1.1	Conclusion Validity	331
	8.1.2	Internal Validity	333
	8.1.3	Construct Validity	334
	8.1.4	External Validity	335
	8.1.5	Essential Validity Threats	337
8.2	Example—Threats to Validity in Fault Prediction System		337
	8.2.1	Conclusion Validity	337
	8.2.2	Internal Validity	339
	8.2.3	Construct Validity	339
	8.2.4	External Validity	340
8.3	Threats and Their Countermeasures		341
Exercises			350
Further Readings			350

9. Reporting Results			353
9.1	Reporting and Presenting Results		353
	9.1.1	When to Disseminate or Report Results?	354
	9.1.2	Where to Disseminate or Report Results?	354

9.1.3   Report Structure...355
        9.1.3.1   Abstract...356
        9.1.3.2   Introduction...357
        9.1.3.3   Related Work..357
        9.1.3.4   Experimental Design 357
        9.1.3.5   Research Methods..358
        9.1.3.6   Research Results...358
        9.1.3.7   Discussion and Interpretation.............................359
        9.1.3.8   Threats to Validity..359
        9.1.3.9   Conclusions 359
        9.1.3.10 Acknowledgment 359
        9.1.3.11 References...359
        9.1.3.12 Appendix 359
9.2   Guidelines for Masters and Doctoral Students.............................359
9.3   Research Ethics and Misconduct..361
    9.3.1   Plagiarism 362
Exercises 362
Further Readings..363

10. **Mining Unstructured Data**...365
10.1   Introduction..365
    10.1.1   What Is Unstructured Data?..366
    10.1.2   Multiple Classifications..366
    10.1.3   Importance of Text Mining..367
    10.1.4   Characteristics of Text Mining..367
10.2   Steps in Text Mining 368
    10.2.1   Representation of Text Documents......................................368
    10.2.2   Preprocessing Techniques 368
        10.2.2.1   Tokenization...370
        10.2.2.2   Removal of Stop Words 370
        10.2.2.3   Stemming Algorithm...371
    10.2.3   Feature Selection 372
    10.2.4   Constructing a Vector Space Model 375
    10.2.5   Predicting and Validating the Text Classifier 377
10.3   Applications of Text Mining in Software Engineering.................378
    10.3.1   Mining the Fault Reports to Predict the Severity
         of the Faults 378
    10.3.2   Mining the Change Logs to Predict the Effort 378
    10.3.3   Analyzing the SRS Document to Classify Requirements into
         NFRs 378
10.4   Example—Automated Severity Assessment of Software Defect
    Reports 379
    10.4.1   Introduction...379
    10.4.2   Data Source 380
    10.4.3   Experimental Design...380
    10.4.4   Result Analysis..382
    10.4.5   Discussion of Results..385

10.4.6   Threats to Validity 387
10.4.7   Conclusion 387
Exercises 387
Further Readings 388

11. Demonstrating Empirical Procedures 391
11.1   Abstract 391
        11.1.1   Basic 391
        11.1.2   Method 392
        11.1.3   Results 392
11.2   Introduction 392
        11.2.1   Motivation 392
        11.2.2   Objectives 392
        11.2.3   Method 393
        11.2.4   Technique Selection 393
        11.2.5   Subject Selection 394
11.3   Related Work 394
11.4   Experimental Design 395
        11.4.1   Problem Definition 395
        11.4.2   Research Questions 395
        11.4.3   Variables Selection 396
        11.4.4   Hypothesis Formulation 397
                  11.4.4.1   Hypothesis Set 397
        11.4.5   Statistical Tests 398
        11.4.6   Empirical Data Collection 399
        11.4.7   Technique Selection 400
        11.4.8   Analysis Process 401
11.5   Research Methodology 401
        11.5.1   Description of Techniques 401
        11.5.2   Performance Measures and Validation Method 404
11.6   Analysis Results 404
        11.6.1   Descriptive Statistics 404
        11.6.2   Outlier Analysis 408
        11.6.3   CFS Results 408
        11.6.4   Tenfold Cross-Validation Results 408
        11.6.5   Hypothesis Testing and Evaluation 416
        11.6.6   Nemenyi Results 419
11.7   Discussion and Interpretation of Results 420
11.8   Validity Evaluation 423
        11.8.1   Conclusion Validity 423
        11.8.2   Internal Validity 423
        11.8.3   Construct Validity 423
        11.8.4   External Validity 423
11.9   Conclusions and Future Work 424
Appendix 425

12. Tools for Analyzing Data...429
    12.1   WEKA..429
    12.2   KEEL...429
    12.3   SPSS..430
    12.4   MATLAB®..430
    12.5   R...431
    12.6   Comparison of Tools..431
    Further Readings...434

Appendix: Statistical Tables...437

References..445

Index..459

Foreword

Dr. Ruchika Malhotra is a leading researcher in the software engineering community in India, whose work has applications and implications across the broad spectrum of software engineering activities and domains. She is also one of the leading scholars in the area of search-based software engineering (SBSE) within this community, a topic that is rapidly attracting interest and uptake within the wider Indian software engineering research and practitioner communities and is also well established worldwide as a widely applicable approach to software product and process optimization.

In this book Dr. Malhotra uses her breadth of software engineering experience and expertise to provide the reader with coverage of many aspects of empirical software engineering. She covers the essential techniques and concepts needed for a researcher to get started on empirical software engineering research, including metrics, experimental design, analysis and statistical techniques, threats to the validity of any research findings, and methods and tools for empirical software engineering research.

As Dr. Malhotra notes in this book, SBSE is one approach to software engineering that is inherently grounded in empirical assessment, since the overall approach uses computational search to optimize software engineering problems. As such, almost all research work on SBSE involves some form of empirical assessment.

Dr. Malhotra also reviews areas of software engineering that typically rely heavily on empirical techniques, such as software repository mining, an area of empirical software engineering research that offers a rich set of insights into the nature of our engineering material and processes. This is a particularly important topic area that exploits the vast resources of data available (in open source repositories, in app stores, and in other electronic resources relating to software projects). Through mining these repositories, we answer fundamentally empirical questions about software systems that can inform practice and software improvement.

The book provides the reader with an introduction and overview of the field and is also backed by references to the literature, allowing the interested reader to follow up on the methods, tools, and concepts described.

Mark Harman
University College London

Preface

Empirical research has become an essential component of software engineering research practice. Empirical research in software engineering—including the concepts, analysis, and applications—is all about designing, collecting, analyzing, assessing, and interpreting empirical data collected from software repositories using statistical and machine-learning techniques. Software practitioners and researchers can use the results obtained from these analyses to produce high quality, low cost, and maintainable software.

Empirical software engineering involves planning, designing, analyzing, assessing, interpreting, and reporting results of validation of empirical data. There is a lack of understanding and level of uncertainty on the empirical procedures and practices in software engineering. The aim of this book is to present the empirical research processes, procedures, and practices that can be implemented in practice by the research community. My several years of experience in the area of empirical software engineering motivated me to write this book.

Universities and software industries worldwide have started realizing the importance of empirical software engineering. Many universities are now offering a full course on empirical software engineering for undergraduate, postgraduate, and doctoral students in the disciplines of software engineering, computer science engineering, information technology, and computer applications.

In this book, a description of the steps followed in the research process in order to carry out replicated and empirical research is presented. Readers will gain practical knowledge about how to plan and design experiments, conduct systematic reviews and case studies, and analyze the results produced by these empirical studies. Hence, the empirical research process will provide the software engineering community the knowledge for conducting empirical research in software engineering.

The book contains a judicious mix of empirical research concepts and real-life case study that makes it ideal for a course and research on empirical software engineering. Readers will also experience the process of developing predictive models (e.g., defect prediction, change prediction) on data collected from source code repositories. The purpose of this book is to introduce students, academicians, teachers, software practitioners and researchers to the research process carried out in the empirical studies in software engineering. This book presents the application of machine-learning techniques and real-life case studies in empirical software engineering, which aims to bridge the gap between research and practice. The book explains the concepts with numerous examples of empirical studies. The main features of this book are as follows:

- Presents empirical processes, the importance of empirical-research ethics in software engineering research, and the types of empirical studies.
- Describes the planning, conducting, and reporting phases of systematic literature reviews keeping in view the challenges encountered in this field.

- Describes software metrics; the most popular metrics given to date are included and explained with the help of examples.
- Provides an in-depth description of experimental design, including research questions formation, literature review, variables description, hypothesis formulation, data collection, and selection of data analysis methods.
- Provides a full chapter on mining data from software repositories. It presents the procedure for extracting data from software repositories such as CVS, SVN, and Git and provides applications of the data extracted from these repositories in the software engineering area.
- Describes methods for analyzing data, hypothesis testing, model prediction, and interpreting results. It presents statistical tests with examples to demonstrate their use in the software engineering area.
- Describes performance measures and model validation techniques. The guidelines for using the statistical tests and performance measures are also provided. It also emphasizes the use of machine-learning techniques in predicting models along with the issues involved with these techniques.
- Summarizes the categories of threats to validity with practical examples. A summary of threats extracted from fault prediction studies is presented.
- Provides guidelines to researchers and doctorate students for publishing and reporting the results. Research misconduct is discussed.
- Presents the procedure for mining unstructured data using text mining techniques and describes the concepts with the help of examples and case studies. It signifies the importance of text-mining procedures in extracting relevant information from software repositories and presents the steps in text mining.
- Presents real-life research-based case studies on software quality prediction models. The case studies are developed to demonstrate the procedures used in the chapters of the book.
- Presents an overview of tools that are widely used in the software industry for carrying out empirical studies.

I take immense pleasure in presenting to you this book and hope that it will inspire researchers worldwide to utilize the knowledge and knowhow contained for newer applications and advancement of the frontiers of software engineering. The importance of feedback from readers is important to continuously improve the contents of the book. I welcome constructive criticism and comments about anything in the book; any omission is due to my oversight. I will appreciatively receive suggestions, which will motivate me to work hard on a next improved edition of the book; as Robert Frost rightly wrote:

> The woods are lovely, dark, and deep,
> But I have promises to keep,
> And miles to go before I sleep,
> And miles to go before I sleep.

Ruchika Malhotra
Delhi Technological University

Acknowledgments

I am most thankful to my father for constantly encouraging me and giving me time and unconditional support while writing this book. He has been a major source of inspiration to me.

This book is a result of the motivation of Yogesh Singh, Director, Netaji Subhas Institute of Technology, Delhi, India. It was his idea that triggered this book. He has been a constant source of inspiration for me throughout my research and teaching career. He has been continuously involved in evaluating and improving the chapters in this book; I will always be indebted to him for his extensive support and guidance.

I am extremely grateful to Mark Harman, professor of software engineering and head of the Software Systems Engineering Group, University College London, UK, for the support he has given to the book in his foreword. He has contributed greatly to the field of software engineering research and was the first to explore the use and relevance of search-based techniques in software engineering.

I am extremely grateful to Megha Khanna, assistant professor, Acharya Narendera Dev College, University of Delhi, India, for constantly working with me in terms of solving examples, preparing case studies, and reading texts. The book would not have been in its current form without her support. My sincere thanks to her.

My heartfelt gratitude is due to Ankita Jain Bansal, assistant professor, Netaji Subhas Institute of Technology, Delhi, India, who worked closely with me during the evolution of the book. She was continuously involved in modifying various portions in the book, especially experimental design procedure and threshold analysis.

I am grateful to Abha Jain, research scholar, Delhi Technological University, Delhi, India, for helping me develop performance measures and text-mining examples for the book. My thanks are also due to Kanishk Nagpal, software engineer, Samsung India Electronics Limited, Delhi, India, who worked closely with me on mining repositories and who developed the DCRS tool provided in Chapter 5.

Thanks are due to all my doctoral students in the Department of Software Engineering, Delhi Technological University, Delhi, India, for motivating me to explore and evolve empirical research concepts and applications in software engineering. Thanks are also due to my undergraduate and postgraduate students at the Department of Computer Science and Software Engineering, Delhi Technological University, for motivating me to study more before delivering lectures and exploring and developing various tools in several projects. Their outlook, debates, and interest have been my main motivation for continuous advancement in my academic pursuit. I also thank all researchers, scientists, practitioners, software developers, and teachers whose insights, opinions, ideas, and techniques find a place in this book.

Thanks also to Rajeev Raje, professor, Department of Computer & Information Science, Indiana University–Purdue University Indianapolis, Indianapolis, for his support and valuable suggestions.

Last, I am thankful to Manju Khari, assistant professor, Ambedkar Institute of Technology, Delhi, India, for her support in gathering some of the material for the further readings sections in the book.

Author

Ruchika Malhotra is an assistant professor in the Department of Software Engineering, Delhi Technological University (formerly Delhi College of Engineering), Delhi, India. She is a Raman Scholar and was awarded the prestigious UGC Raman Postdoctoral Fellowship by the government of India, under which she pursued postdoctoral research in the Department of Computer and Information Science, Indiana University–Purdue University Indianapolis (2014–15), Indiana. She earned her master's and doctorate degrees in software engineering from the University School of Information Technology, Guru Gobind Singh Indraprastha University, Delhi, India. She was an assistant professor at the University School of Information Technology, Guru Gobind Singh Indraprastha University, Delhi, India.

Dr. Malhotra received the prestigious IBM Faculty Award in 2013 and has received the Best Presenter Award in Workshop on Search Based Software Testing, ICSE, 2014, Hyderabad, India. She is an executive editor of *Software Engineering: An International Journal* and is a coauthor of the book, *Object Oriented Software Engineering*.

Dr. Malhotra's research interests are in empirical research in software engineering, improving software quality, statistical and adaptive prediction models, software metrics, the definition and validation of software metrics, and software testing. Her *H*-index as reported by Google Scholar is 17. She has published more than 100 research papers in international journals and conferences, and has been a referee for various journals of international repute in the areas of software engineering and allied fields. She is guiding several doctoral candidates and has guided several undergraduate projects and graduate dissertations. She has visited foreign universities such as Imperial College, London, UK; Indiana University–Purdue University Indianapolis, Indiana; Ball State University, Muncie, Indiana; and Harare Institute of Technology, Zimbabwe. She has served on the technical committees of several international conferences in the area of software engineering (SEED, WCI, ISCON). Dr. Malhotra can be contacted via e-mail at ruchikamalhotra2004@yahoo.com.

1

Introduction

As the size and complexity of software is increasing, software organizations are facing the pressure of delivering high-quality software within a specific time, budget, and available resources. The software development life cycle consists of a series of phases, including requirements analysis, design, implementation, testing, integration, and maintenance. Software professionals want to know which tools to use at each phase in software development and desire effective allocation of available resources. The software planning team attempts to estimate the cost and duration of software development, the software testers want to identify the fault-prone modules, and the software managers seek to know which tools and techniques can be used to reduce the delivery time and best utilize the manpower. In addition, the software managers also desire to improve the software processes so that the quality of the software can be enhanced. Traditionally, the software engineers have been making decisions based on their intuition or individual expertise without any scientific evidence or support on the benefits of a tool or a technique.

Empirical studies are verified by observation or experiment and can provide powerful evidence for testing a given hypothesis (Aggarwal et al. 2009). Like other disciplines, software engineering has to adopt empirical methods that will help to plan, evaluate, assess, monitor, control, predict, manage, and improve the way in which software products are produced. An empirical study of real systems can help software organizations assess large software systems quickly, at low costs. The application of empirical techniques is especially beneficial for large-scale systems, where software professionals need to focus their attention and resources on various activities of the system under development. For example, developing a model for predicting faulty modules allows software organizations to identify faulty portions of source code so that testing activities can be planned more effectively. Empirical studies such as surveys, systematic reviews and experimental studies, help software practitioners to scientifically assess and validate the tools and techniques in software development.

In this chapter, an overview and the types of empirical studies are provided, the phases of the experimental process are described, and the ethics involved in empirical research of software engineering are summarized. Further, this chapter also discusses the key concepts used in the book.

1.1 What Is Empirical Software Engineering?

The initial debate on software as an engineering discipline is over now. It has been realized that without software as an engineering discipline survival is difficult. Engineering compels the development of the product in a scientific, well formed, and systematic manner. Core engineering principles should be applied to produce good quality maintainable software within a specified time and budget. Fritz Bauer coined the term *software engineering* in 1968 at the first conference on software engineering and defined it as (Naur and Randell 1969):

> The establishment and use of sound engineering principles in order to obtain economically developed software that is reliable and works efficiently on real machines.

Software engineering is defined by IEEE Computer Society as (Abren et al. 2004):

> The application of a systematic, disciplined, quantifiable approach to the development, operation and maintenance of software, and the study of these approaches, that is, the application of engineering to software.

The software engineering discipline facilitates the completion of the objective of delivering good quality software to the customer following a systematic and scientific approach. Empirical methods can be used in software engineering to provide scientific evidence on the use of tools and techniques.

Harman et al. (2012a) defined "empirical" as:

> "Empirical" is typically used to define any statement about the world that is related to observation or experience.

Empirical software engineering (ESE) is an area of research that emphasizes the use of empirical methods in the field of software engineering. It involves methods for evaluating, assessing, predicting, monitoring, and controlling the existing artifacts of software development.

ESE applies quantitative methods to the software engineering phenomenon to understand software development better. ESE has been gaining importance over the past few decades because of the availability of vast data sets from open source repositories that contain information about software requirements, bugs, and changes (Meyer et al. 2013).

1.2 Overview of Empirical Studies

Empirical study is an attempt to compare the theories and observations using real-life data for analysis. Empirical studies usually utilize data analysis methods and statistical techniques for exploring relationships. They play an important role in software engineering research by helping to form well-formed theories and widely accepted results. The empirical studies provide the following benefits:

- Allow to explore relationships
- Allow to prove theoretical concepts
- Allow to evaluate accuracy of the models
- Allow to choose among tools and techniques
- Allow to establish quality benchmarks across software organizations
- Allow to assess and improve techniques and methods

Empirical studies are important in the area of software engineering as they allow software professionals to evaluate and assess the new concepts, technologies, tools, and techniques in scientific and proved manner. They also allow improving, managing, and controlling the existing processes and techniques by using evidence obtained from the empirical analysis. The empirical information can help software management in decision making

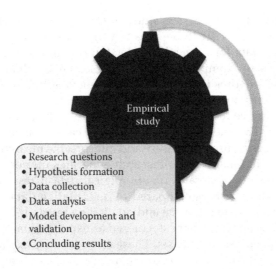

FIGURE 1.1
Steps in empirical studies.

and improving software processes. The empirical studies involve the following steps (Figure 1.1):

- Formation of research questions
- Formation of a research hypothesis
- Gathering data
- Analyzing the data
- Developing and validating models
- Deriving conclusions from the obtained results

Empirical study allows to gather evidence that can be used to support the claims of efficiency of a given technique or technology. Thus, empirical studies help in building a body of knowledge so that the processes and products are improved resulting in high-quality software.

Empirical studies are of many types, including surveys, systematic reviews, experiments, and case studies.

1.3 Types of Empirical Studies

The studies can be broadly classified as quantitative and qualitative. Quantitative research is the most widely used scientific method in software engineering that applies mathematical- or statistical-based methods to derive conclusions. Quantitative research is used to prove or disprove a hypothesis (a concept that has to be tested for further investigation). The aim of a quantitative research is to generate results that are generalizable and unbiased and thus can be applied to a larger population in research. It uses statistical methods to validate a hypothesis and to explore causal relationships.

In qualitative research, the researchers study human behavior, preferences, and nature. Qualitative research provides an in-depth analysis of the concept under investigation and thus uses focused data for research. Understanding a new process or technique in software engineering is an example of qualitative research. Qualitative research provides textual descriptions or pictures related to human beliefs or behavior. It can be extended to other studies with similar populations but generalizations of a particular phenomenon may be difficult. Qualitative research involves methods such as observations, interviews, and group discussions. This method is widely used in case studies.

Qualitative research can be used to analyze and interpret the meaning of results produced by quantitative research. Quantitative research generates numerical data for analysis, whereas qualitative research generates non-numerical data (Creswell 1994). The data of qualitative research is quite rich as compared to quantitative data. Table 1.1 summaries the key differences between quantitative and qualitative research.

The empirical studies can be further categorized as experimental, case study, systematic review, survey, and post-mortem analysis. These categories are explained in the next section. Figure 1.2 presents the quantitative and qualitative types of empirical studies.

1.3.1 Experiment

An experimental study tests the established hypothesis by finding the effect of variables of interest (independent variables) on the outcome variable (dependent variable) using statistical analysis. If the experiment is carried out correctly, the hypothesis is either accepted or rejected. For example, one group uses technique A and the other group uses technique B, which technique is more effective in detecting a larger number of defects? The researcher may apply statistical tests to answer such questions. According to Kitchenham et al. (1995), the experiments are small scale and must be controlled. The experiment must also control the confounding variables, which may affect the accuracy of the results produced by the experiment. The experiments are carried out in a controlled environment and often referred to as controlled experiments (Wohlin 2012).

The key factors involved in the experiments are independent variables, dependent variables, hypothesis, and statistical techniques. The basic steps followed in experimental

TABLE 1.1

Comparison of Quantitative and Qualitative Research

	Quantitative Research	Qualitative Research
General	Objective	Subjective
Concept	Tests theory	Forms theory
Focus	Testing a hypothesis	Examining the depth of a phenomenon
Data type	Numerical	Textual or pictorial
Group	Small	Large and random
Purpose	Predict causal relationships	Describe and interpret concepts
Basis	Based on hypothesis	Based on concept or theory
Method	Confirmatory: established hypothesis is tested	Exploratory: new hypothesis is formed
Variables	Variables are defined by the researchers	Variables may emerge unexpectedly
Settings	Controlled	Flexible
Results	Generalizable	Specialized

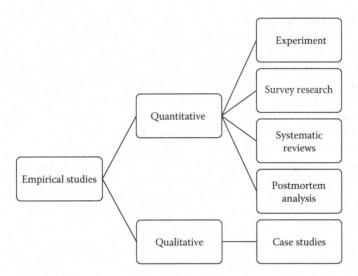

FIGURE 1.2
Types of empirical studies.

FIGURE 1.3
Steps in experimental research.

research are shown in Figure 1.3. The same steps are followed in any empirical study process however the content varies according to the specific study being carried out. In first phase, experiment is defined. The next phase involves determining the experiment design. In the third phase the experiment is executed as per the experiment design. Then, the results are interpreted. Finally, the results are presented in the form of experiment report. To carry out an empirical study, a replicated study (repeating a study with similar settings or methods but different data sets or subjects), or to perform a survey of existing empirical studies, the research methodology followed in these studies needs to be formulated and described.

A controlled experiment involves varying the variables (one or more) and keeping everything else constant or the same and are usually conducted in small or laboratory setting (Conradi and Wang 2003). Comparing two methods for defect detection is an example of a controlled experiment in software engineering context.

1.3.2 Case Study

Case study research represents real-world scenarios and involves studying a particular phenomenon (Yin 2002). Case study research allows software industries to evaluate a tool, method, or process (Kitchenham et al. 1995). The effect of a change in an organization can be studied using case study research. Case studies increase the understanding of the phenomenon under study. For example, a case study can be used to examine whether a unified model language (UML) tool is effective for a given project or not. The initial and new concepts are analyzed and explored by exploratory case studies, whereas the already existing concepts are tested and improvised by confirmatory case studies.

The phases included in the case study are presented in Figure 1.4. The case study design phase involves identifying existing objectives, cases, research questions, and data-collection strategies. The case may be a tool, technology, technique, process, product, individual, or software. Qualitative data is usually collected in a case study. The sources include interviews, group discussions, or observations. The data may be directly or indirectly collected from participants. Finally, the case study is executed, the results obtained are analyzed, and the findings are reported. The report type may vary according to the target audience.

Case studies are appropriate where a phenomenon is to be studied for a longer period of time so that the effects of the phenomenon can be observed. The disadvantages of case studies include difficulty in generalization as they represent a typical situation. Since they are based on a particular case, the validity of the results is questionable.

1.3.3 Survey Research

Survey research identifies features or information from a large scale of a population. For example, surveys can be used when a researcher wants to know whether the use of a particular process has improved the view of clients toward software usability features. This information can be obtained by asking the selected software testers to fill questionnaires. Surveys are usually conducted using questionnaires and interviews. The questionnaires are constructed to collect research-related information.

Preparation of a questionnaire is an important activity and should take into consideration the features of the research. The effective way to obtain a participant's opinion is to get a questionnaire or survey filled by the participant. The participant's feedback and reactions are recorded in the questionnaire (Singh 2011). The questionnaire/survey can be used to detect trends and may provide valuable information and feedback on a particular process, technique, or tool. The questionnaire/survey must include questions concerning the participant's likes and dislikes about a particular process, technique, or tool. The interviewer should preferably handle the questionnaire.

Surveys are classified into three types (Babbie 1990)—descriptive, explorative, and explanatory. Exploratory surveys focus on the discovery of new ideas and insights and are usually conducted at the beginning of a research study to gather initial information. The descriptive survey research is more detailed and describes a concept or topic. Explanatory survey research tries to explain how things work in connections like cause and effect, meaning a researcher wants to explain how things interact or work with each other. For example, while exploring relationship between various independent variables and an

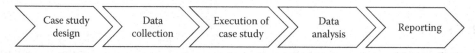

FIGURE 1.4
Case study phases.

outcome variable, a researcher may want to explain why an independent variable affects the outcome variable.

1.3.4 Systematic Reviews

While conducting any study, literature review is an important part that examines the existing position of literature in an area in which the research is being conducted. The systematic reviews are methodically undertaken with a specific search strategy and well-defined methodology to answer a set of questions. The aim of a systematic review is to analyze, assess, and interpret existing results of research to answer research questions. Kitchenham (2007) defines systematic review as:

> A form of secondary study that uses a well-defined methodology to identify, analyze and interpret all available evidence related to a specific research question in a way that is unbiased and (to a degree) repeatable.

The purpose of a systematic review is to summarize the existing research and provide future guidelines for research by identifying gaps in the existing literature. A systematic review involves:

1. Defining research questions.
2. Forming and documenting a search strategy.
3. Determining inclusion and exclusion criteria.
4. Establishing quality assessment criteria.

The systematic reviews are performed in three phases: planning the review, conducting the review, and reporting the results of the review. Figure 1.5 presents the summary of the phases involved in systematic reviews.

In the planning stage, the review protocol is developed that includes the following steps: research questions identification, development of review protocol, and evaluation of review protocol. During the development of review protocol the basic processes in the review are planned. The research questions are formed that address the issues to be

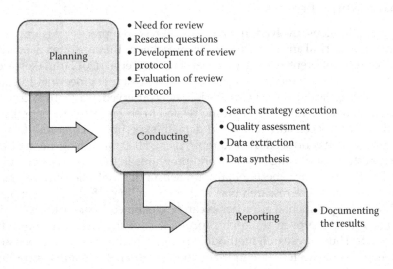

FIGURE 1.5
Phases of systematic review.

answered in the systematic literature review. The development of review protocol involves planning a series of steps—search strategy design, study selection criteria, study quality assessment, data extraction process, and data synthesis process. In the first step, the search strategy is described that includes identification of search terms and selection of sources to be searched to identify the primary studies. The second step determines the inclusion and exclusion criteria for each primary study. In the next step, the quality assessment criterion is identified by forming the quality assessment questionnaire to analyze and assess the studies. The second to last step involves the design of data extraction forms to collect the required information to answer the research questions, and, in the last step, data synthesis process is defined. The above series of steps are executed in the review in the conducting phase. In the final phase, the results are documented. Chapter 2 provides details of systematic review.

1.3.5 Postmortem Analysis

Postmortem analysis is carried out after an activity or a project has been completed. The main aim is to detect how the activities or processes can be improved in the future. The postmortem analysis captures knowledge from the past, after the activity has been completed. Postmortem analysis can be classified into two types: general postmortem analysis and focused postmortem analysis. General postmortem analysis collects all available information from a completed activity, whereas focused postmortem analysis collects information about specific activity such as effort estimation (Birk et al. 2002).

According to Birk et al., in postmortem analysis, large software systems are analyzed to gain knowledge about the good and bad practices of the past. The techniques such as interviews and group discussions can be used for collecting data in postmortem analysis. In the analysis process, the feedback sessions are conducted where the participants are asked whether the concepts told to them have been correctly understood (Birk et al. 2002).

1.4 Empirical Study Process

Before describing the steps involved in the empirical research process, it is important to distinguish between empirical and experimental approaches as they are often used interchangeably but are slightly different from each other. Harman et al. (2012a) makes a distinction between experimental and empirical approaches in software engineering. In experimental software engineering, the dependent variable is closely observed in a controlled environment. Empirical studies are used to define anything related to observation and experience and are valuable as these studies consider real-world data. In experimental studies, data is artificial or synthetic but is more controlled. For example, using 5000 machine-generated instances is an experimental study, and using 20 real-world programs in the study is an empirical study (Meyer et al. 2013). Hence, any experimental approach, under controlled environments, allows the researcher to remove the research bias and confounding effects (Harman et al. 2012a). Both empirical and experimental approaches can be combined in the studies.

Without a sound and proven research process, it is difficult to carry out efficient and effective research. Thus, a research methodology must be complete and repeatable, which, when followed, in a replicated or empirical study, will enable comparisons to be made across various studies. Figure 1.6 depicts the five phases in the empirical study process. These phases are discussed in the subsequent subsections.

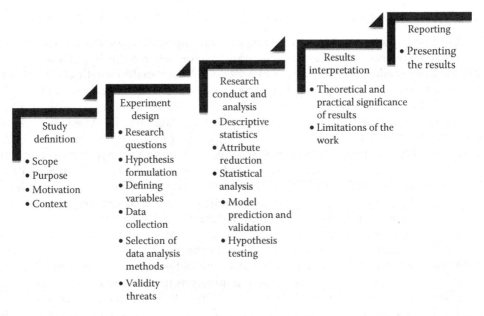

FIGURE 1.6
Empirical study phases.

1.4.1 Study Definition

The first step involves the definition of the goals and objectives of the empirical study. The aim of the study is explained in this step. Basili et al. (1986) suggests dividing the defining phase into the following parts:

- Scope: What are the dimensions of the study?
- Motivation: Why is it being studied?
- Object: What entity is being studied?
- Purpose: What is the aim of the study?
- Perspective: From whose view is the study being conducted (e.g, project manager, customer)?
- Domain: What is the area of the study?

The scope of the empirical study defines the extent of the investigation. It involves listing down the specific goals and objectives of the experiment. The purpose of the study may be to find the effect of a set of variables on the outcome variable or to prove that technique A is superior to technique B. It also involves identifying the underlying hypothesis that is formulated at later stages. The motivation of the experiment describes the reason for conducting the study. For example, the motivation of the empirical study is to analyze and assess the capability of a technique or method. The object of the study is the entity being examined in the study. The entity in the study may be the process, product, or technique. Perspective defines the view from which the study is conducted. For example, if the study is conducted from the tester's point of view then the tester will be interested in planning and allocating resources to test faulty portions of the source code. Two important domains in the study are programmers and programs (Basili et al. 1986).

1.4.2 Experiment Design

This is the most important and significant phase in the empirical study process. The design of the experiment covers stating the research questions, formation of the hypothesis, selection of variables, data-collection strategies, and selection of data analysis methods. The context of the study is defined in this phase. Thus, the sources (university/academic, industrial, or open source) from which the data will be collected are identified. The data-collection process must be well defined and the characteristics of the data must be stated. For example, nature, programming language, size, and so on must be provided. The outcome variables are to be carefully selected such that the objectives of the research are justified. The aim of the design phase should be to select methods and techniques that promote replicability and reduce experiment bias (Pfleeger 1995). Hence, the techniques used must be clearly defined and the settings should be stated so that the results can be replicated and adopted by the industry. The following are the steps carried out during the design phase:

1. Research questions: The first step is to formulate the research problem. This step states the problem in the form of questions and identifies the main concepts and relations to be explored. For example, the following questions may be addressed in empirical studies to find the relationship between software metrics and quality attributes:

    a. What will be the effect of software metrics on quality attributes (such as fault proneness/testing effort/maintenance effort) of a class?

    b. Are machine-learning methods adaptable to object-oriented systems for predicting quality attributes?

    c. What will be the effect of software metrics on fault proneness when severity of faults is taken into account?

2. Independent and dependent variables: To analyze relationships, the next step is to define the dependent and the independent variables. The outcome variable predicted by the independent variables is called the dependent variable. For instance, the dependent variables of the models chosen for analysis may be fault proneness, testing effort, and maintenance effort. A variable used to predict or estimate a dependent variable is called the independent (explanatory) variable.

3. Hypothesis formulation: The researcher should carefully state the hypothesis to be tested in the study. The hypothesis is tested on the sample data. On the basis of the result from the sample, a decision concerning the validity of the hypothesis (acception or rejection) is made.

    Consider an example where a hypothesis is to be formed for comparing a number of methods for predicting fault-prone classes.

    For each method, M, the hypothesis in a given study is the following (the relevant null hypothesis is given in parentheses), where the capital H indicates "hypothesis." For example:

    H–M: M outperform the compared methods for predicting fault-prone software classes (null hypothesis: M does not outperform the compared methods for predicting fault-prone software classes).

4. Empirical data collection: The researcher decides the sources from which the data is to be collected. It is found from literature that the data collected is either from university/academic systems, commercial systems, or open source software. The researcher should state the environment in which the study is performed,

programming language in which the systems are developed, size of the systems to be analyzed (lines of code [LOC] and number of classes), and the duration for which the system is developed.

5. Empirical methods: The data analysis techniques are selected based on the type of the dependent variables used. An appropriate data analysis technique should be selected by identifying its strengths and weaknesses. For example, a number of techniques have been available for developing models to predict and analyze software quality attributes. These techniques could be statistical like linear regression and logistic regression or machine-learning techniques like decision trees, support vector machines, and so on. Apart from these techniques, there are a new set of techniques like particle swarm optimization, gene expression programming, and so on that are called the search-based techniques. The details of these techniques can be found in Chapter 7.

In the empirical study, the data is analyzed corresponding to the details given in the experimental design. Thus, the experimental design phase must be carefully planned and executed so that the analysis phase is clear and unambiguous. If the design phase does not match the analysis part then it is most likely that the results produced are incorrect.

1.4.3 Research Conduct and Analysis

Finally, the empirical study is conducted following the steps described in the experiment design. The experiment analysis phase involves understanding the data by collecting descriptive statistics. The unrelated attributes are removed, and the best attributes (variables) are selected out of a set of attributes (e.g., software metrics) using attribute reduction techniques. After removing irrelevant attributes, hypothesis testing is performed using statistical tests and, on the basis of the result obtained, a decision regarding the acceptance or rebuttal of the hypothesis is made. The statistical tests are described in Chapter 6. Finally, for analyzing the casual relationships between the independent variables and the dependent variable, the model is developed and validated. The steps involved in experiment conduct and analysis are briefly described below.

1. Descriptive statistics: The data is validated for correctness before carrying out the analysis. The first step in the analysis is descriptive statistics. The research data must be suitably reduced so that the research data can be read easily and can be used for further analysis. Descriptive statistics concern development of certain indices or measures to summarize the data. The important statistics measures used for comparing different case studies include mean, median, and standard deviation. The data analysis methods are selected based on the type of the dependent variable being used. Statistical tests can be applied to accept or refute a hypothesis. Significance tests are performed for comparing the predicted performance of a method with other sets of methods. Moreover, effective data assessment should also yield outliers (Aggarwal et al. 2009).

2. Attribute reduction: Feature subselection is an important step that identifies and removes as much of the irrelevant and redundant information as possible. The dimensionality of the data reduces the size of the hypothesis space and allows the methods to operate faster and more effectively (Hall 2000).

3. Statistical analysis: The data collected can be analyzed using statistical analysis by following the steps below.

   a. Model prediction: The multivariate analysis is used for the model prediction. Multivariate analysis is used to find the combined effect of each independent variable on the dependent variable. Based on the results of performance measures, the performance of models predicted is evaluated and the results are interpreted. Chapter 7 describes these performance measures.

   b. Model validation: In systems, where models are independently constructed from the training data (such as in data mining), the process of constructing the model is called training. The subsamples of data that are used to validate the initial analysis (by acting as "blind" data) are called validation data or test data. The validation data is used for validating the model predicted in the previous step.

   c. Hypothesis testing: It determines whether the null hypothesis can be rejected at a specified confidence level. The confidence level is determined by the researcher and is usually less than 0.01 or 0.05 (refer Section 4.7 for details).

1.4.4 Results Interpretation

In this step, the results computed in the empirical study's analysis phase are assessed and discussed. The reason behind the acceptance or rejection of the hypothesis is examined. This process provides insight to the researchers about the actual reasons of the decision made for hypothesis. The conclusions are derived from the results obtained in the study. The significance and practical relevance of the results are defined in this phase. The limitations of the study are also reported in the form of threats to validity.

1.4.5 Reporting

Finally, after the empirical study has been conducted and interpreted, the study is reported in the desired format. The results of the study can be disseminated in the form of a conference article, a journal paper, or a technical report.

The results are to be reported from the reader's perspective. Thus, the background, motivation, analysis, design, results, and the discussion of the results must be clearly documented. The audience may want to replicate or repeat the results of a study in a similar context. The experiment settings, data-collection methods, and design processes must be reported in significant level of detail. For example, the descriptive statistics, statistical tools, and parameter settings of techniques must be provided. In addition, graphical representation should be used to represent the results. The results may be graphically presented using pie charts, line graphs, box plots, and scatter plots.

1.4.6 Characteristics of a Good Empirical Study

The characteristics of a good empirical study are as follows:

1. Clear: The research goals, hypothesis, and data-collection procedure must be clearly stated.

2. Descriptive: The research should provide details of the experiment so that the study can be repeated and replicated in similar settings.

3. Precise: Precision helps to prove confidence in the data. It represents the degree of measure correctness and data exactness. High precision is necessary to specify the attributes in detail.

4. Valid: The experiment conclusions should be valid for a wide range of population.

5. Unbiased: The researcher performing the study should not influence the results to satisfy the hypothesis. The research may produce some bias because of experiment error. The bias may be produced when the researcher selects the participants such that they generate the desired results. The measurement bias may occur during data collection.

6. Control: The experiment design should be able to control the independent variables so that the confounding effects (interaction effects) of variables can be reduced.

7. Replicable: Replication involves repeating the experiment with different data under same experimental conditions. If the replication is successful then this indicates generalizability and validity of the results.

8. Repeatable: The experimenter should be able to reproduce the results of the study under similar settings.

1.5 Ethics of Empirical Research

Researchers, academicians, and sponsors should be aware of research ethics while conducting and funding empirical research in software engineering. The upholding of ethical standards helps to develop trust between the researcher and the participant, and thus smoothens the research process. An unethical study can harm the reputation of the research conducted in software engineering area.

Some ethical issues are regulated by the standards and laws provided by the government. In some countries like the United States, the sponsoring agency requires that the research involving participants must be reviewed by a third-party ethics committee to verify that the research complies with the ethical principles and standards (Singer and Vinson 2001). Empirical research is based on the trust between the participant and the researcher, the ethical information must be explicitly provided to the participants to avoid any future conflicts. The participants must be informed about the risk and ethical issues involved in the research at the beginning of the study. The examples of problems related to ethics that are experienced in industry are given by Becker-Kornstaedt (2001) and summarized in Table 1.2.

TABLE 1.2

Examples of Unethical Research

S. No	Problem
1	Employees misleading the manager to protect himself or herself with the knowledge of the researcher
2	Nonconformance to a mandatory process
3	Revealing identities of the participant or organization
4	Manager unexpectedly joining a group interview or discussion with the participant
5	Experiment revealing identity of the participants of a nonperforming department in an organization
6	Experiment outcomes are used in employee ratings
7	Participants providing information off the record, that is, after interview or discussion is over

The ethical threats presented in Table 1.2 can be reduced by (1) presenting data and results such that no information about the participant and the organization is revealed, (2) presenting different reports to stakeholders, (3) providing findings to the participants and giving them the right to withdraw any time during the research, and (4) providing publication to companies for review before being published. Singer and Vinson (2001) identified that the engineering and science ethics may not be related to empirical research in software engineering. They provided the following four ethical principles:

1. Informed consent: This principle is concerned with subjects participating in the experiment. The subjects or participants must be provided all the relevant information related to the experiment or study. The participants must willingly agree to participate in the research process. The consent form acts as a contract between an individual participant and the researcher.

2. Scientific value: This principle states that the research results must be correct and valid. This issue is critical if the researchers are not familiar with the technology or methodology they are using as it will produce results of no scientific value.

3. Confidentiality: It refers to anonymity of data, participants, and organizations.

4. Beneficence: The research must provide maximum benefits to the participants and protect the interests of the participants. The benefits of the organization must also be protected by not revealing the weak processes and procedures being followed in the departments of the organization.

1.5.1 Informed Content

Informed consent consists of five elements—disclosure, comprehension, voluntariness, consent, and right to withdraw. Disclosure means to provide all relevant details about the research to the participants. This information includes risks and benefits incurred by the participants. Comprehension refers to presenting information in such a manner that can be understood by the participant. Voluntariness specifies that the consent obtained must not be under any pressure or influence and actual consent must be taken. Finally, the subjects must have the right to withdraw from research process at any time. The consent form has the following format (Vinson and Singer 2008):

1. Research title: The title of the project must be included in the consent form.

2. Contact details: The contact details (including ethics contact) will provide the participant information about whom to contact to clarify any questions or issues or complaints.

3. Consent and comprehension: The participant actually gives the consent form in this section stating that they have understood the requirement of the research.

4. Withdrawal: This section states that the participants can withdraw from the research without any penalty.

5. Confidentiality: It states the confidentiality related to the research study.

6. Risks and benefits: This section states the risks and benefits of the research to the participants.

7. Clarification: The participants can ask for any further clarification at any time during the research.

8. Signature: Finally, the participant signs the consent form with the date.

1.5.2 Scientific Value

This ethical issue is concerned with two aspects—relevance of research topic and validity of research results. The research must balance between risks and benefits. In fact, the advantages of the research should outweigh the risks. The results of the research must also be valid. If they are not valid then the results are incorrect and the study has no value to the research community. The reason for invalid results is usually misuse of methodology, application, or tool. Hence, the researchers should not conduct the research for which they are not capable or competent.

1.5.3 Confidentiality

The information shared by the participants should be kept confidential. The researcher should hide the identity of the organization and participant. Vinson and Singer (2008) identified three features of confidentiality—data privacy, participant anonymity, and data anonymity. The data collected must be protected by password and only the people involved in the research should have access to it. The data should not reveal the information about the participant. The researchers should not collect personal information of participant. For example, participant identity must be used instead of the participant name. The participant information hiding is achieved by hiding information from colleagues, professors, and general public. Hiding information from the manager is particularly essential as it may affect the career of the participants. The information must be also hidden from the organization's competitors.

1.5.4 Beneficence

The participants must be benefited by the research. Hence, methods that protect the interest of the participants and do not harm them must be adopted. The research must not pose a threat to the researcher's job, for example, by creating an employee-ranking framework. The revealing of an organization's sensitive information may also bring loss to the company in terms of reputation and clients. For example, if the names of companies are revealed in the publication, the comparison between the processes followed in the companies or potential flaws in the processes followed may affect obtaining contracts from the clients. If the research involves analyzing the process of the organization, the outcome of the research or facts revealed from the research can harm the participants to a significant level.

1.5.5 Ethics and Open Source Software

In the absence of empirical data, data and source code from open source software are being widely used for analysis in research. This poses concerns of ethics, as the open source software is not primarily developed for research purposes. El-Emam (2001) raised two important ethical issues while using open source software namely "informed consent and minimization of harm and confidentiality." Conducting studies that rate the developers or compares two open source software may harm the developer's reputation or the company's reputation (El-Emam 2001).

1.5.6 Concluding Remarks

The researcher must maintain the ethics in the research by careful planning and, if required, consulting ethical bodies that have expertise for guiding them on ethical issues in software engineering empirical research. The main aim of the research involving participants must be to protect the interests of the participants so that they are protected from any harm. Becker-Kornstaedt (2001) suggests that the participant interests can be protected by using techniques such as manipulating data, providing different reports to different stakeholders, and providing the right to withdraw to the participants.

Finally, feedback of the research results must be provided to the participants. The opinion of the participants about the validity of the results must also be asked. This will help in increasing the trust between the researcher and the participant.

1.6 Importance of Empirical Research

Why should empirical studies in software engineering be carried out? The main reason of carrying out an empirical study is to reduce the gap between theory and practice by using statistical tests to test the formed hypothesis. This will help in analyzing, assessing, and improving the processes and procedures of software development. It may also provide guidelines to management for decision making. Thus, without evaluating and assessing new methods, tools, and techniques, their use will be random and effectiveness will be uncertain. The empirical study is useful to researchers, academicians, and the software industry from different perspectives.

1.6.1 Software Industry

The results of ESE must be adopted by the industry. ESE can be used to answer the questions related to practices in industry and can improve the processes and procedures of software development. To match the requirements of the industry, the researcher must ask the following questions while conducting research:

- How does the research aim maps to the industrial problems?
- How can the software practitioners use the research results?
- What are the important problems in the industry?

The predictive models constructed in ESE can be applied to future, similar industrial applications. The empirical research enables software practitioners to use the results of the experiment and ascertain that a set of good processes and procedures are followed during software development. Thus, the empirical study can guide toward determining the quality of the resultant software products and processes. For example, a new technique or technology can be evaluated and assessed. The empirical study can help the software professionals in effectively planning and allocating resources in the initial phases of software development life cycle.

1.6.2 Academicians

While studying or conducting research, academicians are always curious to answer questions that are foremost in their minds. As the academicians dig deeper into their subject or research, the questions tend to become more complex. Empirical research empowers them with a great tool to find an answer by asking or interviewing different stakeholders, by conducting a survey, or by conducting a scientific experiment. Academicians generally make predictions that can be stated in the form of hypotheses. These hypotheses need to be subjected to robust scientific verification or approval. With empirical research, these hypotheses can be tested, and their results can be stated as either being accepted or rejected. Thereafter, based on the result, the academicians can make some generalization or make a conclusion about a particular theory. In other words, a new theory can be generated and some old ones may be disproved. Additionally, sometimes there are practical questions that an academician encounters, empirical research would be highly beneficial in solving them. For example, an academician working in a university may want to find out the most efficient learning approach that yields the best performance among a group of students. The results of the research can be included in the course curricula.

From the academic point of view, high-quality teaching is important for future software engineers. Empirical research can provide management with important information about the use of tools and techniques. The students will further carry forward the knowledge to the software industry and thus improve the industrial practices. The empirical result can support one technique over the other and hence will be very useful in comparing the techniques.

1.6.3 Researchers

From the researchers point of view, the results can be used to provide insight about existing trends and guidelines regarding future research. The empirical study can be repeated or replicated by the researcher in order to establish generalizability of the results to new subjects or data sets.

1.7 Basic Elements of Empirical Research

The basic elements in empirical research are purpose, participants, process, and product. Figure 1.7 presents the four basic elements of empirical research. The purpose defines the reason of the research, the relevance of the topic, specific aims in the form of research questions, and objectives of the research.

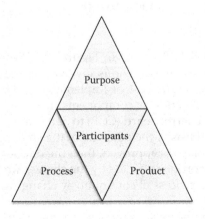

FIGURE 1.7
Elements of empirical research.

Process lays down the way in which the research will be conducted. It defines the sequence of steps taken to conduct a research. It provides details about the techniques, methodologies, and procedures to be used in the research. The data-collection steps, variables involved, techniques applied, and limitations of the study are defined in this step. The process should be followed systematically to produce a successful research.

Participants are the subjects involved in the research. The participants may be interviewed or closely observed to obtain the research results. While dealing with participants, ethical issues in ESE must be considered so that the participants are not harmed in any way.

Product is the outcome produced by the research. The final outcome provides the answer to research questions in the empirical research. The new technique developed or methodology produced can also be considered as a product of the research. The journal paper, conference article, technical report, thesis, and book chapters are products of the research.

1.8 Some Terminologies

Some terminologies that are frequently used in the empirical research in software engineering are discussed in this section.

1.8.1 Software Quality and Software Evolution

Software quality determines how well the software is designed (quality of design), and how well the software conforms to that design (quality of conformance).

In a software project, most of the cost is consumed in making changes rather than developing the software. Software evolution (maintenance) involves making changes in the software. Changes are required because of the following reasons:

1. Defects reported by the customer
2. New functionality requested by the customer
3. Improvement in the quality of the software
4. To adapt to new platforms

The typical evolution process is depicted in Figure 1.8. The figure shows that a change is requested by a stakeholder (anyone who is involved in the project) in the project. The second step requires analyzing the cost of implementing the change and the impact of the change on the related modules or components. It is the responsibility of an expert group known as the change control board (CCB) to determine whether the change must be implemented or not. On the basis of the outcome of the analysis, the CCB approves or disapproves a change. If the change is approved, then the developers implement the change. Finally, the change and the portions affected by the change are tested and a new version of the software is released. The process of continuously changing the software may decrease the quality of the software.

The main concerns during the evolution phase are maintaining the flexibility and quality of the software. Predicting defects, changes, efforts, and costs in the evolution phase

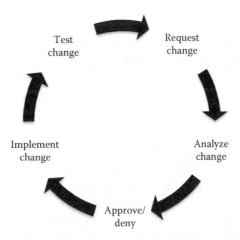

FIGURE 1.8
Software evolution cycle.

is an important area of software engineering research. An effective prediction can lead to decreasing the cost of maintenance by a large extent. This will also lead to high-quality software and hence increasing the modifiability aspect of the software. Change prediction concerns itself with predicting the portions of the software that are prone to changes and will thus add up to the maintenance costs of the software. Figure 1.9 shows the various research avenues in the area of software evolution.

After the detection of the change and nonchange portions in a software, the software developers can take various remedial actions that will reduce the probability of occurrence of changes in the later phases of software development and, consequently, the cost will also reduce exponentially. The remedial steps may involve redesigning or refactoring of modules so that fewer changes are encountered in the maintenance phase. For example, if high value of the coupling metric is the reason for change proneness of a given module. This implies that the given module in question is highly interdependent on other modules. Thus, the module should be redesigned to improve the quality and reduce its probability to be change prone. Similar design corrective actions or other measures can be easily taken once the software professional detects the change-prone portions in a software.

Defect prediction
- What are the defect-prone portions in the maintanence phase?

Change prediction
- What are the change-prone portions in the software?
- How many change requests are expected?

Maintenance costs prediction
- What is the cost of maintaining the software over a period of time?

Maintenance effort prediction
- How much effort will be required to implement a change?

FIGURE 1.9
Prediction during evolution phase.

1.8.2 Software Quality Attributes

Software quality can be measured in terms of attributes. The attribute domains that are required to define for a given software are as follows:

1. Functionality
2. Usability
3. Testability
4. Reliability
5. Maintainability
6. Adaptability

The attribute domains can be further divided into attributes that are related to software quality and are given in Figure 1.10. The details of software quality attributes are given in Table 1.3.

1.8.3 Measures, Measurements, and Metrics

The terms measures, measurements, and metrics are often used interchangeably. However, we should understand the difference among these terms. Pressman (2005) explained this clearly as:

> A measure provides a quantitative indication of the extent, amount, dimension, capacity or size of some attributes of a product or process. Measurement is the act of determining a measure. The metric is a quantitative measure of the degree to which a product or process possesses a given attribute.

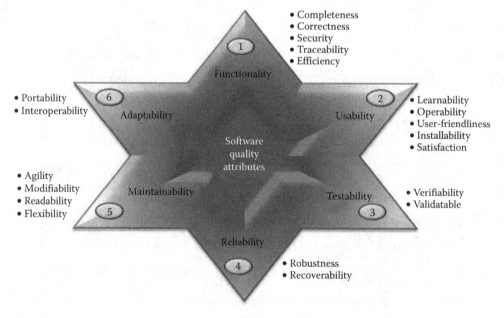

FIGURE 1.10
Software quality attributes.

TABLE 1.3

Software Quality Attributes

Functionality: The degree to which the purpose of the software is satisfied		
1	Completeness	The degree to which the software is complete
2	Correctness	The degree to which the software is correct
3	Security	The degree to which the software is able to prevent unauthorized access to the program data
4	Traceability	The degree to which requirement is traceable to software design and source code
5	Efficiency	The degree to which the software requires resources to perform a software function
Usability: The degree to which the software is easy to use		
1	Learnability	The degree to which the software is easy to learn
2	Operability	The degree to which the software is easy to operate
3	User-friendliness	The degree to which the interfaces of the software are easy to use and understand
4	Installability	The degree to which the software is easy to install
5	Satisfaction	The degree to which the user's feel satisfied with the software
Testability: The ease with which the software can be tested to demonstrate the faults		
1	Verifiability	The degree to which the software deliverable meets the specified standards, procedures, and process
2	Validatable	The ease with which the software can be executed to demonstrate whether the established testing criteria is met
Maintainability: The ease with which the faults can be located and fixed, quality of the software can be improved, or software can be modified in the maintenance phase		
1	Agility	The degree to which the software is quick to change or modify
2	Modifiability	The degree to which the software is easy to implement, modify, and test in the maintenance phase
3	Readability	The degree to which the software documents and programs are easy to understand so that the faults can be easily located and fixed in the maintenance phase
4	Flexibility	The ease with which changes can be made in the software in the maintenance phase
Adaptability: The degree to which the software is adaptable to different technologies and platforms		
1	Portability	The ease with which the software can be transferred from one platform to another platform
2	Interoperability	The degree to which the system is compatible with other systems
Reliability: The degree to which the software performs failure-free functions		
1	Robustness	The degree to which the software performs reasonably under unexpected circumstances
2	Recoverability	The speed with which the software recovers after the occurrence of a failure

Source: Y. Singh and R. Malhotra, *Object-Oriented Software Engineering*, PHI Learning, New Delhi, India, 2012.

For example, a measure is the number of failures experienced during testing. Measurement is the way of recording such failures. A software metric may be the average number of failures experienced per hour during testing.

Fenton and Pfleeger (1996) has defined measurement as:

> It is the process by which numbers or symbols are assigned to attributes of entities in the real world in such a way as to describe them according to clearly defined rules.

Software metrics can be defined as (Goodman 1993): "The continuous application of measurement based techniques to the software development process and its products to supply meaningful and timely management information, together with the use of those techniques to improve that process and its products."

1.8.4 Descriptive, Correlational, and Cause–Effect Research

Descriptive research provides description of concepts. Correlational research provides relation between two variables. Cause–effect research is similar to experiment research in that the effect of one variable on another is found.

1.8.5 Classification and Prediction

Classification predicts categorical outcome variables (ordinal or nominal). The training data is used for model development, and the model can be used for predicting unknown categories of outcome variables. For example, consider a model to classify modules as faulty or not faulty on the basis of coupling and size of the modules. Figure 1.11 represents this example in the form of a decision tree. The tree shows that if the coupling of modules is <8 and the LOC is low then the module is not faulty.

In classification, the classification techniques take training data (comprising of the independent and the dependent variables) as input and generate rules or mathematical formulas that are used by validation data to verify the model predicted. The generated rules or mathematical formulas are used by future data to predict categories of the outcome variables. Figure 1.12 depicts the classification process. Prediction is similar to classification except that the outcome variable is continuous.

1.8.6 Quantitative and Qualitative Data

Quantitative data is numeric, whereas qualitative data is textual or pictorial. Quantitative data can either be discrete or continuous. Examples of quantitative data are LOC, number of faults, number of work hours, and so on. The information obtained by qualitative

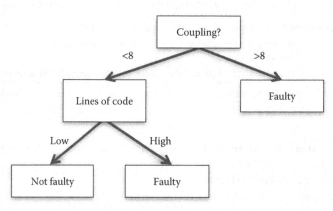

FIGURE 1.11
Example of classification process.

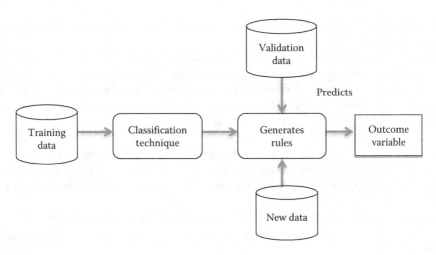

FIGURE 1.12
Steps in classification process.

analysis can be categorized by identifying patterns from the textual information. This can be achieved by reading and analyzing texts and deriving logical categories. This will help organize data in the form of categories. For example, answers to the following questions are presented in the form of categories.

- What makes a good quality system?
  User-friendliness, response time, reliability, security, recovery from failure
- How was the overall experience with the software?
  Excellent, very good, good, average, poor, very poor

Text mining is another way to process qualitative data into useful form that can be used for further analysis.

1.8.7 Independent, Dependent, and Confounding Variables

Variables are measures that can be manipulated or varied in research. There are two types of variables involved in cause–effect analysis, namely, the independent and the dependent variables. They are also known as attributes or features in software engineering research. Figure 1.13 shows that the experimental process analyzes the relationship between the independent and the dependent variables. Independent variables (or predictor variables)

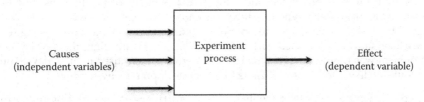

FIGURE 1.13
Independent and dependent variables.

are input variables that are manipulated or controlled by the researcher to measure the response of the dependent variable.

The dependent variable (or response variable) is the output produced by analyzing the effect of the independent variables. The dependent variables are presumed to be influenced by the independent variables. The independent variables are the causes and the dependent variable is the effect. Usually, there is only one dependent variable in the research. Figure 1.13 depicts that the independent variables are used to predict the outcome variable following a systematic experimental process.

Examples of independent variables are lines of source code, number of methods, and number of attributes. Dependent variables are usually measures of software quality attributes. Examples of dependent variable are effort, cost, faults, and productivity. Consider the following research question:

Do software metrics have an effect on the change proneness of a module?

Here, software metrics are the independent variables and change proneness is the dependent variable.

Apart from the independent variables, unknown variables or confounding variables (extraneous variables) may affect the outcome (dependent) variable. Randomization can nullify the effect of confounding variables. In randomization, many replications of the experiments are executed and the results are averaged over multiple runs, which may cancel the effect of extraneous variables in the long course.

1.8.8 Proprietary, Open Source, and University Software

Data-based empirical studies that are capable of being verified by observation or experiment are needed to provide relevant answers. In software engineering empirical research, obtaining empirical data is difficult and is a major concern for researchers. The data collected may be from university/academic software, open source software, or proprietary software.

Undergraduate or graduate students at the university usually develop the university software. To use this type of data, the researchers must ensure that the software is developed by following industrial practices and should document the process of software development and empirical data collection in detail. For example, Aggarwal et al. (2009) document the procedure of data collection as: "All students had at least six months experience with Java language, and thus they had the basic knowledge necessary for this study. All the developed systems were taken of a similar level of complexity and all the developers were made sufficiently familiar with the application they were working on." The study provides a list of the coding standards that were followed by students while developing the software and also provides details about the testing environment as given below by Aggarwal et al. (2009):

> The testing team was constituted under the guidance of senior faculty consisting of a separate group of students who had the prior knowledge of system testing. They were assigned the task of testing systems according to test plans and black-box testing techniques. Each fault was reported back to the development team, since the development environment was representative of real industry environment used in these days. Thus, our results are likely to be generalizable to other development environments.

Open source software is usually a freely available software, developed by many developers from different places in a collaborative manner. For example, Google Chrome, Android operating system, and Linux operating system.

Proprietary software is a licensed software owned by a company. For example, Microsoft Office, Adobe Acrobat, and IBM SPSS are proprietary software. In practice, obtaining data from proprietary software for research validation is difficult as the software companies are usually not willing to share the information about their software systems.

The software developed by the student programmers is generally small and developed by limited number of developers. If the decision is made for collecting and using this type of data in research then the guidelines similar to given above must be followed to promote unbiased and replicated results. These days, open source software repositories are being mined to obtain research data for historical analysis.

1.8.9 Within-Company and Cross-Company Analysis

In within-company analysis, the empirical study collects the data from the old versions/releases of the same software, predicts models, and applies the predicted models to the future versions of the same project. However, in practice, the old data may not be available. In such cases, the data obtained from similar earlier projects developed by different companies are used for prediction in new projects. The process of validating the predicted model using data collected from different projects from which the model has been derived is known as cross-company analysis. For example, He et al. (2012) conducted a study to find the effectiveness of cross-project prediction for predicting defects. They used data collected from different projects to predict models and applied those data on new projects. Figure 1.14 shows that the model (M1) is developed using training data collected from software A, release R1. The next release of software used model M1 to predict the outcome variable. This process is known as within-company prediction, whereas in cross-company prediction, data collected from another software B uses model M1 to predict the outcome variable.

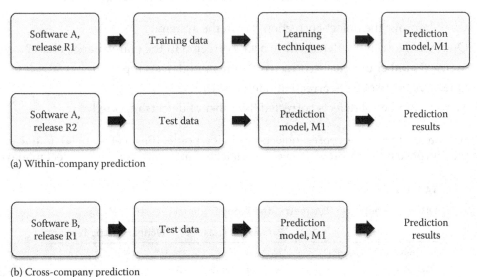

(a) Within-company prediction

(b) Cross-company prediction

FIGURE 1.14
(a) Within-company versus (b) cross-company prediction.

1.8.10 Parametric and Nonparametric Tests

In hypothesis testing, statistical tests are applied to determine the validity of the hypothesis. These tests can be categorized as either parametric or nonparametric. Parametric tests are used for data samples having normal distribution (bell-shaped curve), whereas nonparametric tests are used when the distribution of data samples is highly skewed. If the assumptions of parametric tests are met, they are more powerful as they use more information while computation. The difference between parametric and nonparametric tests is presented in Table 1.4.

1.8.11 Goal/Question/Metric Method

The Goal/Question/Metric (GQM) method was developed by Basili and Weiss (1984) and is a result of their experience, research, and practical knowledge. The GQM method consists of the following three basis elements:

1. Goal
2. Question
3. Metric

In GQM method, measurement is goal-oriented. Thus, first the goals need to be defined that can be measured during the software development. The GQM method defines goals that are transformed into questions and metrics. These questions are answered later to determine whether the goals have been satisfied or not. Hence, GQM method follows top-down approach for dividing goals into questions and mapping questions to metrics, and follows bottom-up approach by interpreting the measurement to verify whether the goals have been satisfied. Figure 1.15 presents the hierarchical view of GQM framework. The figure shows that the same metric can be used to answer multiple questions.

For example, if the developer wants to improve the defect-correction rate during the maintenance phase. The goal, question, and associated metrics are given as:

- Goal: Improve the defect-correction rate in the system.
- Question: How many defects have been corrected in the maintenance phase?
- Metric: Number of defects corrected/Number of defects reported.
- Question: Is the defect-correction rate satisfactory?
- Metric: Number of defects corrected/Number of defects reported.

The goals are defined as purposes, objects, and viewpoints (Basili et al. 1994). In the above example, purpose is "to improve," object is "defects," and viewpoint is "project manager."

TABLE 1.4

Difference between Parametric and Nonparametric Tests

	Parametric Tests	**Nonparametric Tests**
Assumed distribution	Normal	Any
Data type	Ratio or interval	Any
Measures of central tendency	Mean	Median
Example	t-test, ANOVA	Kruskal–Wallis–Wilcoxon test

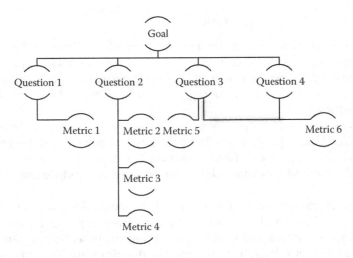

FIGURE 1.15
Framework of GQM.

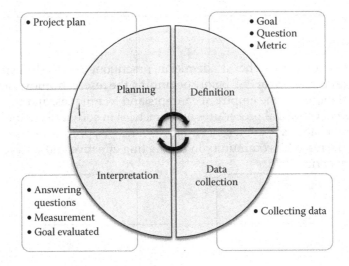

FIGURE 1.16
Phases of GQM.

Figure 1.16 presents the phases of the GQM method. The GQM method has the following four phases:

- Planning: In the first phase, the project plan is produced by recognizing the basic requirements.
- Definition: In this phase goals, questions, and relevant metrics are defined.
- Data collection: In this phase actual measurement data is collected.
- Interpretation: In the final phase, the answers to the questions are provided and the goal's attainment is verified.

1.8.12 Software Archive or Repositories

The progress of the software is managed using software repositories that include source code, documentation, archived communications, and defect-tracking systems. The information contained in these repositories can be used by the researchers and practitioners for maintaining software systems, improving software quality, and empirical validation of data and techniques.

Researchers can mine these repositories to understand the software development, software evolution, and make predictions. The predictions can consist of defects and changes and can be used for planning of future releases. For example, defects can be predicted using historical data, and this information can be used to produce less defective future releases.

The data is kept in various types of software repositories such as CVS, Git, SVN, ClearCase, Perforce, Mercurial, Veracity, and Fossil. These repositories are used for management of software content and changes, including documents, programs, user procedure manuals, and other related information. The details of mining software repositories are presented in Chapter 5.

1.9 Concluding Remarks

It is very important for a researcher, academician, practitioner, and a student to understand the procedures and concepts of ESE before beginning the research study. However, there is a lack of understanding of the empirical concepts and techniques, and the level of uncertainty on the use of empirical procedures and practices in software engineering. The goal of the subsequent chapters is to present empirical concepts, procedures, and practices that can be used by the research community in conducting effective and well-formed research in software engineering field.

Exercises

1.1 What is empirical software engineering? What is the purpose of empirical software engineering?

1.2 What is the importance of empirical studies in software engineering?

1.3 Describe the characteristics of empirical studies.

1.4 What are the five types of empirical studies?

1.5 What is the importance of replicated and repeated studies in empirical software engineering?

1.6 Explain the difference between an experiment and a case study.

1.7 Differentiate between quantitative and qualitative research.

1.8 What are the steps involved in an experiment? What are characteristics of a good experiment?

1.9 What are ethics involved in a research? Give examples of unethical research.

1.10 Discuss the following terms:

  a. Hypothesis testing

  b. Ethics

  c. Empirical research

  d. Software quality

1.11 What are systematic reviews? Explain the steps in systematic review.

1.12 What are the key issues involved in empirical research?

1.13 Compare and contrast classification and prediction process.

1.14 What is GQM method? Explain the phases of GQM method.

1.15 List the importance of empirical research from the perspective of software industries, academicians, and researchers.

1.16 Differentiate between the following:

  a. Parametric and nonparametric tests

  b. Independent, dependent and confounding variables

  c. Quantitative and qualitative data

  d. Within-company and cross-company analysis

  e. Proprietary and open source software

Further Readings

Kitchenham et al. effectively provides guidelines for empirical research in software engineering:

B. A. Kitchenham, S. L. Pfleeger, L. M. Pickard, P. W. Jones, D. C. Hoaglin, K. E. Emam, and J. Rosenberg, "Preliminary guidelines for empirical research in software engineering," *IEEE Transactions on Software Engineering*, vol. 28, pp. 721–734, 2002.

Juristo and Moreno explain a good number of concepts of empirical software engineering:

N. Juristo, and A. N. Moreno, "Lecture notes on empirical software engineering," *Series on Software Engineering and Knowledge Engineering, World Scientific*, vol. 12, 2003.

The basic concept of qualitative research is presented in:

N. Mays, and C. Pope, "Qualitative research: Rigour and qualitative research," *British Medical Journal*, vol. 311, no. 6997, pp. 109–112, 1995.

A. Strauss, and J. Corbin, *Basics of Qualitative Research: Techniques and Procedures for Developing Grounded Theory*, Sage Publications, Thousand Oaks, CA, 1998.

A collection of research from top empirical software engineering researchers focusing on the practical knowledge necessary for conducting, reporting, and using empirical methods in software engineering can be found in:

J. Singer, and D. I. K. Sjøberg, *Guide to Advanced Empirical Software Engineering*, Edited by F. Shull, Springer, Berlin, Germany, vol. 93, 2008.

The detail about ethical issues for empirical software engineering is presented in:

J. Singer, and N. Vinson, "Ethical issues in empirical studies of software engineering," *IEEE Transactions on Software Engineering*, vol. 28, pp. 1171–1180, NRC 44912, 2002.

An overview of empirical observations and laws is provided in:

A. Endres, and D. Rombach, *A Handbook of Software and Systems Engineering: Empirical Observations, Laws, and Theories*, Addison-Wesley, New York, 2003.

Authors present detailed practical guidelines on the preparation, conduct, design, and reporting of case studies of software engineering in:

P. Runeson, M. Host, A. Rainer, and B. Regnell, *Case Study Research in Software Engineering: Guidelines and Examples*, John Wiley & Sons, New York, 2012.

The following research paper provides detailed explanations about software quality attributes:

I. Gorton (ed.), "Software quality attributes," In: *Essential Software Architecture*, Springer, Berlin, Germany, pp. 23–38, 2011.

An in-depth knowledge of prediction is mentioned in:

A. J. Albrecht, and J. E. Gaffney, "Software function, source lines of code, and development effort prediction: A software science validation," *IEEE Transactions on Software Engineering*, vol. 6, pp. 639–648, 1983.

The following research papers provide a brief knowledge of quantitative and qualitative data in software engineering:

A. Rainer, and T. Hall, "A quantitative and qualitative analysis of factors affecting software processes," *Journal of Systems and Software*, vol. 66, pp. 7–21, 2003.
C. B. Seaman, "Qualitative methods in empirical studies of software engineering," *IEEE Transactions on Software Engineering*, vol. 25, pp. 557–572, 1999.

A useful concept of how to analyze qualitative data is presented in:

A. Bryman, and B. Burgess, *Analyzing Qualitative Data*, Routledge, New York, 2002.

Basili explain the major role to controlled experiment in software engineering field in:

V. Basili, *The Role of Controlled Experiments in Software Engineering Research,* Empirical Software Engineering Issues, LNCS 4336, Springer-Verlag, Berlin, Germany, pp. 33–37, 2007.

The following paper presents guidelines for controlling experiments:

A. Jedlitschka, and D. Pfahl, "Reporting guidelines for controlled experiments in software engineering," In *Proceedings of the International Symposium on Empirical Software Engineering Symposium,* IEEE, Noosa Heads, Australia, pp. 95–104, 2005.

A detailed explanation of within-company and cross-company concept with sample case studies may be obtained from:

A. Kitchenham, E. Mendes, and G. H. Travassos, "Cross versus within-company cost estimation studies: A systematic review," *IEEE Transactions on Software Engineering,* vol. 33, pp. 316–329, 2007.

The concept of proprietary, open source, and university software are well explained in the following research paper:

A. MacCormack, J. Rusnak, and C. Y. Baldwin, "Exploring the structure of complex software designs: An empirical study of open source and proprietary code," *Management Science,* vol. 52, pp. 1015–1030, 2006.

The concept of parametric and nonparametric test may be obtained from:

D. G. Altman, and J. M. Bland, "Parametric v non-parametric methods for data analysis," *British Medical Journal,* 338, 2009.

The book by Solingen and Berghout is a classic and a very useful reference, and it gives detailed discussion on the GQM methods:

R. V. Solingen, and E. Berghout, *The Goal/Question/Metric Method: A Practical Guide for Quality Improvement of Software Development,* McGraw-Hill, London, vol. 40, 1999.

A classical report written by Prieto explains the concept of software repositories:

R. Prieto-Díaz, "Status report: Software reusability," *IEEE Software,* vol. 10, pp. 61–66, 1993.

2

Systematic Literature Reviews

Review of existing literature is an essential step before beginning any new research. Systematic reviews (SRs) synthesize the existing research work in such a manner that can be analyzed, assessed, and interpreted to draw meaningful conclusions. The aim of conducting an SR is to gather and interpret empirical evidence from the available research with respect to formed research questions. The benefit of conducting an SR is to summarize the existing trends in the available research, identify gaps in the current research, and provide future guidelines for conducting new research. The SRs also provide empirical evidence in support or opposition of a given hypothesis. Hence, the author of the SR must make all the efforts to provide evidence that support or does not support a given research hypothesis.

In this chapter, guidelines for conducting SRs are given for software engineering researchers and practitioners. The steps to be followed while conducting an SR including planning, conducting and reporting phases are described. The existing high-quality reviews in the areas of software engineering are also presented in this chapter.

2.1 Basic Concepts

SRs are better planned, more rigorous, and thoroughly analyzed as compared to surveys or literature reviews. In this section, we provide an overview of SRs and compare them with traditional surveys.

2.1.1 Survey versus SRs

Literature survey is the process of summarizing, organizing, and documenting the existing research to understand the research carried out in the field. On the other hand, an SR is the process of systematically and critically analyzing the information extracted from the existing research to answer the established research questions. The literature survey only provides the summary of the results of existing literature, whereas an SR opens avenues for new research as it provides future directions for researchers based on thorough analysis of existing literature. Kitchenham (2007) defined SR as:

> A systematic literature review (often referred to as a systematic review) is a means of identifying, evaluating and interpreting all available research relevant to a particular research question, or topic area, or phenomenon of interest.

Glossary of evidence-based medicine (EBM) terms defines SR as (http://ktclearinghouse. ca/cebm/glossary/):

> A summary of the medical literature that uses explicit methods to perform a comprehensive literature search and critical appraisal of individual studies and that uses appropriate statistical techniques to combine these valid studies.

TABLE 2.1

Comparison of Systematic Reviews and Literature Survey

S. No.	Systematic Review	Literature Survey
1	The goal is to identify best practices, strengths and weaknesses of specific techniques, procedures, tools, or methods by combining information from various studies.	The goal is to classify or categorize existing literature.
2	Focused on research questions that assess the techniques under investigation.	Provides an introduction of each paper in literature based on the identified area.
3	Provides a detailed review of existing literature.	Provides a brief overview of existing literature.
4	Extracts technical and useful metadata from the contents.	Extracts general research trends from the studies.
5	Search process is more stringent such that it involves searching references or contacting researchers in the field.	Search process is less stringent.
6	Strong assessment of quality is necessary.	Strong assessment of quality is not necessary.
7	Results are based on high-quality evidence with the aim to answer research questions.	Results only provide summary of existing literature.
8	Often uses statistics to analyze the results.	Does not use statistics to analyze the results.

SRs summarize high-quality research on a specific area. They provide the best available evidence on a particular technique or technology and produce conclusions that can be used by the software practitioners and researchers to select the best available techniques or methodologies. The studies included in the review are known as primary studies and the SRs are known as secondary studies. Table 2.1 presents the summary of difference between SR and literature survey.

2.1.2 Characteristics of SRs

The following are the main characteristics of an SR:

1. It selects high-quality research papers and studies that are relevant, important, and essential, which are summarized in the form of one review paper.

2. It performs a systematic search by forming a search strategy to identify primary studies from the digital libraries. The search strategy is documented so that the readers can analyze the completeness of the process and repeat the same.

3. It forms a valid review protocol and research questions that address the issues to be answered in the SR.

4. It clearly summarizes the characteristics of each selected study, including aims, techniques, and methods used in the studies.

5. It consists of a justified quality assessment criteria for inclusion and exclusion of the studies in the SR so that the effectiveness of each study can be determined.

6. It uses a number of presentation tools for reporting the findings and results of the selected studies to be included in the SR.

7. It identifies gaps in the current findings and highlights future directions.

2.1.3 Importance of SRs

An SR is conducted using scientific methods and minimizes the bias in the studies. The SRs are important as:

1. They gather important empirical evidence on the technique or method being focused in the SR. On the basis of the empirical evidence, the strengths and weaknesses of the technique may be summarized.
2. They identify the gaps in the current research.
3. They report the commonalities and the differences in the primary studies.
4. They provide future guidelines and framework to researchers and practitioners to perform new research.

2.1.4 Stages of SRs

SR consists of a series of steps that are carried out throughout the review process and provides a summary of important issues raised in the study. The stages in the SR enable the researchers to conduct the review in an organized manner. The activities included in the SR are as follows:

1. Planning the review
2. Conducting the review
3. Reporting the review results

The procedure followed in performing the SR is given by Kitchenham et al. (2007). The process is depicted in Figure 2.1. In the first step, the need for the SR is examined and in the second step the research questions are formed that address the issues to be answered in the review. Thereafter, the review protocol is developed that includes the following steps: search strategy design, study selection criteria, study quality assessment criteria, data extraction process, and data synthesis process.

The formation of review protocol consists of a series of stages. In the first step, the search strategy is formed, including identification of search terms and selection of sources to be searched to identify the primary studies. The next step involves determination of relevant studies by setting the inclusion and exclusion criteria for selecting review studies. Thereafter, quality assessment criteria are identified by forming the quality assessment questionnaire to analyze and assess the studies. The second to last stage involves the design of data extraction forms to collect the required information to answer the research questions, and in the final stage, methods for data synthesis are devised. Development of review protocol is an important step in an SR as it reduces the possibility and risk of research bias in the SR. Finally, in the planning stage, the review protocol is evaluated.

The steps planned in the first stage are actually performed in the conducting stage that includes actual collection of relevant studies by applying first the search strategy and then the inclusion and exclusion criteria. Each selected study is ranked according to the quality assessment criteria, and the data extraction and data synthesis steps are followed from only the selected high-quality primary studies. In the final phase, the results of the SR are reported. This step further involves examining, presenting, and verifying the results.

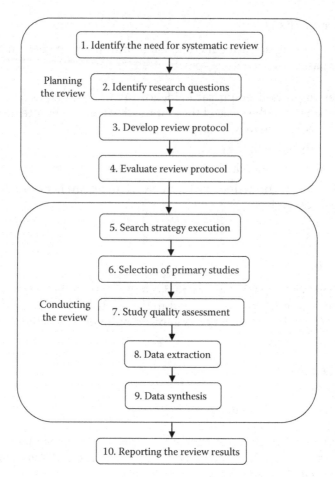

FIGURE 2.1
Systematic review process.

The above stages defined in the SR are iterative and not sequential. For example, the criteria for inclusion and exclusion of primary studies must be developed prior to collecting the studies. The criteria may be refined in the later stages.

2.2 Case Study

Software fault prediction (SFP) involves prediction of classes/modules in a software as faulty or nonfaulty based on the object oriented (OO) metrics for corresponding classes or modules. The identification of faulty or nonfaulty classes/modules enables researchers and practitioners to identify faulty portions in the early phases of software development. These faulty portions need extra attention during software development and the practitioners may focus testing resources on them. There are many techniques such as the statistical and the machine learning (ML) that can be used for classifying a class as faulty or nonfaulty. We conducted an SR of 64 in Malhotra (2015) primary studies from January 1991 to October 2013 for SFP using the ML techniques. The aim of the study is to gather empirical evidence from the literature to facilitate the use of the ML techniques for SFP. The study analyzes and assesses the gathered evidence regarding the use and performance of the ML techniques.

This case study is taken as an example review to explain all the steps in the SR in the subsequent sections and will be referred as systematic review of machine learning techniques (SRML). The detailed results of the case study can be found in Malhotra (2015).

2.3 Planning the Review

Before one begins with the review, it is important and essential to recognize the need for the review. After identifying the need for the SR, the researcher should form the research questions. Subsequently, the researchers must develop, document, and analyze the review protocol. The detailed results of the case study can be found in Malhotra (2015).

2.3.1 Identify the Need for SR

The identification of need for an SR is the most essential and crucial step while performing an SR. For example, Singh et al. (2014) identified the need of a structured review that can provide similarities and differences between results of existing studies on fault proneness. In their study, the summary of the results of the studies that predict fault proneness were provided. Radjenović et al. (2013) observed that many software metrics have been proposed in the literature and many of these metrics have been used for fault prediction. However, finding an appropriate suite of metrics was found to be essential because of the differences in the performance of the metrics. They concluded that there should be more studies that use industrial data sets so that metrics that can be used in the industrial settings can be identified.

To justify the importance of the SR, this step involves the review of all the existing SRs conducted in the same software engineering domain, thus recognizing the existing works and identifying the areas that need to be addressed in the new SR.

The following questions need to be determined before conducting the SR:

1. How many primary studies are available in the software engineering context?
2. What are the strength and weaknesses of the existing SR (if any) in the software engineering context?
3. What is the practical relevance of the proposed SR?
4. How will the proposed SR guide practitioners and researchers?
5. How can the quality of the proposed SR be evaluated?

Checklist is the most common mechanism used for reviewing the quality of the existing SR in the same area. It may also identify the flaws in the existing SR. A checklist may consist of a list of questions to determine the effectiveness of the existing SR. Table 2.2 shows an example of the checklist to assess the quality of an SR. The checklist consists of questions pertaining to the procedures and processes followed during an SR. The existing studies may be rated on a scale of 1–12 so that the quality of each study can be determined.

TABLE 2.2

Checklist for Evaluating Existing SR

S. No.	Questions
1	Is the aim of the review stated?
2	Is the search strategy appropriate?
3	Are the research questions justified?
4	Is the inclusion/exclusion criteria appropriate?
5	Is the quality assessment criteria applied?
6	Are independent reviewers used for quality evaluation of primary studies?
7	Is the data collected from the primary sources in an appropriate manner?
8	Is the data synthesis process effectively carried out?
9	Are the characteristics of the primary studies described?
10	Is any empirical evidence collected from the primary studies to reach a conclusion?
11	Does the review identify gaps in the existing literature?
12	Is the interpretation of the results stated and the guidelines for future research identified?

We may establish a threshold value to identify quality level of the study. If the rating of the existing SR goes below the established threshold value, the quality of the study may be considered as not acceptable and a new SR on the same topic may be conducted.

Thus, if an SR in the same domain with similar aims is located but it was conducted a long time ago, then a new SR adding current studies may be justified. However, if the existing SR is still relevant and is of high quality, then a new SR may not be required.

2.3.2 Formation of Research Questions

The process of formation of the research questions involves identification of relevant issues that need to be answered by the SR. According to Kitchenham (2007), it is the most important activity in any SR. The structure of an SR depends on the content of the research questions formed, and key decisions are based on the questions such as: Which studies to focus? Where to search them? How to assess the quality of these studies? Hence, the research questions must be well formed and constructed after a thorough analysis. The data for answering the identified research questions is collected from the primary studies. While constructing the research questions, the target audience, the tools and techniques to be evaluated, outcomes of the study, and the environment in which the study is conducted (academic or industry) must be determined. Hence, the following things must be kept in mind while forming the research questions:

- Which areas have already been explored in the existing reviews (if any)?
- Which areas are relevant and need to be explored/answered during the proposed SR?
- Are the questions important to the researchers and software practitioners?
- Will the questions assess any similarities in the trends or identify any deviation from the existing trends?

TABLE 2.3

Research Questions for SRML Case Study (Malhotra 2015)

RQ#	Research Questions	Motivation
RQ1	Which ML techniques have been used for SFP?	Identify the ML techniques commonly being used in SFP.
RQ2	What kind of empirical validation for predicting faults is found using the ML techniques found in RQ1?	Assess the empirical evidence obtained.
RQ2.1	Which techniques are used for subselecting metrics for SFP?	Identify techniques reported to be appropriate for selecting relevant metrics.
RQ2.2	Which metrics are found useful for SFP?	Identify metrics reported to be appropriate for SFP.
RQ2.3	Which metrics are found not useful for SFP?	Identify metrics reported to be inappropriate for SFP.
RQ2.4	Which data sets are used for SFP?	Identify data sets reported to be appropriate for SFP.
RQ2.5	Which performance measures are used for SFP?	Identify the measures which can be used for assessing the performance of the ML techniques for SFP.
RQ3	What is the overall performance of the ML techniques for SFP?	Investigate the performance of the ML techniques for SFP.
RQ4	Whether the performance of the ML techniques is better than statistical techniques?	Compare the performance of the ML techniques over statistical techniques for SFP.
RQ5	Are there any ML techniques that significantly outperform other ML techniques?	Assess the performance of the ML techniques over other ML techniques for SFP.
RQ6	What are the strengths and weaknesses of the ML techniques?	Determine the conditions that favor the use of ML techniques.

The following questions address various issues related to SR on the use of the ML techniques for SFP:

- Which ML techniques have been used for SFP?
- Which metrics have been used for SFP?
- What type of data sets have been used for SFP?
- What is the accuracy of the ML techniques for SFP?
- Is the performance of the ML techniques better than the traditional statistical techniques for SFP?

Table 2.3 presents the research questions along with the motivation for SRML. While forming the research questions, the interest of the researchers must be kept in mind. For example, for Masters and PhD student thesis, it is necessary to identify the research relevant to the proposed work so that the current body of knowledge can be formed and the proposed work can be established.

2.3.3 Develop Review Protocol

The development of review protocol is an important step in an SR as it reduces the possibility and risk of research bias in the SR. The development of review protocol involves defining the basic research process and procedures that will be followed during the SR.

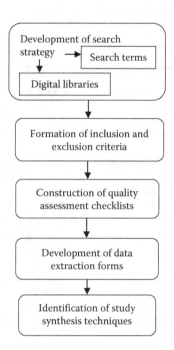

FIGURE 2.2
Steps involved in a review protocol.

In this step, the planning of the search strategy, study selection criteria, quality assessment criteria, data extraction, and data synthesis is carried out.

The purpose of the review must state the options researchers have when deciding which technique or method to adopt in practice. The review protocol is established by frequently holding meetings and group discussions in the group formed comprising of preferably senior members having experience in the area. Hence, this step is iterative and is defined and refined in various iterations. Figure 2.2 shows the steps involved in the development of review protocol.

The first step involves formation of search terms, selection of digital libraries that must be searched, and refinement of search terms. This step allows identification of primary studies that will address the research questions. The initial search terms may be identified by the following steps to form the best suited search string:

- Breaking down the research questions into individual units.
- Using search terms in the titles, keywords, and abstracts of relevant studies.
- Identifying alternative terms and synonyms for the main search terms.

Thereafter, the sophisticated search terms are formed by incorporating alternative terms and synonyms using Boolean expression "OR" and combining main search terms using "AND." The following general search terms were used for identification of primary studies in SRML case study:

Software AND (fault OR defect OR error) AND (proneness OR prone OR prediction OR probability) AND (regression OR ML OR soft computing OR data mining OR classification OR Bayesian network OR neural network [NN] OR decision tree OR support vector machine OR genetic algorithms OR random forest [RF]).

After identifying the search terms, the relevant and important digital portals are to be selected. The portals publishing the journal articles are the right place to search for the relevant studies. The bibliographic databases are also common place of search as they provide title, abstract, and publication source of the study. The selection of digital libraries/portals is very essential, as the number of studies found is dependent on it. Generally, several libraries must be searched to find all the relevant studies that cover the research questions. The selection must not be restricted by the availability of digital portals at the home universities. For example, the following seven electronic digital libraries may be searched for the identification of primary studies:

1. IEEE Xplore
2. ScienceDirect
3. ACM Digital Library
4. Wiley Online Library
5. Google Scholar
6. SpringerLink
7. Web of Science

The reference section of the relevant studies must also be examined/scanned to identify the other relevant studies. The external experts in the areas may also be contacted in this regard.

The next step is to establish the inclusion and exclusion criteria for the SR. The inclusion and exclusion criteria allow the researchers to decide whether to include or exclude the study in the SR. The inclusion and exclusion criteria are based on the research questions. For example, the studies that use data collected from university software developed by student programmers or experiments conducted by students may be excluded from the SR. Similarly, the studies that do not perform any empirical analysis on the techniques and technologies that are being examined in the SR may be excluded. Hence, the inclusion criteria may be specific to the type of tool, technique, or technology being explored in the SR. The data on which the study was conducted or the type of empirical data being used (academia or industry/small, medium, or large sized) may also affect the inclusion criteria.

The following inclusion and exclusion criteria were formed in SRML review:

Inclusion criteria:
- Empirical studies using the ML techniques for SFP.
- Empirical studies combining the ML and non-ML techniques.
- Empirical studies comparing the ML and statistical techniques.

Exclusion criteria:
- Studies without empirical analysis or results of use of the ML techniques for SFP.
- Studies based on fault count as dependent variable.
- Studies using the ML techniques in context other than SFP.
- Similar studies, that is, studies by the same author in conference as well-extended version in journal. However, if the results were different in both the studies, they were retained.
- Studies that only use statistical techniques for SFP.
- Review studies.

The above inclusion and exclusion criteria were applied on each relevant study tested by two researchers independently, and they reached a common decision after detailed discussion. In case of any doubt, full text of a study was reviewed and final decision regarding the inclusion/exclusion of the study was made. Hence, more than one reviewer should check the relevance of a study based on the inclusion and exclusion criteria before a final decision for inclusion or exclusion of a study is made.

The third step in development of a review protocol is to form the quality questionnaire for assessing the relevance and strength of the primary studies. The quality assessment is necessary to investigate and analyze the quality and determine the strength of the studies to be included in final synthesis. It is necessary to limit the bias in the SR and provide guidelines for interpretation of the results.

The assessment criteria must be based on the relevance of a particular study to the research questions and the quality of the processes and methods used in the study. In addition, quality assessment questions must focus on experimental design, applicability of results, and interpretation of results. Some studies may meet the inclusion criteria but may not be relevant with respect to the research design, the way in which data is collected, or may not justify the use of various techniques analyzed. For example, a study on fault proneness may not perform comparative analysis of ML and non-ML techniques.

The quality questionnaire must be constructed by weighing the studies with numerical values. Table 2.4 presents the quality assessment questions for any SR. The studies are rated according to each question and given a score of 1 (yes) if it is satisfactory, 0.5 (partly) if it is moderately satisfactory, and a score of 0 (no) if it is unsatisfactory. The final score is obtained after adding the values assigned to each question. A study could have a maximum score of 10 and a minimum score of 0, if ranked on the basis of quality assessment questions formed in Table 2.4. The studies with low-quality scores may be excluded from the SR or final list of primary studies.

In addition to the questions given in Table 2.4, the following four additional questions were formed in SRML review (see Table 2.5). Hence, a researcher may create specific quality assessment questions with respect to the SR.

The quality score along with the level assigned to the study in the example case study SRML taken in this chapter is given in Table 2.6. The reviewers must decide a threshold value for excluding a study from the SR. For example, studies with quality score >9 were considered for further data extraction and synthesis in SRML review.

TABLE 2.4

Quality Assessment Questions

Q#	Quality Questions	Yes	Partly	No
Q1	Are the aims of the research clearly stated?			
Q2	Are the independent variables clearly defined?			
Q3	Is the data set size appropriate?			
Q4	Is the data-collection procedure clearly defined?			
Q5	Is attributes subselection technique used?			
Q6	Are the techniques clearly defined?			
Q7	Are the results and findings clearly stated?			
Q8	Are the limitations of the study specified?			
Q9	Is the research methodology repeatable?			
Q10	Does the study contribute/add to the literature?			

TABLE 2.5

Additional Quality Assessment Questions for SRML Review

Q#	Quality Questions	Yes	Partly	No
Q11	Are the ML techniques justified?			
Q12	Are the performance measures used to assess the SFP models clearly defined?			
Q13	Is there any comparative analysis conducted among statistical and ML techniques?			
Q14	Is there any comparative analysis conducted among different ML techniques?			

TABLE 2.6

Quality Scores for Quality Assessment questions given in Table 2.4

Quality Score	
$9 \leq \text{score} \leq 10$	Very high
$8 \leq \text{score} \leq 6$	High
$5 \leq \text{score} \leq 4$	Medium
$0 \leq \text{score} \leq 3$	Low

The next step is to construct data extraction forms that will help to summarize the information extracted from the primary studies in view of the research questions. The details of which specific research questions are answered by specific primary study are also present in the data extraction form. Hence, one of the aim of the data extraction is to find which primary study addresses which research question for a given study. In many cases, the data extraction forms will extract the numeric data from the primary studies that will help to analyze the results obtained from these primary studies. The first part of the data extraction card summarizes the author name, title of the primary study, and publishing details, and the second part of the data extraction form contains answers to the research questions extracted from a given primary study. For example, the data set details, independent variables (metrics), and the ML techniques are summarized for the SRML case study (see Figure 2.3).

A team of researchers must collect the information from the primary studies. However, because of the time and resource constraints at least two researchers must evaluate the primary studies to obtain useful information to be included in the data extraction card. The results from these two researchers must then be matched and if there is any disagreement between them, then other researchers may be consulted to resolve these disagreements. The researchers must clearly understand the research questions and the review protocol before collecting the information from the primary studies. In case of Masters and PhD students, their supervisors may collect information from the primary studies and then match their results with those obtained by the students.

The last step involves identification of data synthesis tools and techniques to summarize and interpret the information obtained from the primary studies. The basic objective while synthesizing data is to accumulate and combine facts and figures obtained from the selected primary studies to formulate a response to the research questions. Tables and charts may be used to highlight the similarities and differences between the primary studies. The following

Section I
Reviewer name
Author name
Title of publication
Year of publication
Journal/conference name
Type of study
Section II
Data set used
Independent variables
Feature subselection methods
ML techniques used
Performance measures used
Values of accuracy measures
Strengths of ML techniques
Weaknesses of ML techniques

FIGURE 2.3
Data extraction form.

steps need to be followed before deciding the tools and methods to be used for depicting the results of the research questions:

- Decide which studies to include for answering a particular research question.
- Summarize the information obtained by the primary studies.
- Interpret the information depicted by the answer to the research question.

The effects of the results (performance measures) obtained from the primary studies may be analyzed using statistical measures such as mean, median, and standard deviation (SD).

In addition, the outliers present in the results may be identified and removed using various methods such as box plots. We must also use various tools such as bar charts, scatter plots, forest plots, funnel plots, and line charts to visually present the results of the primary studies in the SR. The aggregation of the results from various studies will allow researchers to provide strong and well-acceptable conclusions and may give strong support in proving a point. The data obtained from these studies may be quantitative (expressed in the form of numerical measures) or qualitative (expressed in the form of descriptive information/texts). For example, the values of performance measures are quantitative in nature, and the strengths and weaknesses of the ML techniques are qualitative in nature.

A detailed description of the methods and techniques that are identified to represent answers to the established research questions in the SRML case study for SFP using the ML techniques are stated as follows:

- To summarize the number of ML techniques used in primary studies the SRML case study will use a visualization technique, that is, a line graph to depict the number of studies pertaining to the ML techniques in each year, and presented a classification taxonomy of various ML techniques with their major categories and subcategories.

The case study also presented a bar chart that shows the total number of studies conducted for each main category of the ML technique and pie charts that depict the distribution of selected studies into subcategories for each ML category.

- The case study will use counting method to find the feature subselection techniques, useful and not useful metrics, and commonly used data sets for SFP. These subparts will be further aided by graphs and pie charts that showcase the distribution of selected primary studies for metrics usage and data set usage. Performance measures will be summarized with the help of a table and a graph.
- The comparison of the result of the primary studies is shown with the help of a table that compares six performance measures for each ML technique. The box plots will be constructed to identify extreme values corresponding to each performance measure.
- A bar chart will be created to depict and analyze the comparison between the performance of the statistical and ML techniques.
- The strengths and weaknesses of different ML techniques for SFP will be summarized in tabular format.

Finally, the review protocol document may consist of the following sections:

1. Background of the review
2. Purpose of the review
3. Contents
   a. Search strategy
   b. Inclusion and exclusion criteria
   c. Study quality assessment criteria
   d. Data extraction
   e. Data synthesis
4. Review evaluation criteria
5. References
6. Appendix

2.3.4 Evaluate Review Protocol

For the evaluation of review protocol a team of independent reviewers must be formed. The team must frequently hold meetings and group discussions to evaluate the completeness and consistency of the review protocol. The evaluation of review protocol involves the confirmation of the following:

1. Development of appropriate search strings that are derived from research questions
2. Adequacy of inclusion and exclusion criteria
3. Completeness of quality assessment questionnaire
4. Design of data extraction forms that address various research questions
5. Appropriateness of data analysis procedures

Masters and PhD students must present the review protocol to their supervisors for the comments and analysis.

2.4 Methods for Presenting Results

The data synthesis provides a summary of the knowledge gained from the existing studies with respect to a specific research question. The appropriate approach for selecting specific technique for qualitative and quantitative synthesis depends on the type of research question being answered. The narrative synthesis and visual representation can be used to conclude the research results of the SR.

2.4.1 Tools and Techniques

The following tools can be used for summarizing and presenting the resultant information:

1. Tabulation: It is the most common approach for representing qualitative and quantitative data. The description of an approach can be summarized in tabular form. The details of study assessment, study design, outcomes of the measure, and the results of the study can be presented in tables. Each table must be referred and interpreted in the results section.

2. Textual descriptions: They are used to highlight the main findings of the studies. The most important findings/outcomes and comparision results must be emphasized in the review and the less important issues should not be overemphasized in the text.

3. Visual diagrams: There are various diagrams that can be used to present and summarize the findings of the study. Meta-analysis is a statistical method to analyze the results of the independent studies so that generalized conclusions can be produced. The outcomes obtained from a given study can be either binary or continuous.

   a. For binary variables the following effects are of interest:

      i. Relative risk (RR, or risk ratio): Risk measures the strength of the relationship between the presence of an attribute and occurrence of an outcome. RR is having the ratio of samples of a positive outcome in two groups included in a study.

      Table 2.7 shows 2×2 contingency table, where $a_{11}, a_{12}, a_{21},$ and a_{22} represent the number of samples in each group with respect to each outcome.

      Table 2.7 can be used to calculate RR and the results are shown below:

$$\text{Risk}_1 = \frac{a_{11}}{a_{11} + a_{12}}, \text{Risk}_2 = \frac{a_{21}}{a_{21} + a_{22}}$$

TABLE 2.7

Contingency Table for Binary Variable

	Outcome Present	Outcome Absent
Group 1	a_{11}	a_{12}
Group 2	a_{21}	a_{22}

$$RR = \frac{Risk_1}{Risk_2}$$

ii. Odds ratio (OR): It measures the strength of the presence or absence of an event. It is the ratio of odds of an outcome in two groups. It is desired that the value is greater than one. The OR is defined as:

$$Odds_1 = \frac{a_{11}}{a_{12}}, Odds_2 = \frac{a_{21}}{a_{22}}$$

$$OR = \frac{Odds_1}{Odds_2}$$

iii. Risk difference: It is also known as measure of absolute effect. It is calculated as the difference between the observed risks (ratio of number of samples present in an individual group with respect to outcome of interest) in the presence of outcome in two groups. The risk difference is given as:

$$Risk\ difference = Risk_1 - Risk_2$$

iv. Area under the ROC curve (AUC): It is obtained from the receiver operating characteristics (ROC) analysis (for details refer Chapter 6) and used to evaluate the models' accuracy by plotting sensitivity and 1-specificity at various cutoff points.

Consider the example given in Table 2.8, it shows the contingency table for classes that are coupled or not coupled in a software with respect to the faulty or nonfaulty binary outcomes.

The values of RR, OR, and risk difference are given below:

$$Risk_1 = \frac{31}{4+31} = 0.885,\ Risk_2 = \frac{4}{4+99} = 0.038$$

$$RR = \frac{0.885}{0.038} = 23.289$$

$$Odds_1 = \frac{31}{4} = 7.75,\ Odds_2 = \frac{4}{99} = 0.04$$

$$OR = \frac{7.75}{0.04} = 193.75$$

$$Risk\ difference = 0.885 - 0.038 = 0.847$$

TABLE 2.8

Example Contingency Table for Binary Variable

	Faulty	Not Faulty	Total
Coupled	31	4	35
Not coupled	4	99	103
Total	35	103	138

b. For continuous variables (variables that do not have any specified range), the following commonly used effects are of interest:

   i. Mean difference: This measure is used when a study reports the same type of outcome and measures them on the same scale. It is also known as "difference of means." It represents the difference between the mean value of each group (Kictenham 2007). Let $\overline{X}_{g1}$ and $\overline{X}_{g2}$ be the mean of two groups (say g1 and g2), which is defined as:

$$\text{Mean difference} = \overline{X}_{g1} - \overline{X}_{g2}$$

   ii. Standardized mean difference: It is used when a study reports the same type of outcome measure but measures it in different ways. For example, the size of a program may be measured by function points or lines of code. Standardized mean difference is defined as the ratio of difference between the means in two groups to the SD of the pooled outcome. Let SD_{pooled} be the SD pooled across groups, SD_{g1} be the SD of one group, SD_{g2} be the SD of another group, and n_{g1} and n_{g2} be the sizes of the two groups. The formula for standardized mean difference is given below:

$$\text{Standardized mean difference} = \frac{\overline{X}_{g1} - \overline{X}_{g2}}{SD_{pooled}}$$

where

$$SD_{pooled} = \sqrt{\frac{(n_{g1} - 1)SD_{g1}^2 + (n_{g2} - 1)SD_{g2}^2}{n_{g1} + n_{g2} - 2}}$$

For example, let $\overline{X}_{g1} = 110$, $\overline{X}_{g2} = 100$, $SD_{g1} = 5$ and $SD_{g2} = 4$, and $n_{g1} = 20$ and $n_{g2} = 20$ of a sample population. Then,

$$SD_{pooled} = \sqrt{\frac{(20-1) \times 5^2 + (20-1) \times 4^2}{20 + 20 - 2}} = 4.527$$

$$\text{Standardized mean difference} = \frac{110 - 100}{4.527} = 2.209$$

Example 2.1

Consider the following data (refer Table 2.9) consisting of an attribute data class that can have binary values true or false, where true represents that the class is data intensive (number of declared variables is high) and false represents that the class is not data intensive (number of declared variables is low). The outcome variable is change that contains "yes" and "no," where "yes" represents presence of change and "no" represents absence of change.

Calculate RR, OR, and risk difference.

Solution

The 2 × 2 contingency table is given in Table 2.10.

$$\text{Risk}_1 = \frac{6}{6+2} = 0.75, \ \text{Risk}_2 = \frac{1}{1+6} = 0.142$$

TABLE 2.9

Sample Data

Data Class	Change
False	No
False	No
True	Yes
False	Yes
True	Yes
False	No
False	No
True	Yes
True	No
False	No
False	No
True	Yes
True	No
True	Yes
True	Yes

TABLE 2.10

Contingency Table for Example Data Given in Table 2.9

Data Class	Change Present	Change Not Present	Total
True	6	2	8
False	1	6	7
Total	7	8	15

$$RR = \frac{0.75}{0.142} = 5.282$$

$$Odds_1 = \frac{6}{2} = 3, \ Odds_2 = \frac{1}{6} = 0.17$$

$$OR = \frac{3}{0.17} = 17.647$$

$$Risk\ difference = 0.75 - 0.142 = 0.608$$

2.4.2 Forest Plots

The forest plot provides a visual assessment of estimate of the overall results of the studies. These overall results may be OR, RR, or AUC in each of the independent studies in the SR or meta-analysis. The confidence interval (CI) of each effect along with overall combined effect at 95% level is computed from the available data. The effects model can be either fixed or random. The fixed effects model assumes that there is a common effect in the

TABLE 2.11

Results of Five Studies

Study	AUC	Standard Error	95% CI
Study 1	0.721	0.025	0.672–0.770
Study 2	0.851	0.021	0.810–0.892
Study 3	0.690	0.008	0.674–0.706
Study 4	0.774	0.017	0.741–0.807
Study 5	0.742	0.031	0.681–0.803
Total (fixed effects)	0.722	0.006	0.709–0.734
Total (random effects)	0.755	0.025	0.705–0.805

studies, whereas in the random effects model there are varied effects in the studies. When heterogeneity is found in the effects, then the random effects model is preferred.

Table 2.11 presents the AUC computed from the ROC analysis, standard error, and upper bound and lower bound of CI. Figure 2.4 depicts the forest plot for five studies using AUC and standard error. Each line represents each study in the SR. The boxes (black-filled squares) depict the weight assigned to each study. The weight is represented as the inverse of the standard error. The lesser the standard error, the more weight is assigned to the study. Hence, in general, weights can be based on the standard error and sample size. The CI is represented through length of lines. The diamond represents summary of combined effects of all the studies, and the edges of the diamond represent the overall effect. The results show the presence of heterogeneity, hence, random effects models is used to analyze the overall accuracy in terms of AUC ranging from 0.69 to 0.85.

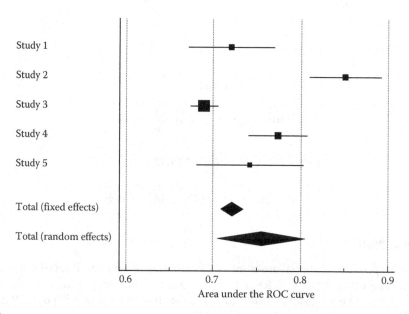

FIGURE 2.4
Forest plots.

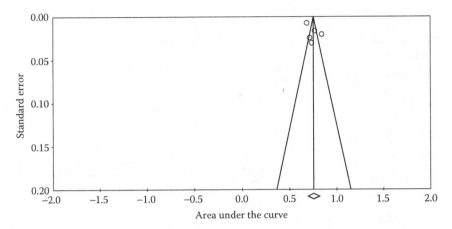

FIGURE 2.5
Funnel plot.

2.4.3 Publication Bias

Publication bias means that the probability of finding studies with positive results is more as compared to the negative results, which are inconclusive. Dickersin et al. found that the possibility of publication of statistically significant results is three times greater than the inconclusive results. The major reason for rejection of a research paper is its inability to produce significant results that can be published. The funnel plot depicts a plot of effect on the horizontal axis and the study size measure (generally standard error) on the vertical axis. The funnel plot can be used to analyze the publication bias and is shown in Figure 2.5. Figure 2.5 presents the plot of effect size against the standard error. If the publication bias is not present, the funnel plot will be like a symmetrical, inverted funnel in which the studies are distributed symmetrically around the combined size of effect. In Figure 2.5, the funnel plot is shown for five studies in which the AUC curve represents the effect size. As shown in funnel plot, all the studies (represented by circles) cluster on the top of the plot, which indicates the presence of the publication bias. In this case, further analysis of the studies lying in the outlying part of the asymmetrical funnel plot is done.

2.5 Conducting the Review

The review protocol is actually put into practice in this phase, including conducting search, selecting primary studies (see Figure 2.6), filling data extraction forms, and data synthesis.

2.5.1 Search Strategy Execution

This step involves comprehensive search of relevant primary studies that meet the search criteria formed from the research questions in the review protocol. The search is performed in the digital libraries identified in the review protocol. The search string may be refined according to the initial results of the search. The studies that are gathered from

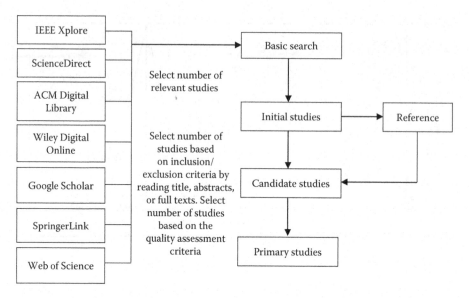

FIGURE 2.6
Search process.

the reference section of the relevant papers must also be included. The multiple copies of the same publications must be removed and the collected publications must be stored in a reference management system, such as Mendeley and JabRef. The list of journals and conferences in which the primary studies have been published must be created. Table 2.12 shows some popular journals and conferences in software engineering.

TABLE 2.12

Popular Journals and Conferences on Software Engineering

Publication Name	Type
IEEE Transactions on Software Engineering	Journal
Journal of Systems and Software	Journal
Empirical Software Engineering	Journal
Information and Software Technology	Journal
IEEE International Symposium on Software Reliability	Conference
International Conference on Predictor Models in Software Engineering (PROMISE)	Conference
International Conference on Software Engineering	Conference
Software Quality Journal	Journal
Automated Software Engineering	Journal
SW Maintenance & Evolution—Research & Practice	Journal
Expert Systems with Applications	Journal
Software Verification, Validation & Testing	Journal
IEEE Software	Journal
Software Practice & Experience	Journal
IET Software	Journal
ACM Transactions on Software Engineering and Methodology	Journal

2.5.2 Selection of Primary Studies

The primary studies are selected based on the established inclusion and exclusion criteria. The selection criteria may be revised during the process of selection, as all the aspects are not apparent in the planning phase. It is advised that two or more researchers must explore the studies to determine their relevance.

The process must begin with the removal of obvious irrelevant studies. The titles, abstracts, or full texts of the collected studies need to be analyzed to identify the primary studies. In some cases, only the title or abstract may be enough to detect the relevance of the study, however, in other cases, the full texts need to be obtained to determine the relevance. Brereton et al. (2007) observed in his study, "The standard of IT and software engineering abstracts is too poor to rely on when selecting primary studies. You should also review the conclusions."

2.5.3 Study Quality Assessment

The studies selected are assigned quality scores based on the quality questions framed in Section 2.3.3. On the basis of the final scores, decision of whether or not to retain the study in the final list of relevant studies is made.

The record of studies that were considered as candidate for selection but were removed after applying thorough inclusion/exclusion criteria must be maintained along with the reasons of rejection.

2.5.4 Data Extraction

After the selection of primary studies, the information from the primary studies is collected in the data extraction forms. The data extraction form was designed during the planning phase and is based on the research questions. The data extraction forms consist of numerical values, weaknesses and strengths of techniques used in studies, CIs, and so on. Brereton et al. (2007) suggested that the following guidelines may be followed during data extraction:

- When large number of primary studies is present, two independent reviewers may be used, one as data collector and the other as a data checker.
- The review protocol and data extraction forms must be clearly understood by the reviewers.

Table 2.13 shows an example of data extraction form collected for SRML case study using research results given by Dejager et al. (2013). A similar form can be made for all the primary studies.

2.5.5 Data Synthesis

The tables and charts are used to summarize the results of the SR. The qualitative results are summarized in tabular form and quantitative results are presented in the form of tables and plots.

In this section, we summarize some of the results obtained by examining the results of SRML case study. Each research question given in Table 2.3 should be answered in

TABLE 2.13

Example Data Extraction Form

Section I	
Reviewer name	Ruchika Malhotra
Author name	Karel Dejaeger, Thomas Verbraken, and Bart Baesens
Title of publication	Toward Comprehensible Software Fault Prediction Models Using Bayesian Network Classifiers
Year of publication	2013
Journal/conference name	*IEEE Transactions on Software Engineering*
Type of the study	Research paper

Section II				
Data set used	NASA data sets (JM1, MC1, KC1, PC1, PC2, PC3, PC4, PC5), Eclipse			
Independent variables	Static code measures (Halstead and McCabe)			
Feature subselection method	Markov Blanket			
ML techniques used	Naïve Bayes, Random Forest			
Performance measures used	AUC, *H*-measure			
Values of accuracy measures (AUC)	Data	RF		NB
	JM1	0.74	0.74	0.69	0.69
	KC1	0.82	0.8	0.8	0.81
	MC1	0.92	0.92	0.81	0.79
	PC1	0.84	0.81	0.77	0.85
	PC2	0.73	0.66	0.81	0.79
	PC3	0.82	0.78	0.77	0.78
	PC4	0.93	0.89	0.79	0.8
	PC5	0.97	0.97	0.95	0.95
	Ecl 2.0a	0.82	0.82	0.8	0.79
	Ecl 2.1a	0.75	0.73	0.74	0.74
	Ecl 3.0a	0.77	0.77	0.76	0.86
Strengths (Naïve Bayes)	It is easy to interpret and construct			
	Computationally efficient			
Weaknesses (Naïve Bayes)	Performance of model is dependent on attribute selection technique used			
	Unable to discard irrelevant attributes			

TABLE 2.14

Distribution of Studies Across ML Techniques
Based on Classification

Method	# of Studies	Percent
Decision tree	31	47.7
NN	17	26.16
Support vector machine	18	27.7
Bayesian learning	31	47.7
Ensemble learning	12	18.47
Evolutionary algorithm	8	12.31
Rule-based learning	5	7.7
Misc.	16	24.62

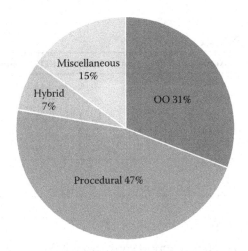

FIGURE 2.7
Primary study distribution according to the metrics used.

the results section by using visual diagrams and tables. For example, Table 2.14 presents the number of studies covering various ML techniques. There are various ML techniques available in the literature such as decision tree, NNs, support vector machine, and bayesian learning. The table shows that 31 studies analyzed decision tree techniques, 17 studies analyzed NN techniques, 18 studies examined support vector machines, and so on. Similarly, the software metrics are divided into various categories in the SRML case study—OO, procedural, hybrid, and miscellaneous. Figure 2.7 depicts the percentage of studies examining each category of metrics, such as 31% of studies examine OO metrics. The pie chart shows that the procedural metrics are most commonly used metrics with 47% of the total primary studies.

The results of the ML techniques that were assessed in at least 5 out of 64 selected primary studies are provided using frequently used performance measures in the 64 primary studies. The results showed that accuracy, *F*-measure, precision, recall, and AUC are the most frequently used performance measures in the selected primary studies. Tables 2.15 and 2.16 present the minimum, maximum, mean, median, and SD values for the selected performance measures. The results are shown for RF and NN techniques.

TABLE 2.15

Results of RF Technique

RF	Accuracy	Precision	Recall	AUC	Specificity
Minimum	55.00	59.00	62.00	0.66	64.3
Maximum	93.40	78.90	100.00	1.00	80.7
Mean	75.63	70.63	81.35	0.83	72.5
Median	75.94	71.515	80.25	0.82	72.5
SD	15.66	7.21	12.39	0.09	11.6

TABLE 2.16

Results of NN Technique

MLP	Accuracy	Precision	Recall	ROC	Specificity
Minimum	64.02	2.20	36.00	0.54	61.60
Maximum	93.44	76.55	98.00	0.95	79.06
Mean	82.23	52.36	69.11	0.78	70.29
Median	83.46	65.29	71.70	0.77	71.11
SD	9.44	27.57	12.84	0.09	5.27

2.6 Reporting the Review

The last step in the SR is to prepare a report consisting of the results of the review and distributing it to the target audience. The results of the SR may be reported in the following:

- Journal or conferences
- Technical report
- PhD thesis

The detailed reporting of the results of the SR is very important and critical so that academicians can have an idea about the quality of the study. The detailed reporting consists of the review protocol, inclusion/exclusion criteria, list of primary studies, list of rejected studies, quality scores assigned to studies, and raw data pertaining to the primary studies, for example, number of research questions addressed by the primary studies and so on should be reported. The review results are generally longer than the normal original study. However, the journals may not permit publication of long SR. Hence, the details may be kept in appendix and stored in electronic form. The details in the form of technical report may also be published online.

Table 2.17 presents the format and contents of the SR. The table provides the contents along with its detailed description. The strengths and limitations of the SR must also be discussed along with the explanation of its effect on the findings.

TABLE 2.17

Format of an SR Report

Section	Subsections	Description	Comments
Title	–		The title should be short and informative.
Authors Details	–		–
Abstract	Background	What is the relevance and importance of the SR?	It allows the researchers to gain insight about the importance, addressed areas, and main findings of the study.
	Method	What are the tools and techniques used to perform the SR?	
	Results	What are the major findings obtained by the SR?	
	Conclusions	What are the main implications of the results and guidelines for the future research?	

(Continued)

TABLE 2.17 (*Continued*)

Format of SR Report

Section	Subsections	Description	Comments
Introduction		What is the motivation and need of the SR?	It will provide the justification for the need of the SR. It also presents the overview of an existing SR.
Method	Research Questions	What are the areas to be addressed during the SR?	The review methods must be based on the review protocol.
			This is the most important part of the SR.
	Search Strategy	What are the relevant studies found during the SR?	It identifies the initial list of relevant studies using the keywords and searching the digital portals.
	Study Selection	What is the inclusion/exclusion criterion for selecting the studies?	It describes the criteria for including and excluding the studies in the SR.
	Quality Assessment Criteria	What are the quality assessment questions that need to be evaluated?	The rejected studies along with the reason of the rejection need to be maintained.
		How will the scores be assigned to the studies?	
		Which studies have been rejected?	
	Data Extraction	What should be the format of the data extraction forms?	The data extraction forms are used to summarize the information from the primary studies.
	Data Synthesis	Which tools are used to present the results of the analysis?	The tools and techniques used to summarize the results of the research are presented in this section.
Results	Description of Primary Studies	What are the primary sources of the selected primary studies?	It summarizes the description of the primary studies.
	Answers to Research Questions	What are the findings of the areas to be explored?	It presents the detailed findings of the SR by addressing the research questions.
			Qualitative findings of the research are summarized in tabular form and quantitative findings are depicted through tables and plots.
	Discussions	What are the applications and meaning of the findings?	It provides the similarities and differences in the results of the primary studies so that the results can be generalized.
			It discusses the risks and effects of the summarized studies.
			The main strengths and weaknesses of the techniques used in the primary studies are summarized in this section.
Threats to Validity		What are the threats to the validity of the results?	The main limitations of the SR are presented in this section.
Conclusions	Summary of Current Trends	What are the implications of the findings for the researchers and practitioners?	It summarizes the main findings and its implications for the practitioners.
	Future Directions	What are the guidelines for future research?	
References		–	It provides references to the primary studies, rejected studies, and referred studies.
Appendix		–	The appendix can present the quality scores assigned to each primary study and the number of research questions addressed by each study.

2.7 SRs in Software Engineering

There are many SRs conducted in software engineering. Table 2.18 summarizes few of them with author details, year and review topics, the number of studies reviewed (study size), whether quality assessment of the studies was performed (QA used), data synthesis methods, and conclusions.

TABLE 2.18

Systematic Reviews in Software Engineering

Authors	Year	Research Topics	Study Size	QA Used	Data Synthesis Methods	Conclusions
Kitchenham et al.	2007	Cost estimation models, cross-company data, within-company data	10	Yes	Tables	• Strict quality control on data collection is not sufficient to ensure that a cross-company model performs as well as a within-company model. • Studies where within-company predictions were better than cross-company predictions employed smaller within-company data sets, smaller number of projects in the cross-company models, and smaller databases.
Jørgensen and Shepperd	2007	Cost estimation	304	No	Tables	• Increase the breadth of the search for relevant studies. • Search manually for relevant papers within a carefully selected set of journals. • Conduct more studies on the estimation methods commonly used by the software industry. • Increase the awareness of how properties of the data sets impact the results when evaluating estimation methods.
Stol et al.	2009	Open source software (OSS)–related empirical research	63	No	Pie charts, bar charts, tables	• Most research is done on OSS communities. • Most studies investigate projects in the "system" and "internet" categories. • Among research methods used, case study, survey, and quantitative analysis are the most popular.

(Continued)

TABLE 2.18 (*Continued*)

Systematic Reviews in Software Engineering

Authors	Year	Research Topics	Study Size	QA Used	Data Synthesis Methods	Conclusions
Riaz et al.	2009	Software maintainability prediction	15	Yes	Tables	• Maintainability prediction models are based on algorithmic techniques. • Most commonly used predictors are based on size, complexity, and coupling. • Prediction techniques, accuracy measures, and cross-validation methods are not much used for validating prediction models. • Most commonly used maintainability metric employed an ordinal scale and is based on expert judgment.
Hauge et al.	2010	OSS, organizations	112	No	Bar charts, tables	• Practitioners should use the opportunities offered by OSS. • Researchers should conduct more empirical research on the topics important to organizations.
Afzal et al.	2009	Search-based software testing, meta-heuristics	35	Yes	Tables, figures	• Meta-heuristic search techniques (including simulated annealing, tabu search, genetic algorithms, ant colony methods, grammatical evolution, genetic programming, and swarm intelligence methods) are applied for nonfunctional testing of execution time, quality of service, security, usability, and safety.
Wen et al.	2012	Effort estimation, machine learning	84	Yes	Narrative synthesis, tables, pie charts, box plots	• Models predicted using ML methods is close to acceptable level. • Accuracy of ML models is better than non-ML models. • Case-based reasoning and artificial NN methods are more accurate than decision trees.

(*Continued*)

TABLE 2.18 (*Continued*)

Systematic Reviews in Software Engineering

Authors	Year	Research Topics	Study Size	QA Used	Data Synthesis Methods	Conclusions
Catal	2011	Fault prediction, machine learning, and statistical-based approaches	90	No	Theoretical	• Most of the studies used method-level metrics. • Most studies used ML techniques. • Naïve Bayes is a robust machine-learning algorithm.
Radjenović et al.	2013	Fault prediction, software metrics	106	Yes	Line chart, bubble chart	• OO metrics were used nearly twice as often as traditional source code metrics and process metrics. • OO metrics predict better models as compared to size and complexity metrics.
Ding et al.	2014	Software documentation, knowledge-based approach	60	Yes	Tables, line graph, bar charts, bubble chart	• Knowledge capture and representation is the widely used approach in software documentation. • Knowledge retrieval and knowledge recovery approaches are useful but still need to be evaluated.
Malhotra	2015	Fault prediction, ML technique	64	Yes	Tables, line charts, bar charts	• ML techniques show acceptable prediction capability for estimating software Fault Proneness • ML techniques outperformed Logistic regression technique for software fault models predictions • Random forest was superior as compared to all the other ML techniques

Exercises

2.1 What is an SR? Why do we really need to perform SR?

2.2 a. Discuss the advantages of SRs.

   b. Differentiate between a survey and an SR.

2.3 Explain the characteristics and importance of SRs.

TABLE 2.12.1

Contingency Table from Study on change Prediction

	Change Prone	Not Change Prone	Total
Coupled	14	12	26
Not coupled	16	22	38
Total	30	34	64

2.4 a. What are the search strategies available for selecting primary studies? How will you select the digital portals for searching primary studies?

b. What is the criteria for forming a search string?

2.5 What is the criteria for determining the number of researchers for conducting the same steps in an SR?

2.6 What is the purpose of quality assessment criteria? How will you construct the quality assessment questions?

2.7 Why identification of the need for an SR is considered the most important step in planning the review?

2.8 How will you decide on the tools and techniques to be used during the data synthesis?

2.9 What is publication bias? Explain the purpose of funnel plots in detecting publication bias?

2.10 Explain the steps in SRs with the help of an example case study.

2.11 Define the following terms:

a. RR

b. OR

c. Risk difference

d. Standardized mean difference

e. Mean difference

2.12 Given the contingency table for all classes that are coupled or not coupled in a software with respect to a dichotomous variable change proneness, calculate the RR, OR, and risk difference (Table 2.12.1).

Further Readings

A classic study that describes empirical results in software engineering is given by:

L. M. Pickarda, B. A. Kitchenham, and P. W. Jones, "Combining empirical results in software engineering," *Information and Software Technology*, vol. 40, no. 14, pp. 811–821, 1998.

A detailed survey that summarizes approaches that mine software repositories in the context of software evolution is given in:

> H. Kagdi, M. L. Collard, and J. I. Maletic, "A survey and taxonomy of approaches for mining software repositories in the context of software evolution," *Journal of Software Evolution and Maintenance: Research and Practice*, vol. 19, no. 2, pp. 77–131, 2007.

The guidelines for preparing the review protocols are given in:

> "Guidelines for preparation of review protocols," The Campbell Corporation, http://www.campbellcollaboration.org.

A review on the research synthesis performed in SRs is given in:

> D. S. Cruzes, and T. Dybå, "Research synthesis in software engineering: A tertiary study," *Information and Software Technology*, vol. 53, no. 5, pp. 440–455, 2011.

For details on meta-analysis, see the following publications:

> M. Borenstein, L. V. Hedges, J. P. T. Higgins, and H. R. Rothstein, *Introduction to Meta-Analysis*, Wiley, Chichester, 2009.
>
> R. DerSimonian, and N. Laird, "Meta-analysis in clinical trials," *Controlled Clinical Trials*, vol. 7, no. 3, pp. 177–188, 1986.
>
> J. P. T. Higgins, and S. Green, *Cochrane Handbook for Systematic Reviews of Interventions Version 5.1.0*, The Cochrane Collaboration, 2011. Available from www.cochrane-handbook.org.
>
> J. P. Higgins, S. G. Thompson, J. J. Deeks, and D. G. Altman, "Measuring inconsistency in meta-analyses," *British Medical Journal*, vol. 327, no. 7414, pp. 557–560, 2003.
>
> N. Mantel, and W. Haenszel, "Statistical aspects of the analysis of data from the retrospective analysis of disease," *Journal of the National Cancer Institute*, vol. 22, no. 4, pp. 719–748, 1959.
>
> A. Petrie, J. S. Bulman, and J. F. Osborn, "Further statistics in dentistry. Part 8: Systematic reviews and meta-analyses," *British Dental Journal*, vol. 194, no. 2, pp. 73–78, 2003.
>
> K. Ried, "Interpreting and understanding meta-analysis graphs: A practical guide," *Australian Family Physician*, vol. 35, no. 8, pp. 635–638, 2006.

For further understanding on forest and funnel plots, see the following publications:

> J. Anzures-Cabrera, and J. P. T. Higgins, "Graphical displays for meta-analysis: An overview with suggestions for practice," *Research Synthesis Methods*, vol. 1, no. 1, pp. 66–80, 2010.
>
> A. G. Lalkhen, and A. McCluskey, "Statistics V: Introduction to clinical trials and systematic reviews," *Continuing Education in Anaesthesia, Critical Case and Pain*, vol. 18, no. 4, pp. 143–146, 2008.
>
> R. J. Light, and D. B. Pillemer, *Summing Up: The Science of Reviewing Research*, Harvard University Press, Cambridge, 1984.

J. L. Neyeloff, S. C. Fuchs, and L. B. Moreira, "Meta-analyses and Forest plots using a Microsoft excel spreadsheet: Step-by-step guide focusing on descriptive data analysis," *British Dental Journal Research Notes*, vol. 5, no. 52, pp. 1–6, 2012.

An effective meta-analysis of a number of high-quality defect prediction studies is provided in:

M. Shepperd, D. Bowes, and T. Hall, "Researcher bias: The use of machine learning in software defect prediction," *IEEE Transactions on Software Engineering*, vol. 40, no. 6, pp. 603–616, 2014.

3

Software Metrics

Software metrics are used to assess the quality of the product or process used to build it. The metrics allow project managers to gain insight about the progress of software and assess the quality of the various artifacts produced during software development. The software analysts can check whether the requirements are verifiable or not. The metrics allow management to obtain an estimate of cost and time for software development. The metrics can also be used to measure customer satisfaction. The software testers can measure the faults corrected in the system, and this decides when to stop testing.

Hence, the software metrics are required to capture various software attributes at different phases of the software development. Object-oriented (OO) concepts such as coupling, cohesion, inheritance, and polymorphism can be measured using software metrics. In this chapter, we describe the measurement basics, software quality metrics, OO metrics, and dynamic metrics. We also provide practical applications of metrics so that good-quality systems can be developed.

3.1 Introduction

Software metrics can be used to adequately measure various elements of the software development life cycle. The metrics can be used to provide feedback on a process or technique so that better or improved strategies can be developed for future projects. The quality of the software can be improved using the measurements collected by analyzing and assessing the processes and techniques being used.

The metrics can be used to answer the following questions during software development:

1. What is the size of the program?
2. What is the estimated cost and duration of the software?
3. Is the requirement testable?
4. When is the right time to stop testing?
5. What is the effort expended during maintenance phase?
6. How many defects have been corrected that are reported during maintenance phase?
7. How many defects have been detected using a given activity such as inspections?
8. What is the complexity of a given module?
9. What is the estimated cost of correcting a given defect?
10. Which technique or process is more effective than the other?
11. What is the productivity of persons working on a project?
12. Is there any requirement to improve a given process, method, or technique?

The above questions can be addressed by gathering information using metrics. The information will allow software developer, project manager, or management to assess, improve, and control software processes and products during the software development life cycle.

3.1.1 What Are Software Metrics?

Software metrics are used for monitoring and improving various processes and products in software engineering. The rationale arises from the notion that "you cannot control what you cannot measure" (DeMarco 1982). The most essential and critical issues involved in monitoring and controlling various artifacts during software development can be addressed by using software metrics. Goodman (1993) defined software metrics as:

> The continuous application of measurement based techniques to the software development process and its products to supply meaningful and timely management information, together with the use of those techniques to improve that process and its products.

The above definition provides all the relevant details. Software metrics should be collected from the initial phases of software development to measure the cost, size, and effort of the project. Software metrics can be used to ascertain and monitor the progress of the software throughout the software development life cycle.

3.1.2 Application Areas of Metrics

Software metrics can be used in various domains. One of the key applications of software metrics is estimation of cost and effort. The cost and effort estimation models can be derived using the historical data and can be applied in the early phases of software development.

Software metrics can be used to measure the effectiveness of various activities or processes such as inspections and audits. For example, the project managers can use the number of defects detected by inspection technique to assess the effectiveness of the technique. The processes can be improved and controlled by analyzing the values of metrics. The graphs and reports provide indications to the software developers and they can decide in which direction to move.

Various software constructs such as size, coupling, cohesion, or inheritance can be measured using software metrics. The alarming values (thresholds) of the software metrics can be computed and based on these values and then the required corrective actions can be taken by the software developers to improve the quality of the software.

One of the most important areas of application of software metrics is the prediction of software quality attributes. There are many quality attributes proposed in the literature such as maintainability, testability, usability, and reliability. The benefits of developing the quality models is that they can be used by software developers, project managers, and management personnel in the early phases of software development for resource allocation and identification of problematic areas.

Testing metrics can be used to measure the effectiveness of the test suite. These metrics include the number of statements, percentage of statement coverage, number of paths covered in a program graph, number of independent paths in a program graph, and percentage of branches covered.

Software metrics can also be used to provide meaningful and timely information to the management. The software quality, process efficiency, and people productivity can be computed using the metrics. Hence, this information will help the management in making effective decisions. The effective application of metrics can improve the quality of the software and produce software within the budget and on time. The contributions of software metrics in building good-quality system are provided in Section 3.9.1.

3.1.3 Characteristics of Software Metrics

A metric is only relevant if it is easily understood, calculated, valid, and economical:

1. Quantitative: The metrics should be expressible in values.
2. Understandable: The way of computing the metric must be easy to understand.
3. Validatable: The metric should capture the same attribute that it is designed for.
4. Economical: It should be economical to measure a metric.
5. Repeatable: The values should be same if measured repeatedly, that is, can be consistently repeated.
6. Language independent: The metrics should not depend on any language.
7. Applicability: The metric should be applicable in the early phases of software development.
8. Comparable: The metric should correlate with another metric capturing the same feature or concept.

3.2 Measurement Basics

Software metrics should preserve the empirical relations corresponding to numerical relations for real-life entities. For example, for "taller than" empirical relation, ">" would be an appropriate numeric relation. Figure 3.1 shows the steps of defining measures. In the first step, the characteristics for representing real-life entities should be identified. In the third step, the empirical relations for these characteristics are identified. The third step

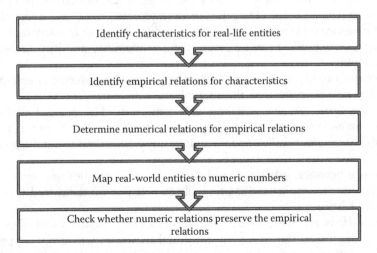

FIGURE 3.1
Steps in software measurement.

determines the numerical relations corresponding to the empirical relations. In the next step, real-world entities are mapped to numeric numbers, and in the last step, we determine whether the numeric relations preserve the empirical relation.

3.2.1 Product and Process Metrics

The entities in software engineering can be divided into two different categories:

1. Process: The process is defined as the way in which the product is developed.
2. Product: The final outcome of following a given process or a set of processes is known as a product. The product includes documents, source codes, or artifacts that are produced during the software development life cycle.

The process uses the product produced by an activity, and a process produces products that can be used by another activity. For example, the software design document is an artifact produced from the design phase, and it serves as an input to the implementation phase. The effectiveness of the processes followed during software development is measured using the process metrics. The metrics related to products are known as product metrics. The efficiency of the products is measured using the product metrics.

The process metrics can be used to

1. Measure the cost and duration of an activity.
2. Measure the effectiveness of a process.
3. Compare the performance of various processes.
4. Improve the processes and guide the selection of future processes.

For example, the effectiveness of the inspection activity can be measured by computing costs and resources spent on it and the number of defects detected during the inspection activity. By assessing whether the number of faults found outweighs the costs incurred during the inspection activity or not, the project managers can decide about the effectiveness of the inspection activity.

The product metrics are used to measure the effectiveness of deliverables produced during the software development life cycle. For example, size, cost, and effort of the deliverables can be measured. Similarly, documents produced during the software development (SRS, test plans, user guides) can be assessed for readability, usability, understandability, and maintainability.

The process and product metrics can further be classified as internal or external attributes. The internal attribute concerns with the internal structure of the process or product. The common internal attributes are size, coupling, and complexity. The external attributes concern with the behavior aspects of the process or product. The external attributes such as testability, understandability, maintainability, and reliability can be measured using the process or product metrics.

The difference between attributes and metrics is that metrics are used to measure a given attribute. For example, size is an attribute that can be measured through lines of source code (LOC) metric.

The internal attributes of a process or product can be measured without executing the source code. For instance, the examples of internal attributes are number of paths, number of branches, coupling, and cohesion. External attributes include quality attributes of the system. They can be measured by executing the source code such as the number of failures,

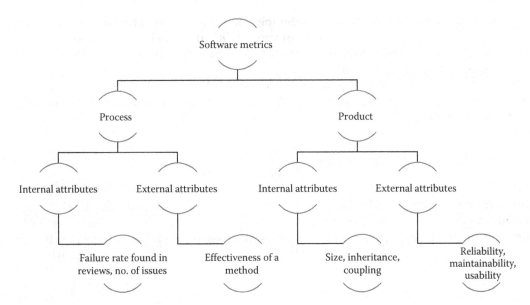

FIGURE 3.2
Categories of software metrics.

response time, and navigation easiness of an item. Figure 3.2 presents the categories of software metrics with examples at the lowest level in the hierarchy.

3.2.2 Measurement Scale

The data can be classified into two types—metric (continuous) and nonmetric (categorical). Metric data is of continuous type that represents the amount of magnitude of a given entity. For example, the number of faults in a class or number of LOC added or deleted during maintenance phase. Table 3.1 shows the LOC added and deleted for the classes A, B, and C.

Nonmetric data is of discrete or categorical type that is represented in the form of categories or classes. For example, weather is sunny, cloudy, or rainy. Metric data can be measured on interval, ratio, or absolute scale. The interval scale is used when the interpretation of difference between values is same. For example, difference between 40°C and 50°C is same as between 70°C and 80°C. In interval scale, one value cannot be represented as a multiple of other value as it does not have an absolute (true) zero point. For example, if the temperature is 20°C, it cannot be said to be twice hotter than when the temperature was 10°C. The reason is that on Fahrenheit scale, 10°C is 50 and 20°C is 68. Hence, ratios cannot be computed on measures with interval scale.

Ratio scales provide more precision as they have absolute zero points and one value can be expressed as a multiple of other. For example, with weight 200 pounds A is twice

TABLE 3.1

Example of Metrics Having Continuous Scale

Class#	LOC Added	LOC Deleted
A	34	5
B	42	10
C	17	9

heavier than B with weight 100 pounds. Simple counts are represented by absolute scale. The example of simple counts is number of faults, LOC, and number of methods. In absolute type of scale, the descriptive statistics such as mean, median, and standard deviation can be applied to summarize data.

Nonmetric type of data can be measured on nominal or ordinal scales. Nominal scale divides metric into classes, categories, or levels without considering any order or rank between these classes. For example, Change is either present or not present in a class.

$$\text{Change} = \begin{cases} 0, \text{ no change present} \\ 1, \text{ change present} \end{cases}$$

Another example of nominal scale is programming languages that are used as labels for different categories. In ordinal scale, one category can be compared with the other category in terms of "higher than," "greater than," or "lower than" relationship. For example, the overall navigational capability of a web page can be ranked into various categories as shown below:

$$\text{What is the overall navigational capability of a webpage?} = \begin{cases} 1, \text{ excellent} \\ 2, \text{ good} \\ 3, \text{ medium} \\ 4, \text{ bad} \\ 5, \text{ worst} \end{cases}$$

Table 3.2 summarizes the differences between measurement scales with examples.

TABLE 3.2

Summary of Measurement Scales

Measurement Scale	Characteristics	Statistics	Operations	Transformation	Examples
Interval	• =, <, > • Ratios not allowed • Arbitrary zero point	Mode, mean, median, interquartile range,	Addition and subtraction	$M = xM' + y$	Temperatures, date, and time
Ratio	• Absolute zero point	variance, standard	All arithmetic operations	$M = xM'$	Weight, height, and length
Absolute	• Simple count values	deviation	All arithmetic operations	$M = M'$	LOC
Nominal	• Order not considered	Frequencies	None	One-to-one mapping	Fault proneness (0—not present, 1—present)
Ordinal	• Order or rank considered • Monotonic increasing function (=, <, >)	Mode, median, interquartile range	None	Increasing function $M(x) > M(y)$	Programmer capability levels (high, medium, low), severity levels (critical, high, medium, low)

Example 3.1

Consider the count of number of faults detected during inspection activity:

1. What is the measurement scale for this definition?
2. What is the measurement scale if number of faults is classified between 1 and 5, where 1 means very high, 2 means high, 3 means medium, 4 means low, and 5 means very low?

Solution:

1. The measurement scale of the number of faults is absolute as it is a simple count of values.
2. Now, the measurement scale is ordinal since the variable has been converted to be categorical (consists of classes), involving ranking or ordering among categories.

3.3 Measuring Size

The purpose of size metrics is to measure the size of the software that can be taken as input by the empirical models to further estimate the cost and effort during the software development life cycle. Hence, the measurement of size is very important and crucial to the success of the project. The LOC metric is the most popular size metric used in the literature for estimation and prediction purposes during the software development. The LOC metric can be counted in various ways. The source code consists of executable lines and unexecutable lines in the form of blank and comment lines. The comment lines are used to increase the understandably and readability of the source code.

The researchers may measure only the executable lines, whereas some may like to measure the LOC with comment lines to analyze the understandability of the software. Hence, the researcher must be careful while selecting the method for counting LOC. Consider the function to check greatest among three numbers given in Figure 3.3.

The function "find-maximum" in Figure 3.3 consists of 20 LOC, if we simply count the number of LOC.

Most researchers and programmers exclude blank lines and comment lines as these lines do not consume any effort and only give the illusion of high productivity of the staff that is measured in terms of LOC/person month (LOC/PM). The LOC count for the function shown in Figure 3.2 is 16 and is computed after excluding the blank and comment lines. The value is computed following the definition of LOC given by Conte et al. (1986):

> A line of code is any line of program text that is not a comment or blank line, regardless of the number of statements or fragments of statements on the line. This specifically includes all lines containing program headers, declarations, and executable and non-executable statements.

In OO software development, the size of software can be calculated in terms of classes and the attributes and functions included in the classes. The details of OO size metrics can be found in Section 3.5.6.

```
/*This function checks greatest amongst three numbers*/
int find-maximum (int i, int j, int k)
        {
                int max;
/*compute the greatest*/
                if(i>j)
                        {
                        if (i<k)
                                max=i;
                        else
                                max=k;
                        }
                        else if (j>k)
                                max=j;
                        else
                                max=k;

/*return the greatest number*/
                        return (max);
        }
```

FIGURE 3.3
Operation to find greatest among three numbers.

3.4 Measuring Software Quality

Maintaining software quality is an essential part of the software development and thus all aspects of software quality should be measured. Measuring quality attributes will guide the software professionals about the quality of the software. Software quality must be measured throughout the software development life cycle phases.

3.4.1 Software Quality Metrics Based on Defects

Defect is defined by IEEE/ANSI as "an accidental condition that causes a unit of the system to fail to function as required" IEEE/ANSI (Standard 982.2). A failure occurs when a fault executes and more than one failure may be associated with a given fault. The defect-based metrics can be classified at product and process levels. The difference of the two terms fault and the defect is unclear from the definitions. In practice, the difference between the two terms is not significant and these terms are used interchangeably. The commonly used product metrics are defect density and defect rate that are used for measuring defects. In the subsequent chapters, we will use the terms fault and defect interchangeably.

3.4.1.1 Defect Density

Defect density metric can be defined as the ratio of the number of defects to the size of the software. Size of the software is usually measured in terms of thousands of lines of code (KLOC) and is given as:

$$\text{Defect density} = \frac{\text{Number of defects}}{\text{KLOC}}$$

The number of defects measure counts the defects detected during testing or by using any verification technique.

Defect rate can be measured as the defects encountered over a period of time, for instance per month. The defect rate may be useful in predicting the cost and resources that will be utilized in the maintenance phase of software development. Defect density during testing is another effective metric that can be used during formal testing. It measures the defect density during the formal testing after completion of the source code and addition of the source code to the software library. If the value of defect density metric during testing is high, then the tester should ask the following questions:

1. Whether the software is well designed or developed?
2. Whether the testing technique is effective in defect detection?

If the reason for high number of defects is the first one then the software should be thoroughly tested to detect the high number of defects. However, if the reason for high number of defects is the second one, it implies that the quality of the system is good because of the presence of fewer defects.

3.4.1.2 Phase-Based Defect Density

It is an extension of defect density metric where instead of calculating defect density at system level it is calculated at various phases of the software development life cycle, including verification techniques such as reviews, walkthroughs inspections, and audits before the validation testing begins. This metric provides an insight about the procedures and standards being used during the software development. Some organizations even set "alarming values" for these metrics so that the quality of the software can be assessed and monitored, thus appropriate remedial actions can be taken.

3.4.1.3 Defect Removal Effectiveness

Defect removal effectiveness (DRE) is defined as:

$$DRE = \frac{\text{Defects removed in a given life cycle phase}}{\text{Latent defects}}$$

For a given phase in the software development life cycle, latent defects are not known. Thus, they are calculated as the estimation of the sum of defects removed during a phase and defects detected later. The higher the value of the DRE, the more efficient and effective is the process followed in a particular phase. The ideal value of DRE is 1. The DRE of a product can also be calculated by:

$$DRE = \frac{D_B}{D_B + D_A}$$

where:
D_B depicts the defects encountered before software delivery
D_A depicts the defects encountered after software delivery

3.4.2 Usability Metrics

The ease of use, user-friendliness, learnability, and user satisfaction can be measured through usability for a given software. Bevan (1995) used MUSIC project to measure usability attributes. There are a number of performance measures proposed in this project and the metrics are defined on the basis of these measures. The task effectiveness is defined as follows:

$$\text{Task effectiveness} = \frac{1}{100} \times (\text{quantity} \times \text{quality})\%$$

where:

Quantity is defined as the amount of task completed by a user

Quality is defined as the degree to which the output produced by the user satisfies the targets of a given task

Quantity and quality measures are expressed in percentages. For example, consider a problem of proofreading an eight-page document. Quantity is defined as the percentage of proofread words, and quality is defined as the percentage of the correctly proofread document. Suppose quantity is 90% and quality is 70%, then task effectiveness is 63%.

The other measures of usability defined in MUSIC project are (Bevan 1995):

$$\text{Temporal efficiency} = \frac{\text{Effectiveness}}{\text{Task time}}$$

$$\text{Productive peroid} = \frac{\text{Task time} - \text{unproductive time}}{\text{Task time}} \times 100$$

$$\text{Relative user efficiency} = \frac{\text{User efficiency}}{\text{Expert efficiency}} \times 100$$

There are various measures that can be used to measure the usability aspect of the system and are defined below:

1. Time for learning a system
2. Productivity increase by using the system
3. Response time

In testing web-based applications, usability can be measured by conducting a survey based on the questionnaire to measure the satisfaction of the customer. The expert having knowledge must develop the questionnaire. The sample size should be sufficient enough to build the confidence level on the survey results. The results are rated on a scale. For example, the difficulty level is measured for the following questions in terms of very easy, easy, difficult, and very difficult. The following questions may be asked in the survey:

- How the user is able to easily learn the interface paths in a webpage?
- Are the interface titles understandable?
- Whether the topics can be found in the 'help' easily or not?

The charts, such as bar charts, pie charts, scatter plots, and line charts, can be used to depict and assess the satisfaction level of the customer. The satisfaction level of the customer must be continuously monitored over time.

3.4.3 Testing Metrics

Testing metrics are used to capture the progress and level of testing for a given software. The amount of testing done is measured by using the test coverage metrics. These metrics can be used to measure the various levels of coverage, such as statement, path, condition, and branch, and are given below:

1. The percentage of statements covered while testing is defined by statement coverage metric.
2. The percentage of branches covered while testing the source code is defined by branch coverage metric.
3. The percentage of operations covered while testing the source code is defined by operation coverage metric.
4. The percentage of conditions covered (both for true and false) is evaluated using condition coverage metric.
5. The percentage of paths covered in a control flow graph is evaluated using condition coverage metric.
6. The percentage of loops covered while testing a program is evaluated using loop coverage metric.
7. All the possible combinations of conditions are covered by multiple coverage metrics.

NASA developed a test focus (TF) metric defined as the ratio of the amount of effort spent in finding and removing "real" faults in the software to the total number of faults reported in the software. The TF metric is given as (Stark et al. 1992):

$$TF = \frac{\text{Number of STRs fixed and closed}}{\text{Total number of STRs}}$$

where:
STR is software trouble report

The fault coverage metric (FCM) is given as:

$$FCM = \frac{\text{Number of faults addressed} \times \text{severity of faults}}{\text{Total number of faults} \times \text{severity of faults}}$$

Some of the basic process metrics used to measure testing are given below:

1. Number of test cases designed
2. Number of test cases executed
3. Number of test cases passed
4. Number of test cases failed
5. Test case execution time
6. Total execution time
7. Time spent for the development of a test case
8. Testing effort
9. Total time spent for the development of test cases

On the basis of above direct measures, the following additional testing-related metrics can be computed to derive more useful information from the basic metrics as given below.

1. Percentage of test cases executed
2. Percentage of test cases passed
3. Percentage of test cases failed
4. Actual execution time of a test case/estimated execution time of a test case
5. Average execution time of a test case

3.5 OO Metrics

Because of growing size and complexity of software systems in the market, OO analysis and design principles are being used by organizations to produce better designed, high–quality, and maintainable software. As the systems are being developed using OO software engineering principles, the need for measuring various OO constructs is increasing.

Features of OO paradigm (programming languages, tools, methods, and processes) provide support for many quality attributes. The key concepts of OO paradigm are: classes, objects, attributes, methods, modularity, encapsulation, inheritance, and polymorphism (Malhotra 2009). An object is made up of three basic components: an identity, a state, and a behavior (Booch 1994). The identity distinguishes two objects with same state and behavior. The state of the object represents the different possible internal conditions that the object may experience during its lifetime. The behavior of the object is the way the object will respond to a set of received messages.

A class is a template consisting of a number of attributes and methods. Every object is the instance of a class. The attributes in a class define the possible states in which an instance of that class may be. The behavior of an object depends on the class methods and the state of the object as methods may respond differently to input messages depending on the current state. Attributes and methods are said to be encapsulated into a single entity. Encapsulation and data hiding are key features of OO languages.

The main advantage of encapsulation is that the values of attributes remain private, unless the methods are written to pass that information outside of the object. The internal working of each object is decoupled from the other parts of the software thus achieving modularity. Once a class has been written and tested, it can be distributed to other programmers for reuse in their own software. This is known as reusability. The objects can be maintained separately leading to easier location and fixation of errors. This process is called maintainability.

The most powerful technique associated to OO methods is the inheritance relationship. If a class B is derived from class A. Class A is said to be a base (or super) class and class B is said to be a derived (or sub) class. A derived class inherits all the behavior of its base class and is allowed to add its own behavior.

Polymorphism (another useful OO concept) describes multiple possible states for a single property. Polymorphism allows programs to be written based only on the abstract interfaces of the objects, which will be manipulated. This means that future extension in the form of new types of objects is easy, if the new objects conform to the original interface.

Nowadays, the software organizations are focusing on software process improvement. This demand led to new/improved approaches in software development area, with perhaps the most promising being the OO approach. The earlier software metrics (Halstead, McCabe, LOCs) were aimed at procedural-oriented languages. The OO paradigm includes new concepts. Therefore, a number of OO metrics to capture the key concepts of OO paradigm have been proposed in literature in the last two decades.

3.5.1 Popular OO Metric Suites

There are a number of OO metric suites proposed in the literature. These metric suites are summarized below. Chidamber and Kemerer (1994) defined a suite of six popular metrics. This suite has received widest attention for predicting quality attributes in literature. The metrics summary along with the construct they are capturing is provided in Table 3.3.

Li and Henry (1993) assessed the Chidamber and Kemerer metrics given in Table 3.3 and provided a metric suite given in Table 3.4.

Bieman and Kang (1995) proposed two cohesion metrics loose class cohesion (LCC) and tight class cohesion (TCC).

Lorenz and Kidd (1994) proposed a suite of 11 metrics. These metrics address size, coupling, inheritance, and so on and are summarized in Table 3.5.

Briand et al. (1997) proposed a suite of 18 coupling metrics. These metrics are summarized in Table 3.6. Similarly, Tegarden et al. (1995) have proposed a large suite of metrics based on variable, object, method and system level. The detailed list can be found in Henderson-Sellers (1996). Lee et al. (1995) has given four metrics, one for measuring cohesion and three metrics for measuring coupling (see Table 3.7).

The system-level polymorphism metrics are measured by Benlarbi and Melo (1999). These metrics are used to measure static and dynamic polymorphism and are summarized in Table 3.8.

TABLE 3.3

Chidamber and Kemerer Metric Suites

Metric	Definition	Construct Being Measured
CBO	It counts the number of other classes to which a class is linked.	Coupling
WMC	It counts the number of methods weighted by complexity in a class.	Size
RFC	It counts the number of external and internal methods in a class.	Coupling
LCOM	Lack of cohesion in methods	Cohesion
NOC	It counts the number of immediate subclasses of a given class.	Inheritance
DIT	It counts the number of steps from the leaf to the root node.	Inheritance

TABLE 3.4

Li and Henry Metric Suites

Metric	Definition	Construct Being Measured
DAC	It counts the number of abstract data types in a class.	Coupling
MPC	It counts a number of unique send statements from a class to another class.	Coupling
NOM	It counts the number of methods in a given class.	Size
SIZE1	It counts the number of semicolons.	Size
SIZE2	It is the sum of number of attributes and methods in a class.	Size

TABLE 3.5

Lorenz and Kidd Metric Suites for measuring Inheritance

Metric	Definition
NOP	It counts the number of immediate parents of a given class.
NOD	It counts the number of indirect and direct subclasses of a given class.
NMO	It counts the number of methods overridden in a class.
NMI	It counts the number of methods inherited in a class.
NMA	It counts the number of new methods added in a class.
SIX	Specialization index

TABLE 3.6

Briand et al. Metric Suites

IFCAIC	These coupling metrics count the number of interactions between classes.
ACAIC	These metrics distinguish the relationship between the classes (friendship, inheritance,
OCAIC	none), different types of interactions, and the locus of impact of the interaction.
FCAEC	The acronyms for the metrics indicates what interactions are counted:
DCAEC	
OCAEC	• The first or first two characters indicate the type of coupling relationship between classes:
IFCMIC	
ACMIC	A: ancestor, D: descendents, F: friend classes, IF: inverse friends (classes that declare a given class A as their friend), O: others, implies none of the other relationships.
DCMIC	• The next two characters indicate the type of interaction:
FCMEC	
DCMEC	CA: There is a class–attribute interaction if class x has an attribute of type class y.
OCMEC	CM: There is a class–method interaction if class x consist of a method that has parameter of type class y.
IFMMIC	
AMMIC	MM: There is a method–method interaction if class x calls method of another class y, or class x has a method of class y as a parameter.
OMMIC	• The last two characters indicate the locus of impact:
FMMEC	
DMMEC	IC: Import coupling, counts the number of other classes called by class x.
OMMEC	EC: Export coupling, count number of other classes using class y.

TABLE 3.7

Lee et al. Metric Suites

Metric	Definition	Construct Being Measured
ICP	Information flow-based coupling	Coupling
IHICP	Information flow-based inheritance coupling	Coupling
NIHICP	Information flow-based noninheritance coupling	Coupling
ICH	Information-based cohesion	Cohesion

Yap and Henderson-Sellers (1993) have proposed a suite of metrics to measure cohesion and reuse in OO systems. Aggarwal et al. (2005) defined two reusability metrics namely function template factor (FTF) and class template factor (CTF) that are used to measure reuse in OO systems. The relevant metrics summarized in tables are explained in subsequent sections.

TABLE 3.8

Benlarbi and Melo Polymorphism Metrics

Metric	Definition
SPA	It measures static polymorphism in ancestors.
DPA	It measures dynamic polymorphism in ancestors.
SP	It is the sum of SPA and SPD metrics.
DP	It is the sum of DPA and DPD metrics.
NIP	It measures polymorphism in noninheritance relations.
OVO	It measures overloading in stand-alone classes.
SPD	It measures static polymorphism in descendants.
DPD	It measures dynamic polymorphism in descendants.

3.5.2 Coupling Metrics

Coupling is defined as the degree of interdependence between modules or classes. It is measured by counting the number of other classes called by a class, during the software analysis and design phases. It increases complexity and decreases maintainability, reusability, and understandability of the system. Thus, interclass coupling must be kept to a minimum. Coupling also increases amount of testing effort required to test classes (Henderson-Sellers 1996). Thus, the aim of the developer should be to keep coupling between two classes as minimum as possible.

Information flow metrics represent the amount of coupling in the classes. Fan-in and fan-out metrics indicate the number of classes collaborating with the other classes:

1. Fan-in: It counts the number of other classes calling class X.
2. Fan-out: It counts the number of classes called by class X.

Figure 3.4 depicts the values for fan-in and fan-out metrics for classes A, B, C, D, E, and F of an example system. The values of fan-out should be as low as possible because of the fact that it increases complexity and maintainability of the software.

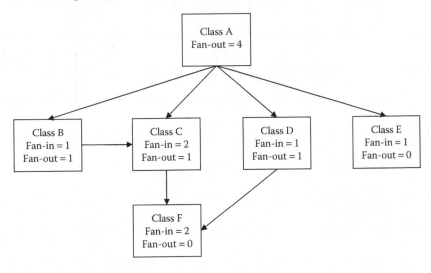

FIGURE 3.4
Fan-in and fan-out metrics

Chidamber and Kemerer (1994) defined coupling as:

> Two classes are coupled when methods declared in one class use methods or instance variables of the other classes.

This definition also includes coupling based on inheritance. Chidamber and Kemerer (1994) defined coupling between objects (CBO) as "the count of number of other classes to which a class is coupled." The CBO definition given in 1994 includes inheritance-based coupling. For example, consider Figure 3.5, three variables of other classes (class B, class C, and class D) are used in class A, hence, the value of CBO for class A is 3. Similarly, classes D, F, G, and H have the value of CBO metric as zero.

Li and Henry (1993) used data abstraction technique for defining coupling. Data abstraction provides the ability to create user-defined data types called abstract data types (ADTs). Li and Henry defined data abstraction coupling (DAC) as:

$$DAC = \text{number of ADTs defined in a class}$$

In Figure 3.5, class A has three ADTs (i.e., three nonsimple attributes). Li and Henry defined another coupling metric known as message passing coupling (MPC) as "number of unique send statements in a class." Hence, if three different methods in class B access the same method in class A, then MPC is 3 for class B, as shown in Figure 3.6.

Chidamber and Kemerer (1994) defined response for a class (RFC) metric as a set of methods defined in a class and called by a class. It is given by RFC = |RS|, where RS, the response set of the class, is given by:

$$RS = I_i \cup all\, j\left\{E_{ij}\right\}$$

where:
I_i = set of all methods in a class (total i)
$R_i = \{R_{ij}\}$ = set of methods called by M_i

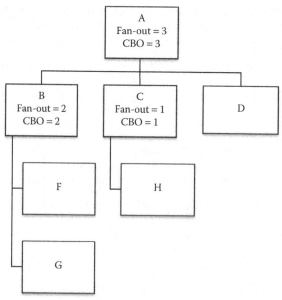

FIGURE 3.5
Values of CBO metric for a small program.

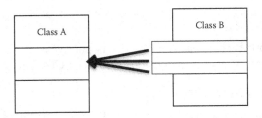

FIGURE 3.6
Example of MPC metric.

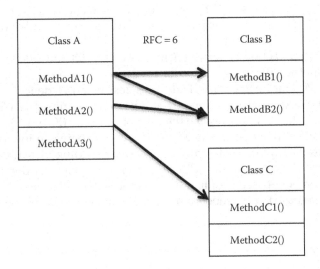

FIGURE 3.7
Example of RFC metric.

For example, in Figure 3.7, RFC value for class A is 6, as class A has three methods of its own and calls 2 other methods of class B and one of class C.

A number of coupling metrics with respect to OO software have been proposed by Briand et al. (1997). These metrics take into account the different OO design mechanisms provided by the C++ language: friendship, classes, specialization, and aggregation. These metrics may be used to guide software developers about which type of coupling affects the maintenance cost and reduces reusability. Briand et al. (1997) observed that the coupling between classes could be divided into different facets:

1. Relationship: It signifies the type of relationship between classes—friendship, inheritance, or other.
2. Export or import coupling (EC/IC): It determines the number of classes calling class A (export) and the number of classes called by class A (import).
3. Type of interaction: There are three types of interactions between classes—class–attribute (CA), class–method (CM), and method–method (MM).
    i. CA interaction: If there are nonsimple attributes declared in a class, the type of interaction is CA. For example, consider Figure 3.8, there are two nonsimple

attributes in class A, B1 of type class B and C1 of type class C. Hence, any changes in class B or class C may affect class A.

ii. CM interaction: If the object of class A is passed as parameter to method of class B, then the type of interaction is said to be CM. For example, as shown in Figure 3.8, object of class B, B1, is passed as parameter to method M1 of class A, thus the interaction is of CM type.

iii. MM interaction: If a method M_i of class K_i calls method M_j of class K_j or if the reference of method M_i of class K_i is passed as an argument to method M_j of class K_j, then there is MM type of interaction between class K_i and class K_j. For example, as shown in Figure 3.8, the method M2 of class B calls method M1 of class A, hence, there is a MM interaction between class B and class A. Similarly, method B1 of class B type is passed as reference to method M3 of class C.

The metrics for CM interaction type are IFCMIC, ACMIC, OCMIC, FCMEC, DCMEC, and OCMEC. In these metrics, the first one/two letters denote the type of relationship (IF denotes inverse friendship, A denotes ancestors, D denotes descendant, F denotes friendship, and O denotes others). The next two letters denote the type of interaction (CA, CM, MM) between classes. Finally, the last two letters denote the type of coupling (IC or EC).

Lee et al. (1995) acknowledged the need to differentiate between inheritance-based and noninheritance-based coupling by proposing the corresponding measures: noninheritance information flow-based coupling (NIH-ICP) and information flow-based inheritance coupling (IH-ICP). Information flow-based coupling (ICP) metric is defined as the sum of NIH-ICP and IH-ICP metrics and is based on method invocations, taking polymorphism into account.

```
class A
{
B B1; // Nonsimple attributes
C C1;
public:
void M1(B B1)
{
}
};
class B
{
public:
void M2()
{
A A1;
A1.M1();// Method of class A called
}
};
class C
{
void M3(B::B1) //Method of class B passed as parameter
{
}
};
```

FIGURE 3.8
Example for computing type of interaction.

3.5.3 Cohesion Metrics

Cohesion is a measure of the degree to which the elements of a module are functionally related to each other. The cohesion measure requires information about attribute usage and method invocations within a class. A class that is less cohesive is more complex and is likely to contain more number of faults in the software development life cycle. Chidamber and Kemerer (1994) proposed lack of cohesion in methods (LCOM) metric in 1994. The LCOM metric is used to measure the dissimilarity of methods in a class by taking into account the attributes commonly used by the methods.

The LCOM metric calculates the difference between the number of methods that have similarity zero and the number of methods that have similarly greater than zero. In LCOM, similarity represents whether there is common attribute usage in pair of methods or not. The greater the similarly between methods, the more is the cohesiveness of the class. For example, consider a class consisting of four attributes (A1, A2, A3, and A4). The method usage of the class is given in Figure 3.9.

There are few problems related to LCOM metric, proposed by Chidamber and Kemerer (1994), which were addressed by Henderson-Sellers (1996) as given below:

1. The value of LCOM metric was zero in a number of real examples because of the presence of dissimilarity among methods. Hence, although a high value of LCOM metric suggests low cohesion, the zero value does not essentially suggest high cohesion.

2. Chidamber and Kemerer (1994) gave no guideline for interpretation of value of LCOM. Thus, Henderson-Sellers (1996) revised the LCOM value. Consider m methods accessing a set of attributes Di $(i = 1,...,n)$. Let $\mu(D_i)$ be the number of methods that access each datum. The revised LCOM1 metric is given as follows:

$$\text{LCOM1} = \frac{(1/N)\sum_{i=1}^{n} \mu(D_i) - m}{1 - m}$$

$M1 = \{A1, A2, A3, A4\}$
$M2 = \{A1, A2\}$
$M3 = \{A3\}$
$M4 = \{A3, A4\}$
$M5 = \{A2\}$
$M1 \cap M2 = 1$
$M1 \cap M3 = 1$
$M1 \cap M4 = 1$
$M1 \cap M5 = 1$
$M2 \cap M3 = 0$
$M2 \cap M4 = 0$
$M2 \cap M5 = 1$
$M3 \cap M4 = 1$
$M3 \cap M5 = 0$
$M4 \cap M5 = 0$
LCOM $= 4 - 6$, Hence, LCOM $= 0$

FIGURE 3.9
Example of LCOM metric.

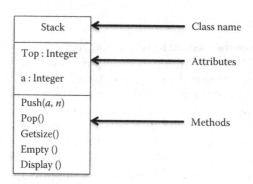

FIGURE 3.10
Stack class.

The approach by Bieman and Kang (1995) to measure cohesion was based on that of Chidamber and Kemerer (1994). They proposed two cohesion measures—TCC and LCC. TCC metric is defined as the percentage of pairs of directly connected public methods of the class with common attribute usage. LCC is the same as TCC, except that it also considers indirectly connected methods. A method M1 is indirectly connected with method M3, if method M1 is connected to method M2 and method M2 is connected to method M3. Hence, transitive closure of directly connected methods is represented by indirectly connected methods. Consider the class stack shown in Figure 3.10.

Figure 3.11 shows the attribute usage of methods. The pair of public functions with common attribute usage is given below:

{(empty, push), (empty, pop), (empty, display), (getsize, push), (getsize, pop), (push, pop), (push, display), (pop, display)}

Thus, TCC for stack class is as given below:

$$TCC(Stack) = \frac{8}{10} \times 100 = 80\%$$

The methods "empty" and "getsize" are indirectly connected, since "empty" is connected to "push" and "getsize" is also connected to "push." Thus, by transitivity, "empty" is connected to "getsize." Similarly "getsize" is indirectly connected to "display."
LCC for stack class is as given below:

$$LCC(Stack) = \frac{10}{10} \times 100 = 100\%$$

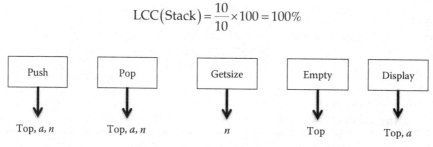

FIGURE 3.11
Attribute usage of methods of class stack.

Lee et al. (1995) proposed information flow-based cohesion (ICH) metric. ICH for a class is defined as the weighted sum of the number of invocations of other methods of the same class, weighted by the number of parameters of the invoked method.

3.5.4 Inheritance Metrics

The inheritance represents parent–child relationship and is measured in terms of number of subclasses, base classes, and depth of inheritance hierarchy by many authors in the literature. Inheritance represents form of reusability. Chidamber and Kemerer (1994) defined depth of inheritance tree (DIT) metric as maximum number of steps from class to root node in a tree. Thus, in case concerning multiple inheritance, the DIT will be counted as the maximum length from the class to the root of the tree. Consider Figure 3.12, DIT for class D and class F is 2.

The average inheritance depth (AID) is calculated as (Yap and Henderson-Sellers 1993):

$$AID = \frac{\sum \text{depth of each class}}{\text{Total number of classes}}$$

In Figure 3.11, the depth of subclass D is 2 ([2 + 2]/2).

The AID of overall inheritance structure is: 0(A) + 1(B) + 1(C) + 2(D) + 0(E) + 1.5(F) + 0(G) = 5.5. Finally, dividing by total number of classes we get 5.5/6 = 0.92.

Chidamber and Kemerer (1994) yet proposed another metric, number of children (NOC), which counts the number of immediate subclasses of a given class in an inheritance hierarchy. A class with more NOC requires more testing. In Figure 3.12, class B has 1 and class C has 2 subclasses. Lorenz and Kidd (1994) proposed number of parents (NOP) metric that counts the number of direct parent classes for a given class in inheritance hierarchy. For example, class D has NOP value of 2. Similarly, Lorenz and Kidd (1994) also developed number of descendants (NOD) metric. The NOD metric defines the number of direct and indirect subclasses of a class. In Figure 3.12, class E has NOD value of 3 (C, D, and F). Tegarden et al. (1992) define number of ancestors (NA) as the number of indirect and direct parent classes of a given class. Hence, as given in Figure 3.12, NA(D) = 4 (A, B, C, and E).

Other inheritance metrics defined by Lorenz and Kidd include the number of methods added (NMA), number of methods overridden (NMO), and number of methods inherited (NMI). NMO counts number of methods in a class with same name and signature as in its

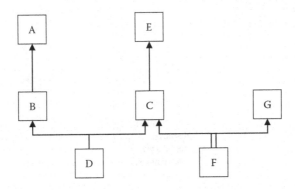

FIGURE 3.12
Inheritance hierarchy.

parent class. NMA counts the number of new methods (neither overridden nor inherited) added in a class. NMI counts number of methods inherited by a class from its parent class. Finally, Lorenz and Kidd (1994) defined specialization index (SIX) using DIT, NMO, NMA, and NMI metrics as given below:

$$SIX = \frac{NMO \times DIT}{NMO + NMA + NMI}$$

Consider the class diagram given in Figure 3.13. The class employee inherits class person. The class employee overrides two functions, addDetails() and display(). Thus, the value of NMO metric for class student is 2. Two new methods is added in this class (getSalary() and compSalary()). Hence, the value of NMA metric is 2.

Thus, for class Employee, the value of NMO is 2, NMA is 2, and NMI is 1 (getEmail()). For the class Employee, the value of SIX is:

$$SIX = \frac{2 \times 1}{2 + 2 + 1} = \frac{2}{5} = 0.4$$

The maximum number of levels in the inheritance hierarchy that are below the class are measured through class to leaf depth (CLD). The value of CLD for class Person is 1.

3.5.5 Reuse Metrics

An OO development environment supports design and code reuse, the most straight-forward type of reuse being the use of a library class (of code), which perfectly suits the

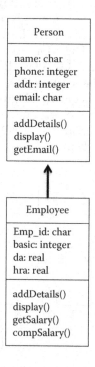

FIGURE 3.13
Example of inheritance relationship.

requirements. Yap and Henderson-Sellers (1993) discuss two measures designed to evaluate the level of reuse possible within classes. The reuse ratio (U) is defined as:

$$U = \frac{\text{Number of superclasses}}{\text{Total number of classes}}$$

Consider Figure 3.13, the value of U is 1/2. Another metric is specialization ratio (S), and is given as:

$$S = \frac{\text{Number of subclasses}}{\text{Number of superclasses}}$$

In Figure 3.13, Employee is the subclass and Person is the parent class. Thus, S = 1.

Aggarwal et al. (2005) proposed another set of metrics for measuring reuse by using generic programming in the form of templates. The metric FTF is defined as ratio of number of functions using function templates to total number of functions as shown below:

$$FTF = \frac{\text{Number of functions using function templates}}{\text{Total number of functions}}$$

Consider a system with methods $F_1,\ldots,F_n$. Then,

$$FTF = \frac{\sum_{i-1}^{n} uses\_FT(F_i)}{\sum_{i-1}^{n} F}$$

where:

$$uses\_FT(F_i) \begin{cases} 1, \text{ iff function uses function template} \\ 0, \text{ otherwise} \end{cases}$$

In Figure 3.14, the value of metric FTF = (1/3).

The metric CTF is defined as the ratio of number of classes using class templates to total number of classes as shown below:

$$CTF = \frac{\text{Number of classes using class templates}}{\text{Total number of classes}}$$

```
void method1(){
.........}
template<class U>
void method2(U &a, U &b){
.........}
void method3(){
.........}
```

FIGURE 3.14
Source code for calculation of FTF metric.

```
class X{
.....};
template<class U, int size>
class Y{
U ar1[size];
....};
```

FIGURE 3.15
Source code for calculating metric CTF.

Consider a system with classes $C_1,\ldots,C_n$. Then,

$$CTF = \frac{\sum_{i-1}^{n} uses\_CT(C_i)}{\sum_{i-1}^{n} C_i}$$

where:

$$uses\_CT(C_i) \begin{cases} 1 \text{iff class uses class template} \\ 0 \text{ otherwise} \end{cases}$$

In Figure 3.15, the value of metric $CTF = \dfrac{1}{2}$.

3.5.6 Size Metrics

There are various conventional metrics applicable to OO systems. The traditional LOC metric measures the size of a class (refer Section 3.3). However, the OO paradigm defines many concepts that require additional metrics that can measure them. Keeping this in view, many OO metrics have been proposed in the literature. Chidamber and Kemerer (1994) developed weighted methods per class (WMC) metric as count of number of methods weighted by complexities and is given as:

$$WMC = \sum_{i=1}^{n} C_i$$

where:
$M_1,\ldots M_n$ are methods defined in class K_1
$C_1,\ldots C_n$ are the complexities of the methods

Lorenz and Kidd defined number of attributes (NOA) metric given as the sum of number of instance variables and number of class variables. Li and Henry (1993) defined number of methods (NOM) as the number of local methods defined in a given class. They also defined two other size metrics—namely, SIZE1 and SIZE2. These metrics are defined below:

SIZE1 = number of semicolons in a class
SIZE2 = sum of NOA and NOM

3.6 Dynamic Software Metrics

The dynamic behavior of the software is captured though dynamic metrics. The dynamic metrics related to coupling, cohesion, and complexity have been proposed in literature. The difference between static and dynamic metrics is presented in Table 3.9 (Chhabra and Gupta 2010).

3.6.1 Dynamic Coupling Metrics

Yacoub et al. (1999) developed a set of metrics for measuring dynamic coupling—namely, export object coupling (EOC) and import object coupling (IOC). These metrics are based on executable code. EOC metric calculates the percentage of ratio of number of messages sent from one object o_1 to another object o_2 to the total number of messages exchanged between o_1 and o_2 during the execution of a scenario. IOC metric calculates percentage of ratio of number of messages sent from object o_2 to o_1 to the total number of messages exchanged between o_1 and o_2 during the execution of a scenario. For example, four messages are sent from object o_1 to object o_2 and three messages are sent from object o_2 to object o_1, $EOC(o_1) = 4/7 \times 100$ and $IOC(o_1) = 3/7 \times 100$.

Arisholm et al. (2004) proposed a suite of 12 dynamic IC and EC metrics. There are six metrics defined at object level and six defined at class level. The first two letters of the metric describe the type of coupling—import or export. The EC represents the number of other classes calling a given class. IC represents number of other classes called by a class. The third letter signifies object or class, and the last letter signifies the strength of coupling (D—dynamic messages, M—distinct method, C—distinct classes). Mitchell and Power developed dynamic coupling metric suite summarized in Table 3.10.

3.6.2 Dynamic Cohesion Metrics

Mitchell and Power (2003, 2004) proposed extension of Chidamber and Kemerer's LCOM metric as dynamic LCOM. They proposed two variations of LCOM metric: runtime simple LCOM (RLCOM) and runtime call-weighted LCOM (RWLCOM). RLCOM considers instance variables accessed at runtime. RWLCOM assigns weights to each instance variable by the number of times it is accessed at runtime.

TABLE 3.9

Difference between static and dynamic metrics

S. No.	Static Metrics	Dynamic Metrics
1	Collected without execution of the program	Collected at runtime execution
2	Easy to collect	Difficult to collect
3	Available in the early phases of software development	Available in later phases of software development
4	Less accurate as compared to dynamic metrics	More accurate
5	Inefficient in dealing with dead code and OO concepts such as polymorphism and dynamic binding	Efficient in dealing with all OO concepts

TABLE 3.10

Mitchell and Power Dynamic Coupling Metric Suite

Metric	Definition
Dynamic coupling between objects	This metric is same as Chidamber and Kemerer's CBO metric, but defined at runtime.
Degree of dynamic coupling between two classes at runtime	It is the percentage of ratio of number of times a class A accesses the methods or instance variables of another class B to the total no of accesses of class A.
Degree of dynamic coupling within a given set of classes	The metric extends the concept given by the above metric to indicate the level of dynamic coupling within a given set of classes.
Runtime import coupling between objects	Number of classes assessed by a given class at runtime.
Runtime export coupling between objects	Number of classes that access a given class at runtime.
Runtime import degree of coupling	Ratio of number of classes assessed by a given class at runtime to the total number of accesses made.
Runtime export degree of coupling	Ratio of number of classes that access a given class at runtime to the total number of accesses made.

3.6.3 Dynamic Complexity Metrics

Determining the complexity of a program is important to analyze its testability and maintainability. The complexity of the program may depend on the execution environment. Munson and Khoshgoftar (1993) proposed dynamic complexity metric.

3.7 System Evolution and Evolutionary Metrics

Software evolution aims at incorporating and revalidating the probable significant changes to the system without being able to predict a priori how user requirements will evolve. The current system release or version can never be said to be complete and continues to evolve. As it evolves, the complexity of the system will grow unless there is a better solution available to solve these issues.

The main objectives of software evolution are ensuring the reliability and flexibility of the system. During the past 20 years, the life span of a system could be on average 6–10 years. However, it was recently found that a system should be evolved once every few months to ensure it is adapted to the real-world environment. This is because of the rapid growth of World Wide Web and Internet resources that make it easier for users to find related information.

The idea of software evolution leads to open source development as anybody could download the source codes and, hence, modify it. The positive impact in this case is that a number of new ideas would be discovered and generated that aims to improve the quality of system with a variety of choices. However, the negative impact is that there is no copyright if a software product has been published as open source.

Over time, software systems, programs as well as applications, continue to develop. These changes will require new laws and theories to be created and justified. Some models would also require additional aspects in developing future programs. Innovations, improvements, and additions lead to unexpected form of software effort in maintenance phase. The maintenance issues also would probably change to adapt to the evolution of the future software.

Software process and development are an ongoing process that has a never-ending cycle. After going through learning and refinements, it is always an arguable issue when it comes to efficiency and effectiveness of the programs.

A software system may be analyzed by the following evolutionary and change metrics (suggested by Moser et al. 2008), which may prove helpful in understanding the evolution and release history of a software system.

3.7.1 Revisions, Refactorings, and Bug-Fixes

The metrics related to refactoring and bug-fixes are defined below:

- *Revisions:* Number of revisions of a software repository file
- *Refactorings:* Number of times a software repository file has been refactored
- *Bug-fixes:* Number of times a file has been associated with bug-fixing
- *Authors:* Number of distinct or different authors who have committed or checked in a software repository

3.7.2 LOC Based

The LOC-based evolution metrics are described as:

- *LOC added:* Sum total of all the lines of code added to a file for all of its revisions in the repository
- *Max LOC added:* Maximum number of lines of code added to a file for all of its revisions in the repository
- *Average LOC added:* Average number of lines of code added to a file for all of its revisions in the repository
- *LOC deleted:* Sum total of all the lines of code deleted from a file for all of its revisions in the repository
- *Max LOC deleted:* Maximum number of lines of code deleted from a file for all of its revisions in the repository
- *Average LOC deleted:* Average number of lines of code deleted from a file for all of its revisions in the repository

3.7.3 Code Churn Based

- *Code churn:* Sum total of (difference between added lines of code and deleted lines of code) for a file, considering all of its revisions in the repository
- *Max code churn:* Maximum code churn for all of the revisions of a file in the repository
- *Average code churn:* Average code churn for all of the revisions of a file in the repository

3.7.4 Miscellaneous

The other related evolution metrics are:

- *Max change set:* Maximum number of files that are committed or checked in together in a repository
- *Average change set:* Average number of files that are committed or checked in together in a repository
- *Age:* Age of repository file, measured in weeks by counting backward from a given release of a software system
- *Weighted Age:* Weighted Age of a repository file is given as:

$$\frac{\sum_{i=1}^{N} \text{Age}(i) \times \text{LOC added}(i)}{\sum \text{LOC added}(i)}$$

where:

> i is a revision of a repository file and N is the total number of revisions for that file

3.8 Validation of Metrics

Several researchers recommend properties that software metrics should posses to increase their usefulness. For instance, Basili and Reiter suggest that metrics should be sensitive to externally observable differences in the development environment, and must correspond to notions about the differences between the software artifacts being measured (Basili and Reiter 1979). However, most recommended properties tend to be informal in the evaluation of metrics. It is always desirable to have a formal set of criteria with which the proposed metrics can be evaluated. Weyuker (1998) has developed a formal list of properties for software metrics and has evaluated number of existing software metrics against these properties. Although many authors (Zuse 1991, Briand et al. 1999b) have criticized this approach, it is still a widely known formal, analytical approach.

Weyuker's (1988) first four properties address how sensitive and discriminative the metric is. The fifth property requires that when two classes are combined their metric value should be greater than the metric value of each individual class. The sixth property addresses the interaction between two programs/classes. It implies that the interaction between program/class A and program/class B is different than the interaction between program/class C and program/class B given that the interaction between program/class A and program/class C. The seventh property requires that a measure be sensitive to statement order within a program/class. The eighth property requires that renaming of variables does not affect the value of a measure. Last property states that the sum of the metric values of a program/class could be less than the metric value of the program/class when considered as a whole (Henderson-Sellers 1996). The applicability of only the properties for OO metrics are given below:

Let u be the metric of program/class P and Q

Property 1: This property states that

$$(\exists P),(\exists Q)\big[u(p)\neq u(Q)\big]$$

It ensures that no measure rates all program/class to be of same metric value.

Property 2: Let c be a nonnegative number. Then, there are finite numbers of program/class with metric c. This property ensures that there is sufficient resolution in the measurement scale to be useful.

Property 3: There are distinct program/class P and Q such that $u(p)=u(Q)$.

Property 4: For OO system, two programs/classes having the same functionality could have different values.

$$(\exists P)(\exists Q)\big[P\equiv Q\,\text{and}\,u(P)\neq(Q)\big]$$

Property 5: When two programs/classes are concatenated, their metric should be greater than the metrics of each of the parts.

$$(\forall P)(\forall Q)\big[u(P)\leq u(P+Q)\,\text{and}\,u(Q)\leq u(P+Q)\big]$$

Property 6: This property suggests nonequivalence of interaction. If there are two program/class bodies of equal metric value which, when separately concatenated to a same third program/class, yield program/class of different metric value.

For program/class P, Q, R

$$(\exists P)(\exists Q)(\exists R)\big[u(P)=u(Q)\,\text{and}\,u(P+R)\neq u(Q+R)\big]$$

Property 7: This property is not applicable for OO metrics (Chidamber and Kemerer 1994).

Property 8: It specifies that "if P is a renaming of Q, then $u(P)=u(Q)$."

Property 9: This property is not applicable for OO metrics (Chidamber and Kemerer 1994).

3.9 Practical Relevance

Empirical assessment of software metrics is important to ensure their practical relevance in the software organizations. Such analysis is of high practical relevance and especially beneficial for large-scale systems, where the experts need to focus their attention and resources to problem areas in the system under development. In the subsequent section, we describe the role of metrics in research and industry. We also provide the approach for calculating metric thresholds.

3.9.1 Designing a Good Quality System

During the entire life cycle of a project, it is very important to maintain the quality and to ensure that it does not deteriorate as a project progresses through its life cycle. Thus, the project manager must monitor quality of the system on a continuous basis. To plan and control quality, it is very important to understand how the quality can be measured. Software metrics are widely used for measuring, monitoring, and evaluating the quality of a project. Various software metrics have been proposed in the literature to assess the software quality attributes such as change proneness, fault proneness, maintainability of a class or module, and so on. A large portion of empirical research has been involved with the development and evaluation of the quality models for procedural and OO software.

Software metrics have found a wide range of applications in various fields of software engineering. As discussed, some of the familiar and common uses of software metrics are scheduling the time required by a project, estimating the budget or cost of a project, estimating the size of the project, and so on. These parameters can be estimated at the early phases of software development life cycle, and thus help software managers to make judicious allocation of resources. For example, once the schedule and budget has been decided upon, managers can plan in advance the amount of person-hours (effort) required. Besides this, the design of software can be assessed in the industry by identifying the out of range values of the software metrics. One way to improve the quality of the system is to relate structural attribute measures intended to quantify important concepts of a given software, such as the following:

- Encapsulation
- Coupling
- Cohesion
- Inheritance
- Polymorphism

to external quality attributes such as the following:

- Fault proneness
- Maintainability
- Testing effort
- Rework effort
- Reusability
- Development effort

The ability to assess quality of software in the early phases of the software life cycle is the main aim of researchers so that structural attribute measures can be used for predicting external attribute measures. This would greatly facilitate technology assessment and comparisons.

Researchers are working hard to investigate the properties of software measures to understand the effectiveness and applicability of the underlying measures. Hence, we need to understand what these measures are really capturing, whether they are really different, and whether they are useful indicators of quality attributes of interest? This will build a body of evidence, and present commonalities and differences across various studies. Finally, these empirical studies will contribute largely in building good quality systems.

3.9.2 Which Software Metrics to Select?

The selection of software metrics (independent variables) in the research is a crucial decision. The researcher must first decide on the domain of the metrics. After deciding the domain, the researcher must decide the attributes to capture in the domain. Then, the popular and widely used software metrics suite available to measure the constructs is identified from the literature. Finally, a decision on the selection must be made on software metrics. The criterion that can be used to select software metrics is that the selected software metrics must capture all the constructs, be widely used in the literature, easily understood, fast to compute, and computationally less expensive. The choice of metric suite heavily depends on the goals of the research. For instance, in quality model prediction, OO metrics proposed by Chidamber and Kemerer (1994) are widely used in the empirical studies.

In cases where multiple software metrics are used, the attribute reduction techniques given in Section 6.2 must be applied to reduce them, if model prediction is being conducted.

3.9.3 Computing Thresholds

As seen in previous sections, there are a number of metrics proposed and there are numerous tools to measure them (see Section 5.8.3). Metrics are widely used in the field of software engineering to identify problematic parts of the software that need focused and careful attention. A researcher can also keep a track of the metric values, which will allow to identify benchmarks across organizations. The products can be compared or rated, which will allow to assess their quality. In addition to this, threshold values can be defined for the metrics, which will allow the metrics to be used for decision making. Bender (1999) defined threshold as "Breakpoints that are used to identify the acceptable risk in classes." In other words, a threshold can be defined as a benchmark or an upper bound such that the values greater than a threshold value are considered to be problematic, whereas the values lower are considered to be acceptable.

During the initial years, many authors have derived threshold values based on their experience and, thus, those values are not universally accepted. For example, McCabe (1976) defined a value of 10 as threshold for the cyclomatic complexity metric. Similarly, for the maintainability index metric, 65 and 85 are defined as thresholds (Coleman et al. 1995). Since these values are based on intuition or experience, it is not possible to generalize results using these values. Besides the thresholds based on intuition, some authors defined thresholds using mean (μ) and standard deviation (σ). For example, Erni and Lewerentz (1996) defined the minimum and maximum values of threshold as $T = \mu + \sigma$ and $T = \mu - \sigma$, respectively. However, this methodology did not gain popularity as it used the assumption that the metrics should be normally distributed, which is not applicable always. French (1999) used Chebyshev's inequality theorem (not restricted to normal distribution) in addition to mean (μ) and standard deviation (σ) to derive threshold values. According to French, a threshold can be defined as $T = \mu + k \times \sigma$ (k = number of standard deviations). However, this methodology was also not used much as it was restricted to only two-tailed symmetric distributions, which is not justified.

A statistical model (based on logistic regression) to calculate the threshold values was suggested by Ulm (1991). Benlarbi et al. (2000) and El Emam et al. (2000b) estimated the threshold values of a number of OO metrics using this model. However, they found that there was no statistical difference between the two models: the model built using the thresholds and the model built without using the thresholds. Bender (1999) working in the epidemiological field found that the proposed threshold model by Ulm (1991) has some drawbacks. The model assumed that the probability of fault in a class is constant when a metric value is below the threshold, and the fault probability increases according to the logistic function, otherwise. Bender (1999) redefined the threshold effects as an acceptable risk level. The proposed threshold methodology was recently used by Shatnawi (2010) to identify the threshold values of various OO metrics. Besides this, Shatnawi et al. (2010) also investigated the use of receiver operating characteristics (ROCs) method to identify threshold values. The detailed explanation of the above two methodologies is provided in the below sub sections (Shatnawi 2006). Malhotra and Bansal (2014a), evaluated the threshold approach proposed by Bender (1999) for fault prediction.

3.9.3.1 Statistical Model to Compute Threshold

The Bender (1999) method known as value of an acceptable risk level (VARL) is used to compute the threshold values, where the acceptable risk level is given by a probability Po (e.g., Po = 0.05 or 0.01). For the classes with metrics values below VARL, the risk of a fault occurrence is lower than the probability (Po). In other words, Bender (1999) has suggested that the value of Po can be any probability, which can be considered as the acceptable risk level.

The detailed description of VARL is given by the formula for VARL as follows (Bender 1999):

$$VARL = \frac{1}{\beta}\left[\ln\left(\frac{Po}{1-Po}\right)-\alpha\right]$$

where:
α is a constant
β is the estimated coefficient
Po is the acceptable risk level

In this formula, α and β are obtained using the standard logistic regression formula (refer Section 7.2.1). This formula will be used for each metric individually to find its threshold value.

For example, consider the following data set (Table A.8 in Appendix I) consisting of the metrics (independent variables): LOC, DIT, NOC, CBO, LCOM, WMC, and RFC. The dependent variable is fault proneness. We calculate the threshold values of all the metrics using the following steps:

Step 1: Apply univariate logistic regression to identify significant metrics.
The formula for univariate logistic regression is:

$$P = \frac{e^{g(x)}}{1+e^{g(x)}}$$

where:

$$g(x) = \alpha + \beta x$$

where:
x is the independent variable, that is, an OO metric
α is the Y-intercept or constant
β is the slope or estimated coefficient

Table 3.11 shows the statistical significance (sig.) for each metric. The "sig." parameter provides the association between each metric and fault proneness. If the "sig."

TABLE 3.11

Statistical Significance of Metrics

Metric	Significance
WMC	**0.013**
CBO	**0.01**
RFC	**0.003**
LOC	**0.001**
DIT	0.296
NOC	0.779
LCOM	**0.026**

value is below or at the significance threshold of 0.05, then the metric is said to be significant in predicting fault proneness (shown in bold). Only for significant metrics, we calculate the threshold values. It can be observed from Table 3.11 that DIT and NOC metrics are insignificant, and thus are not considered for further analysis.

Step 2: Calculate the values of constant and coefficient for significant metrics.

For significant metrics, the values of constant (α) and coefficient (β) using univariate logistic regression are calculated. These values of constant and coefficient will be used in the computation of threshold values. The coefficient shows the impact of the independent variable, and its sign shows whether the impact is positive or negative. Table 3.12 shows the values of constant (α) and coefficient (β) of all the significant metrics.

Step 3: Computation of threshold values.

We have calculated the threshold values (VARL) for the metrics that are found to be significant using the formula given above. The VARL values are calculated for different values of Po, that is, at different levels of risks (between Po = 0.01 and Po = 0.1). The threshold values at different values of Po (0.01, 0.05, 0.08, and 0.1) for all the significant metrics are shown in Table 3.13. It can be observed that the threshold values of all the metrics change significantly as Po changes. This shows that Po plays a significant role in calculating threshold values. Table 3.13 shows that at risk level 0.01 and 0.05, VARL values are out of range (i.e., negative values) for all of the metrics. At Po = 0.1, the threshold values are within the observation range of all the metrics. Hence, in this example, we say that Po = 0.1 is the appropriate risk level and the threshold values (at Po = 0.1) of WMC, CBO, RFC, LOC, and LOCM are 17.99, 14.46, 52.37, 423.44, and 176.94, respectively.

TABLE 3.12

Constant (α) and Coefficient (β) of Significant Metrics

Metric	Coefficient (β)	Constant (α)
WMC	0.06	−2.034
CBO	0.114	−2.603
RFC	0.032	−2.629
LOC	0.004	−2.648
LCOM	0.004	−1.662

TABLE 3.13

Threshold Values on the basis of Logistic Regression Method

Metrics	VARL at 0.01	VARL at 0.05	VARL at 0.08	VARL at 0.1
WMC	−42.69	−15.17	−6.81	17.99
CBO	−17.48	−2.99	1.41	14.46
RFC	−61.41	−9.83	5.86	52.37
LOC	−486.78	−74.11	51.41	423.44
LCOM	−733.28	−320.61	−195.09	176.94

3.9.3.2 Usage of ROC Curve to Calculate the Threshold Values

Shatnawi et al. (2010) calculated threshold values of OO metrics using ROC curve. To plot the ROC curve, we need to define two variables: one binary (i.e., 0 or 1) and another continuous. Usually, the binary variable is the actual dependent variable (e.g., fault proneness or change proneness) and the continuous variable is the predicted result of a test. When the results of a test fall into one of the two obvious categories, such as change prone or not change prone, then the result is a binary variable (1 if the class is change prone, 0 if the class is not change prone) and we have only one pair of sensitivity and specificity. But, in many situations, making a decision in binary is not possible and, thus, the decision or result is given in probability (i.e., probability of correct prediction). Thus, the result is a continuous variable. In this scenario, different cutoff points are selected that make each predicted value (probability) as 0 or 1. In other words, different cutoff points are used to change the continuous variable into binary. If the predicted probability is more than the cutoff then the probability is 1, otherwise it is 0. In other words, if the predicted probability is more than the cutoff then the class is classified as change prone, otherwise it is classified as not change prone.

The procedure of ROC curves is explained in detail in Section 7.5.6, however, we summarize it here to explain the concept. This procedure is carried for various cutoff points, and values of sensitivity and 1-specificity is noted at each cutoff point. Thus, using the (sensitivity, 1-specificity) pairs, the ROC curve is constructed. In other words, ROC curves display the relationship between sensitivity (true-positive rate) and 1-specificity (false-positive rate) across all possible cutoff values. We find an optimal cutoff point, the cutoff point where balance between sensitivity and specificity is provided. This optimal cutoff point is considered as the threshold value for that metric. Thus, threshold value (optimal cutoff point) is obtained for each metric.

For example, consider the data set shown in Table A.8 (given in Appendix I). We need to calculate the threshold values for all the metrics with the help of ROC curve. As discussed, to plot ROC curve, we need a continuous variable and a binary variable. In this example, the continuous variable will be the corresponding metric and the binary variable will be "fault." Once ROC curve is constructed, the optimal cutoff point where sensitivity equals specificity is found. This cutoff point is the threshold of that metric. The thresholds (cutoff points) of all the metrics are given in Table 3.14. When the ROC curve, is constructed the optimal cutoff point is found to be 62. Thus, the threshold value of LOC is 62. This means that if a class has LOC value >62, it is more prone to faults (as our dependent variable in this example is fault proneness) as compared to other classes. Thus, focused attention can be laid on such classes and judicious allocation of resources can be planned.

TABLE 3.14

Threshold Values or the basis of
ROC Curve Method

Metric	Threshold Value
WMC	7.5
DIT	1.5
NOC	0.5
CBO	8.5
RFC	43
LCOM	20.5
LOC	304.5

3.9.4 Practical Relevance and Use of Software Metrics in Research

From the research point of view, the software metrics have a wide range of applications, which help to design a better and much improved quality system:

1. Using software metrics, the researcher can identify change/fault-prone classes that
   a. Enables software developers to take focused preventive actions that can reduce maintenance costs and improve quality.
   b. Helps software managers to allocate resources more effectively. For example, if we have 26% testing resources, then we can use these resources in testing top 26% of classes predicted to be faulty/change prone.
2. Among a large set of software metrics (independent variables), we can find a suitable subset of metrics using various techniques such as correlation-based feature selection, univariate analysis, and so on. These techniques help in reducing the number of independent variables (termed as "data dimensionality reduction"). Only the metrics that are significant in predicting the dependent variable are considered. Once the metrics found to be significant in detecting faulty/change-prone classes are identified, software developers can use them in the early phases of software development to measure the quality of the system.
3. Another important application is that once one knows the metrics being captured by the models, and then such metrics can be used as quality benchmarks to assess and compare products.
4. Metrics also provide an insight into the software, as well as the processes used to develop and maintain it.
5. There are various metrics that calculate the complexity of the program. For example, McCabe metric helps in assessing the code complexity, Halstead metrics helps in calculating different measurable properties of software (programming effort, program vocabulary, program length, etc.), Fan-in and Fan-out metrics estimate maintenance complexity, and so on. Once the complexity is known, more complex programs can be given focused attention.
6. As explained in Section 3.9.3, we can calculate the threshold values of different software metrics. By using threshold values of the metrics, we can identify and focus on the classes that fall outside the acceptable risk level. Hence, during the

project development and progress, we can scrutinize the classes and prepare alternative design structures wherever necessary.

7. Evolutionary algorithms such as genetic algorithms help in solving the optimization problems and require the fitness function to be defined. Software metrics help in defining the fitness function (Harman and Clark 2004) in these algorithms.

8. Last, but not the least, some new software metrics that help to improve the quality of the system in some way can be defined in addition to the metrics proposed in the literature.

3.9.5 Industrial Relevance of Software Metrics

The software design measurements can be used by the software industry in multiple ways: (1) Software designers can use them to obtain quality benchmarks to assess and compare various software products (Aggarwal et al. 2009). (2) Managers can use software metrics in controlling and auditing the quality of the software during the software development life cycle. (3) Software developers can use the software metrics to identify problematic areas and use source code refactoring to improve the internal quality of the software. (4) Software testers can use the software metrics in effective planning and allocation of testing and maintenance resources (Aggarwal et al. 2009). In addition to this, various companies can maintain a large database of software metrics, which allow them to compare a specific company's application software with the rest of the industry. This gives an opportunity to relatively measure that software against its competitors. Comparing the planned or projected resource consumption, code completion, defect rates, and milestone completions against the actual consumption as the work progresses can make an assessment of progress of the software. If there are huge deviations from the expectation, then the managers can take corrective actions before it is too late. Also, to compare the process productivity (can be derived from size, schedule time, and effort [person-months]) of projects completed in a company within a given year against that of projects completed in previous years, the software metrics on the projects completed in a given year can be compared against the projects completed in the previous years. Thus, it can be seen that software metrics contribute in a great way to software industry.

Exercises

3.1 What are software metrics? Discuss the various applications of metrics.

3.2 Discuss categories of software metrics with the help of examples of each category.

3.3 What are categorical metric scales? Differentiate between nominal scale and ordinal scale in the measurements and also discuss both the concepts with examples.

3.4 What is the role and significance of Weyuker's properties in software metrics.

3.5 Define the role of fan-in and fan-out in information flow metrics.

3.6 What are various software quality metrics? Discuss them with examples.

3.7 Define usability. What are the various usability metrics? What is the role of customer satisfaction?

3.8 Define the following metrics:

a. Statement coverage metric

b. Defect density

c. FCMs

3.9 Define coupling. Explain Chidamber and Kemerer metrics with examples.

3.10 Define cohesion. Explain some cohesion metrics with examples.

3.11 How do we measure inheritance? Explain inheritance metrics with examples.

3.12 Define the following metrics:

a. CLD

b. AID

c. NOC

d. DIT

e. NOD

f. NOA

g. NOP

h. SIX

3.13 What is the purpose and significance of computing the threshold of software metrics?

3.14 How can metrics be used to improve software quality?

3.15 Consider that the threshold value of CBO metric is 8 and WMC metric is 15. What does these values signify? What are the possible corrective actions according to you that a developer must take if the values of CBO and WMC exceed these values?

3.16 What are the practical applications of software metrics? How can the metrics be helpful to software organizations?

3.17 What are the five measurement scales? Explain their properties with the help of examples.

3.18 How are the external and internal attributes related to process and product metrics?

3.19 What is the difference between process and product metrics?

3.20 What is the relevance of software metrics in research?

Further Readings

An in-depth study of eighteen different categories of software complexity metrics was provided by Zuse, where he tried to give basic definition for metrics in each category:

H. Zuse, *Software Complexity: Measures and Methods*, Walter De Gryter, Berlin, Germany, 1991.

Fenton's book on software metrics is a classic and useful reference as it provides in-depth discussions on measurement and key concepts related to metrics:

N. Fenton, and S. Pfleeger, *Software Metrics: A Rigorous & Practical Approach*, PWS Publishing Company, Boston, MA, 1997.

The traditional Software Science metrics proposed by Halstead are listed in:

H. Halstead, *Elements of Software Science*, Elsevier North-Holland, Amsterdam, the Netherlands, 1977.

Chidamber and Kemerer (1991) proposed the first significant OO design metrics. Then, another paper by Chidamber and Kemerer defined and validated the OO metrics suite in 1994. This metrics suite is widely used and has obtained widest attention in empirical studies:

S. Chidamber, and C. Kemerer, "A metrics suite for object-oriented design," *IEEE Transactions on Software Engineering*, vol. 20, no. 6, pp. 476–493, 1994.

Detailed description on OO metrics can be obtained from:

B. Henderson-Sellers, *Object Oriented Metrics: Measures of Complexity*, Prentice Hall, Englewood Cliffs, NJ, 1996.

M. Lorenz, and J. Kidd, *Object-Oriented Software Metrics*, Prentice Hall, Englewood Cliffs, NJ, 1994.

The following paper explains various OO metric suites with real-life examples:

K.K. Aggarwal, Y. Singh, A. Kaur, and R. Malhotra,"Empirical study of object-oriented metrics," *Journal of Object Technology*, vol.5, no. 8, pp. 149–173, 2006.

Other relevant publications on OO metrics can be obtained from:

www.acis.pamplin.vt.edu/faculty/tegarden/wrk-pap/ooMETBIB.PDF

Complete list of bibliography on OO metrics is provided at:

"Object-oriented metrics: An annotated bibliography," http://dec.bournemouth.ac.uk/ESERG/bibliography.html.

4

Experimental Design

After the problem is defined, the experimental design process begins. The study must be carefully planned and designed to draw useful conclusions from it. The formation of a research question (RQ), selection of variables, hypothesis formation, data collection, and selection of data analysis techniques are important steps that must be carefully carried out to produce meaningful and generalized conclusions. This would also facilitate the opportunities for repeated and replicated studies.

The empirical study involves creation of a hypothesis that is tested using statistical techniques based on the data collected. The model may be developed using multivariate statistical techniques or machine learning techniques. The steps involved in the experimental design are presented to ensure that proper steps are followed for conducting an empirical study. In the absence of a planned analysis, a researcher may not be able to draw well-formed and valid conclusions. All the activities involved in empirical design are explained in detail in this chapter.

4.1 Overview of Experimental Design

Experimental design is a very important activity, which involves laying down the background of the experiment in detail. This includes understanding the problem, identifying goals, developing various RQ, and identifying the environment. The experimental design phase includes eight basic steps as shown in Figure 4.1. In this phase, an extensive survey is conducted to have a complete overview of all the work done in literature till date. Besides this, the research is formally stated, including a null hypothesis and an alternative hypothesis. The next step in design phase is to determine and define the variables. The variables are of two types: dependent and independent variables. In this step, the variables are identified and defined. The measurement scale should also be defined. This imposes restrictions on the type of data analysis method to be used. The environment in which the experiment will be conducted is also determined, for example, whether the experiment will use data obtained from industry, open source, or university. The procedure for mining data from software repositories is given in Chapter 5. Finally, the data analysis methods to be used for performing the analysis are selected.

4.2 Case Study: Fault Prediction Systems

An example of empirical study is taken to illustrate the experimental process and various empirical concepts. The study will continue in Chapters 6 through 8, wherever required, to help in explaining the concepts. The empirical study is based on predicting severity levels

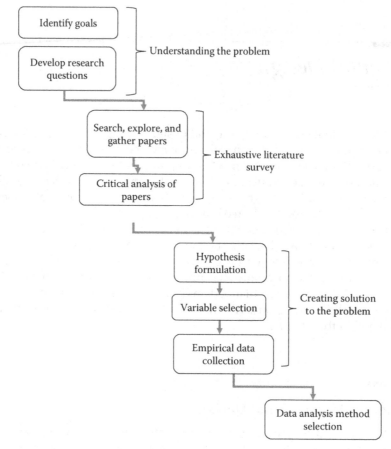

FIGURE 4.1
Steps in experimental design.

of fault and has been published in Singh et al. (2010). Hereafter, the study will be referred to as fault prediction system (FPS). The objective, motivation, and context of the study are described below.

4.2.1 Objective of the Study

The aim of the work is to find the relationship between object-oriented (OO) metrics and fault proneness at different severity levels of faults.

4.2.2 Motivation

The study predicts an important quality attribute, fault proneness during the early phases of software development. Software metrics are used for predicting fault proneness. The important contribution of this study is taking into account of the severity of faults during fault prediction. The value of severity quantifies the impact of the fault on the software operation. The IEEE standard (1044–1993, IEEE 1994) states, "Identifying the severity of an anomaly is a mandatory category as is identifying the project schedule, and project cost impacts of any possible solution for the anomaly." All the failures are not of the same type; they may vary in the impact that they may cause. For example, a failure caused by a fault may lead to a whole system crash or an inability to open a file (El Emam et al. 1999; Aggarwal et al. 2009). In this example, it can be seen that the former failure is more severe than the latter. Lack of determination of severity of faults is one of the main criticisms of the approaches to fault prediction in the study by Fenton and Neil (1999). Therefore, there is a need to develop prediction models that can be used to identify classes that are prone to have serious faults. The software practitioners can use the model predicted with respect to high severity of faults to focus the testing on those parts of the system that are likely to cause serious failures. In this study, the faults are categorized with respect to all the severity levels given in the NASA data set to improve the effectiveness of the categorization and provide meaningful, correct, and detailed analysis of fault data. Categorizing the faults according to different severity levels helps prioritize the fixing of faults (Afzal 2007). Thus, the software practitioners can deal with the faults that are at higher priority first, before dealing with the faults that are comparatively of lower priority. This would allow the resources to be judiciously allocated based on the different severity levels of faults. In this work, the faults are categorized into three levels: high severity, medium severity, and low severity.

Several regression (such as linear and logistic regression [LR]) and machine learning techniques (such as decision tree [DT] and artificial neural network [ANN]) have been proposed in the literature. There are few studies that are using machine learning techniques for fault prediction using OO metrics. Most of the prediction models in the literature are built using statistical techniques. There are many machine learning techniques, and there is a need to compare the results of various machine learning techniques as they give different results. ANN and DT methods have seen an explosion of interest over the years and are being successfully applied across a wide range of problem domains such as finance, medicine, engineering, geology, and physics. Indeed, these methods are being introduced to solve the problems of prediction, classification, or control (Porter 1990; Eftekhar 2005; Duman 2006; Marini 2008). It is natural for software practitioners and potential users to wonder, "Which classification technique is best?," or more realistically, "What methods tend to work well for a given type of data set?" More data-based empirical studies, which are capable of being verified by observation, or experiments are needed. Today, the evidence gathered through these empirical studies is considered to be the most powerful support possible for testing a given hypothesis (Aggarwal et al. 2009). Hence, conducting empirical studies to compare regression and machine learning techniques is necessary to build an adequate body of knowledge to draw strong conclusions leading to widely accepted and well-formed theories.

4.2.3 Study Context

This study uses the public domain data set KC1 obtained from the NASA metrics data program (MDP) (NASA 2004; PROMISE 2007). The independent variables used in the study are various OO metrics proposed by Chidamber and Kemerer (1994), and the dependent variable is fault proneness. The performance of the predicted models is evaluated using receiver operating characteristic (ROC) analysis.

4.2.4 Results

The results show that the area under the curve (measured from the ROC analysis) of models predicted using high-severity faults is low compared with the area under the curve of the model predicted with respect to medium- and low-severity faults.

4.3 Research Questions

The first step in the experimental design is to formulate the RQs. This step states the problem in the form of questions, and identifies the main concepts and relations to be explored.

4.3.1 How to Form RQ?

Most essential aspect of research is formulating an RQ that is clear, simple, and easy to understand. In other words, the scientific process begins after defining RQs. The first question that comes to our mind when doing a research is "What is the need to conduct research (about a particular topic)?" The existing literature can provide answers to questions of researchers. If the questions are not yet answered, the researcher intends to answer those questions and carry forward the research. Thus, this fills the required "gap" by finding the solution to the problem.

A research problem can be defined as a condition that can be studied or investigated through the collection and analysis of data having theoretical or practical significance. Research problem is defined as a part of research for which the researcher is continuously thinking about and wants to find a solution for it. The RQs are extracted from the problem, and the researcher may ask the following questions before framing the RQs:

- What issues need to be addressed in the study?
- Who can benefit from the analysis?
- How can the problem be mapped to realistic terms and measures?
- How can the problem be quantified?
- What measures should be taken to control the problem?
- Are there any unique scenarios for the problem?
- Any expected relationship between causes and outcomes?

Hence, the RQs must fill the gap between existing literature and current work and must give some new perspective to the problem. Figure 4.2 depicts the context of the RQs. The RQ may be formed according to the research types given below:

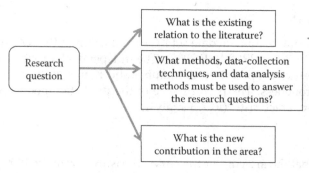

FIGURE 4.2
Context of research questions.

1. Causal relationships: It determines the causal relationships between entities. Does coupling cause increase in fault proneness?
2. Exploratory research: This research type is used to establish new concepts and theories. What are the experiences of programmers using unified modeling language (UML) tool?
3. Explanatory research: This research type provides explanation of the given theories. Why do developers fail to develop good requirement document?
4. Descriptive research: It describes underlying mechanisms and events. How does inspection technique actually work?

Some examples of RQs are as follows:

- Is inspection technique more effective than the walkthrough method in detecting faults?
- Which software development life cycle model is more successful in the software industry?
- Is the new approach effective in reducing testing effort?
- Can search-based techniques be applied to software engineering problems?
- What is the best approach for testing a software?
- Which test data generation technique is effective in the industry?
- What are the important attributes that affect the maintainability of the software?
- Is effort dependent on the programming language, developer's experience, or size of the software?
- Which metrics can be used to predict software faults at the early phases of software development?

4.3.2 Characteristics of an RQ

The following are the characteristics of a good RQ:

1. Clear: The reader who may not be an expert in the given topic should understand the RQs. The questions should be clearly defined.
2. Unambiguous: The use of vague statements that can be interpreted in multiple ways should be avoided while framing RQs. For example, consider the following RQ:

   Are OO metrics significant in predicting various quality attributes?

   The above statement is very vague and can lead to multiple interpretations. This is because a number of quality attributes are present in the literature. It is not clear which quality attribute one wants to consider. Thus, the above vague statement can be redefined in the following way. In addition, the OO metrics can also be specified.

   Are OO metrics significant in predicting fault proneness?
3. Empirical focus: This property requires generating data to answer the RQs.
4. Important: This characteristic requires that answering an RQ adds significant contribution to the research and that there will be beneficiaries.

5. Manageable: The RQ should be answerable, that is, it should be feasible to answer.
6. Practical use: What is the practical application of answering the RQ? The RQ must be of practical importance to the software industry and researchers.
7. Related to literature: The RQ should relate to the existing literature. It should fill gaps in the existing literature.
8. Ethically neutral: The RQ should be ethically neutral. The problem statement should not contain the words "should" or "ought". Consider the following example:

Should the techniques, peer reviews, and walkthroughs be used for verification in contrast to using inspection?

The above statement is said to be not ethically neutral, as it appears that the researcher is favoring the techniques, peer reviews, and walkthroughs in contrast to inspection. This should not be the situation and our question should appear to be neutral by all means.

It could be restated scientifically as follows:

What are the strengths and weaknesses of various techniques available for verification, that is, peer review, walkthrough, and inspection? Which technique is more suitable as compared to other in a given scenario?

Finally, the research problem must be stated in either a declarative or interrogative form. The examples of both the forms are given below:

Declarative form: The present study focuses on predicting change-prone parts of the software at the early stages of software development life cycle. Early prediction of change-prone classes will lead to saving lots of resources in terms of money, manpower, and time. For this, consider the famous Chidamber and Kemerer metrics suite and determine the relationship between metrics and change proneness.

Interrogative form: What are the consequences of predicting the change-prone parts of the software at the early stages of software development life cycle? What is the relationship between Chidamber and Kemerer metrics and change proneness?

4.3.3 Example: RQs Related to FPS

The empirical study given in Section 4.2 addresses some RQs, which it intends to answer. The formulation of such RQs will help the authors to have a clear understanding of the problem and also help the readers to have a clear idea of what the study intends to discover. The RQs are stated below:

- RQ1: Which OO metrics are related to fault proneness of classes with regard to high-severity faults?
- RQ2: Which OO metrics are related to fault proneness of classes with regard to medium-severity faults?
- RQ3: Which OO metrics are related to fault proneness of classes with regard to low-severity faults?
- RQ4: Is the performance of machine learning techniques better than the LR method?

4.4 Reviewing the Literature

Once the research problem is clearly understood and stated, the next step in the initial phases of the experiment design is to conduct an extensive literature review. A literature review identifies the related and relevant research and determines the position of the work being carried out in the specified field.

4.4.1 What Is a Literature Review?

According to Bloomberg and Volpe (2008), literature review is defined as:

> An imaginative approach to searching and reviewing the literature includes having a broad view of the topic; being open to new ideas, methods and arguments; "playing" with different ideas to see whether you can make new linkages; and following ideas to see where they may lead.

The main aim of the research is to contribute toward a better understanding of the concerned field. A literature review analyzes a body of literature related to a research topic to have a clear understanding of the topic, what has already been done on the topic, and what are the key issues that need to be addressed. It provides a complete overview of the existing work in the field. Figure 4.3 depicts various questions that can be answered while conducting a literature review.

The literature review involves collection of research publications (articles, conference paper, technical reports, book chapters, journal papers) on a particular topic. The aim is to gather ideas, views, information, and evidence on the topic under investigation.

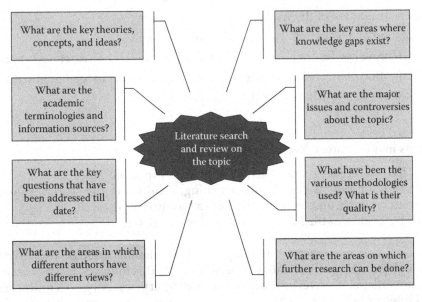

FIGURE 4.3
Key questions while conducting a review.

The purpose of the literature review is to effectively perform analysis and evaluation of literature in relation to the area being explored. The major benefit of the literature review is that the researcher becomes familiar with the current research before commencing his/ her own research in the same area.

The literature review can be carried out by two aspects. The research students perform the review to gain idea about the relevant materials related to their research so that they can identify the areas where more work is required. The literature review carried out as a part of the experimental design is related to the second aspect. The aim is to examine whether the research area being explored is worthwhile or not. For example, search-based techniques have shown the predictive capabilities in various areas where classification problem was of complex nature. But till date, mostly statistical techniques have been explored in software engineering-related problems. Thus, it may be worthwhile to explore the performance capability of search-based techniques in software engineering-related problems. The second aspect of the literature review concerns with searching and analyzing the literature after selecting a research topic. The aim is to gather idea about the current work being carried out by the researcher, whether it has created new knowledge and adds value to the existing research. This type of literature review supports the following claims made by the researcher:

- The research topic is essential.
- The researcher has added some new knowledge to the existing literature.
- The empirical research supports or contradicts the existing results in the literature.

The goals of conducting a literature review are stated as follows:

1. Increase in familiarity with the previous relevant research and prevention from duplication of the work that has already been done.
2. Critical evaluation of the work.
3. Facilitation of development of new ideas and thoughts.
4. Highlighting key findings, proposed methodologies, and research techniques.
5. Identification of inconsistencies, gaps, and contradictions in the literature.
6. Extraction of areas where attention is required.

4.4.2 Steps in a Literature Review

There are four main steps that need to be followed in a literature review. These steps involve identifying digital portals for searching, conducting the search, analyzing the most relevant research, and using the results in the current research.

The four basic steps in the literature review are as follows:

1. Develop search strategy: This step involves identification of digital portals, research journals, and formation of search string. This involves survey of scholarly journal articles, conference articles, proceeding articles, books, technical reports, and Internet resources in various research-related digital portals such as:
   a. IEEE
   b. Springer

c. ScienceDirect/Elsevier

d. Wiley

e. ACM

f. Google Scholar

Before searching in digital portals, the researchers need to identify the most credible research journals in the related areas. For example, in the area of software engineering, some of the important journals in which search can be done are: *Software: Practice and Experience, Software Quality Journal, IEEE Transactions on Software Engineering, Information and Software Technology, Journal of Computer Science and Technology, ACM Transactions on Software Engineering Methodology, Empirical Software Engineering, IEEE Software Maintenance, Journal of Systems and Software, and Software Maintenance and Evolution.*

Besides searching the journals and portals, various educational books, scientific monograms, government documents and publications, dissertations, gray literature, and so on that are relevant to the concerned topic or area of research should be explored. Most importantly, the bibliographies and reference lists of the materials that are read need to be searched. These will give the pointers to more articles and can also be a good estimate about how much have been read on the selected topic of research.

After the digital portals and Internet resources have been identified, the next step is to form the search string. The search string is formed by using the key terms from the selected topic in the research. The search string is used to search the literature from the digital portal.

2. Conduct the search: This step involves searching the identified sources by using the formed search string. The abstracts and/or full texts of the research papers should be obtained for reading and analysis.

3. Analyze the literature: Once the research papers relevant to the research topic have been obtained, the abstract should be read, followed by the introduction and conclusion sections. The relevant sections can be identified and read by the section headings. In case of books, the index must be scanned to obtain an idea about the relevant topics. The materials that are highly relevant in terms of making the greatest contribution in the related research or the material that seems the most convincing can be separated. Finally, a decision about reading the necessary content must be made.

The strengths, drawbacks, and omissions in the literature review must be identified on the basis of the evidence present in the papers. After thoroughly and critically analyzing the literature, the differences of the proposed work from the literature must be highlighted.

4. Use the results: The results obtained from the literature review must then be summarized for later comparison with the results obtained from the current work.

4.4.3 Guidelines for Writing a Literature Review

A literature review should have an introduction section, followed by the main body and the conclusion section. The "introduction" section explains and establishes the importance of the subject under concern. It discusses the kind of work that is done on the concerned topic of research, along with any controversies that may have been encountered by different authors. The "body" contains and focuses on the main idea behind each paper in the review. The relevance of the papers cited should be clearly stated in this section of the review. It is not important to simply restate what the other authors have said, but instead our main aim should be to critically evaluate each paper. Then, the conclusion should be provided that summarizes what the literature says. The conclusion summarizes all the evidence presented and shows its significance. If the review is an introduction to our own research, it indicates how the previous research has lead to our own research focusing and highlighting on the gaps in the previous research (Bell 2005). The following points must be covered while writing a literature review:

- Identify the topics that are similar in multiple papers to compare and contrast different authors' view.
- Group authors who draw similar conclusions.
- Group authors who are in disagreement with each other on certain topics.
- Compare and contrast the methodologies proposed by different authors.
- Show how the study is related to the previous studies in terms of the similarities and the differences.
- Highlight exemplary studies and gaps in the research.

The above-mentioned points will help to carry out effective and meaningful literature review.

4.4.4 Example: Literature Review in FPS

A summary of studies in the literature is presented in Table 4.1. The studies closest to the FPS study are discussed below with key differences.

Zhou and Leung (2006) validated the same data set as in this study to predict fault proneness of models with respect to two categories of faults: high and low. They categorized faults with severity rating 1 as high-severity faults and faults with other severity levels as low-severity faults. They did not consider the faults that originated from design and commercial off-the-shelf (COTS)/operating system (OS). The approach in FPS differs from Zhou and Leung (2006) as this work categorized faults into three severity levels: high, medium, and low. The medium-severity level of faults is more severe than low-severity level of faults. Hence, the classes having faults of medium-severity level must be given more attention compared with the classes with low-severity level of faults. In the study conducted by Zhou and Leung, the classes were not categorized into medium- and low-severity level of faults. Further, the faults produced from the design were not taken into account. The FPS study also analyzed two different machine learning techniques (ANN and DT) for predicting fault proneness of models and evaluated the performance of these models using ROC analysis. Pai and Bechta Dugan (2007) used the same data set using a Bayesian approach to find the relationship of software product metrics to fault content and fault proneness. They did not categorize faults at different severity levels and mentioned that a natural extension to their analysis is severity classification using Bayesian network models. Hence, their work is not comparable with the work in the FPS study.

TABLE 4.1

Literature Review of Fault Prediction Studies

Studies	Empirical Data Collection			Statistical Techniques		
	Language	Environment	Independent Variables	Univariate Analysis	Multivariate Analysis	Predicted Model Evaluation
Basili et al. (1996)	C++	University environment, 180 classes	C&K metrics, 3 code metrics	LR	LR	Contingency table, correctness, completeness
Abreu and Melo (1996)	C++	University environment, UMD: 8 systems	MOOD metrics	Pearson correlation	Linear least square	R^2
Binkley and Schach (1998)	C++, Java	4 case studies, CCS: 113 classes, 82K SLOC, 29 classes, 6K SLOC	CDM, DIT, NOC, NOD, NCIM, NSSR, CBO	Spearman rank correlation	–	–
Harrison et al. (1999)	C++	University environment, SEG1: 16 classes SEG2: 22 classes SEG3: 27 classes	DIT, NOC, NML, NMO	Spearman rho	–	–
Benlarbi and Melt (1999)	C++	LALO: 85 classes, 40K SLOC	OVO, SPA, SPD, DPA, DPD, CHNL, C&K metrics, part of coupling metrics	LR	LR	–
El Emam et al. (2001a)	Java	V0.5: 69 classes, V0.6: 42 classes	Coupling metrics, C&K metrics	LR	LR	R^2, leave one-out cross-validation
El Emam et al. (2001b)	C++	Telecommunication framework: 174 classes	Coupling metrics, DIT	LR	LR	R^2
Tang et al. (1999)	C++	System A: 20 classes System B: 45 classes System C: 27 classes	C&K metrics (without LCOM)	LR	–	–
Briand et al. (2000)	C++	University environment, UMD: 180 classes	Suite of coupling metrics, 49 metrics	LR	LR	R^2, 10 cross-validation, correctness, completeness

(Continued)

TABLE 4.1 (Continued)

Literature Review of Fault Prediction Studies

	Empirical Data Collection			Statistical Techniques		
Studies	Language	Environment	Independent Variables	Univariate Analysis	Multivariate Analysis	Predicted Model Evaluation
Glasberg et al. (2000)	Java	145 classes	NOC, DIT ACAIC, OCAIC, DCAEC, OCAEC	LR	LR	R^2, leave one-out cross-validation, ROC curve, cost-saving model
El Emam et al. (2000a)	C++, Java	Telecommunication framework: 174 classes, 83 classes, 69 classes of Java system	C&K metrics, NOM, NOA	LR	—	—
Briand et al. (2001)	C++	Commercial system, LALO: 90 classes, 40K SLOC	Suite of coupling metrics, OVO, SPA, SPD, DPA, DPD, NIP, SP, DP, 49 metrics	LR	LR	R^2, 10 cross-validation, correctness, completeness
Cartwright and Shepperd (2000)	C++	32 classes, 133K SLOC	ATTRIB, STATES, EVENT, READS, WRITES, DIT, NOC	Linear regression	Linear regression	—
Briand and Wüst (2002)	Java	Commercial system, XPOSE & JWRITER: 144 classes	Polymorphism metrics, C&K	LR	LR, Mars	10 cross-validation, correctness, completeness
Yu et al. (2002)	Java	123 classes, 34K SLOC	C&K metrics, Fan-in, WMC	OLS+LDA	—	—
Subramanyam and Krishnan (2003)	C++, Java		C&K metrics	OLS	OLS	—
Gyimothy et al. (2005)	C++	Mozilla v1.6: 3,192 classes	C&K metrics, LCOMN, LOC	LR, linear regression, NN, DT	LR, linear regression, NN, DT	10 cross-validation, correctness, completeness
Aggarwal et al. (2006a, 2006b)	Java	University environment, 136 classes	Suite of coupling metrics	LR	LR	10 cross-validation, correctness, completeness
Arisholm and Briand (2006)	Java	XRadar and JHawk	C&K metrics	LR	LR	10 cross-validation, sensitivity, specificity

(Continued)

TABLE 4.1 (*Continued*)

Literature Review of Fault Prediction Studies

	Empirical Data Collection			Statistical Techniques		
Studies	Language	Environment	Independent Variables	Univariate Analysis	Multivariate Analysis	Predicted Model Evaluation
Yuming and Hareton (2006)	C++	NASA data set, 145 classes	C&K metrics	LR, ML	LR, ML	Correctness, completeness
Kanmani et al. (2007)	C++	Library management S/w system developed by students, 1,185classes	C&K and Briand metrics: Total 64 (10 cohesion, 18 inheritance, 29 coupling, and 7 size)	PC method	LDA, LR, NN (BPN and PNN)	Type I and II error, correctness, completeness, efficiency, effectiveness
Aggarwal et al. (2009)	Java	12 systems developed by undergraduate at the University School of Information Technology (USIT)	52 metrics (26 coupling, 7 cohesion, 11 inheritance, and 8 size) by C&K (1991, 1994), Li and Henry (1993), Lee et al. (1995), Briand et al. (1999a), Hitz and Montazeri (1995), Bieman and Kang (1995), Tegarden et al. (1995), Henderson-Sellers (1996), Lorenz and Kidd (1994), Lake and Cook (1994)	LR	LR	9 cross-validation
Singh et al. (2010)	C++	Public domain data set KC1 from NASA MDP, 145 classes, 2,107 methods, 40K LOC	C&K metrics, LOC	LR, ML (NN, DT)	LR, ML (NN,DT)	Sensitivity, specificity, precision, completeness
Singh et al. (2009)	C++	Public domain data set KC1 from the NASA MDP, 145 classes, 2107 methods, 40K LOC	C&K metrics, LOC	ML (SVM)	ML (SVM)	Sensitivity, specificity, precision, completeness
Malhotra et al. (2010)	C++	Public domain data set KC1 from the NASA MDP, 145 classes, 2,107 methods, 40K LOC	C&K metrics, LOC	ML (SVM)	ML (SVM)	Sensitivity, specificity, precision, completeness
Zhou et al. (2010)	Java	Three major releases, 2.0, 2.1, and 3.0, with sizes 796, 988, and 1,306K SLOC, respectively	10 metrics by C&K, Michura and Capretz (2005), Etzkorn et al. (1999), Olague et al. (2008), Lorenz and Kidd (1994), Briand et al. (2001)	LR and ML (NB, KStar, Adtree)	LR and ML (NB, KStar, Adtree)	Accuracy, sensitivity, specificity, precision, F-measure

(*Continued*)

TABLE 4.1 (*Continued*)

Literature Review of Fault Prediction Studies

Studies	Empirical Data Collection		Independent Variables	Statistical Techniques		
	Language	Environment		Univariate Analysis	Multivariate Analysis	Predicted Model Evaluation
Di Martino et al. (2011)	Java	Versions 4.0, 4.2, and 4.3 of the jEdit system	C&K, NPM, LOC	–	Combination of GA+SVM, LR, C4.5, NB, MLP, KNN, and RF	Precision, accuracy, recall, *F*-measure
Azar and Vybihad (2011)	Java	8 open source software systems	22 metrics by Henderson-Sellers (2007), Barnes and Swim (1993), Coppick and Cheatham (1992), C&K	–	ACO, C4.5, random guessing	Accuracy
Malhotra and Singh (2011)	–	Open source data set Arc, 234 classes	C&K and QMOOD metrics	LR	LR and ML (ANN, RF, LB, AB, NB, KStar, Bagging)	Sensitivity, specificity, precision
Malhotra and Jain (2012)	Java	Apache POI, 422 classes	MOOD, QMOOD, C&K (19 metrics)	LR	LR, MLP (RF, Bagging, MLP, SVM, genetic algorithm)	Sensitivity, specificity, precision

Source:　Compiled from multiple sources.

–implies that feature not examined.

LR: logistic regression, LDA: linear discriminant analysis, ML: machine learning, OLS: ordinary least square linear regression, PC: principal component analysis, NN: neural network, BPN: back propagation neural network, PPN: probabilistic neural network, DT: decision tree, MLP: multilayer perceptron, SVM: support vector machine, RF: random forest, GA+SVM: combination of genetic algorithm and support vector machine, NB: naïve Bayes, KNN: *k*-nearest neighbor, C4.5: decision tree, ACO: ant colony optimization, Adtree: alternating decision tree, AB: adaboost, LB: logitboost, CHNL: class hierarchy nesting level: NCIM: number of classes inheriting a method, NSSR: number of subsystems-system relationship: NPM, number of public methods: LCOMN, lack of cohesion on methods allowing negative value. Related to metrics: C&K: Chidamber and Kemerer, MOOD: metrics for OO design, QMOOD: quality metrics for OO design.

4.5 Research Variables

Before the detailed experiment design begins, the relevant independent and dependent variables have to be selected.

4.5.1 Independent and Dependent Variables

The variables used in an experiment can be divided into two types: dependent variable and independent variable. While conducting a research, the dependent and the independent variables that are used in the study need to be defined.

In an empirical study, the independent variable is a variable that can be changed or varied to see its effect on the dependent variable. In other words, a dependent variable is a variable that has effect on the independent variables and can be controlled. Thus, the dependent variable is "dependent" on the independent variable. As the experimenter changes the independent variable, the change in the dependent variable is observed. The selection of variables also involves selecting the measurement scale. Any of these variables may be discrete or continuous. A binary variable has only two values. For example, whether a component is faulty or is not faulty. A continuous variable has many values (refer to Chapter 3 for details). In software product metric validation, continuous variables are usually counts. A count is characterized by being a non-negative integer, and hence is a continuous variable. Usually, in empirical studies in software engineering, there is one dependent variable. Figure 4.4 depicts the relationship between the dependent and independent variable. Table 4.2 states the key differences between the independent and dependent variables.

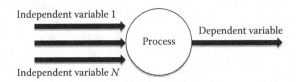

FIGURE 4.4
Relationship between dependent and independent variables.

TABLE 4.2

Differences between Dependent and Independent Variables

Independent Variable	Dependent Variable
Variable that is varied, changed, or manipulated.	It is not manipulated. The response or outcome that is measured when the independent variable is varied.
It is the presumed cause.	It is the presumed effect.
Independent variable is the antecedent.	Dependent variable is the consequent.
Independent variable refers to the status of the "cause," which leads to the changes in the status of the dependent variable.	Dependent variable refers to the status of the "outcome" in which the researcher is interested.
Also known as explanatory or predictor variable.	Also known as response or predictor or target variable.
For example, various metrics that can be used to measure various software constructs.	For example, whether a module is faulty or not.

4.5.2 Selection of Variables

The selection of appropriate variables is not easy and generally based on the domain knowledge of the researcher. In research, the variables are identified according to the things being measured. When exploring new topics, for selection of variables a researcher must carefully analyze the research problem and identify the variables affecting the dependent variable. However, in case of explored topics, the literature review can also help in identification of variables. Hence, the selection is based on the researcher's experience, information obtained from existing published research, and judgment or advice obtained from the experts in the related areas. In fact, the independent variables are selected most of the time on the basis of information obtained from the published empirical work of a researcher's own as well as from other researchers. Hence, the research problem must be thoroughly and carefully analyzed to identify the variables.

4.5.3 Variables Used in Software Engineering

The independent variables can be different software metrics proposed in existing studies. There are different types of metrics, that is, product-related metrics and process-related metrics. Under product-related metrics, there are class level, method level, component level, and file level metrics. All these metrics can be utilized as independent variables.

For example, software metrics such as volume, lines of code (LOC), cyclomatic complexity, and branch count can be used as independent variables.

The variable describing the quality attributes of classes to be predicted is called dependent variable. A variable used to explain a dependent variable is called independent variable. The binary dependent variables of the models can be fault proneness, change proneness, and so on, whereas, the continuous dependent variables can be testing effort, maintenance effort, and so on.

Fault proneness is defined as the probability of fault detection in a class. Change proneness is defined as the probability of a class being changed in future. Testing effort is defined as LOC changed or added throughout the life cycle of the defect per class. Maintenance effort is defined as LOC changed per class in its maintenance history. The quality attributes are somewhat interrelated, for example, as the fault proneness of a class increase so will the testing effort required to correct the faults in the class.

4.5.4 Example: Variables Used in the FPS

The independent variables are various OO metrics proposed by Chidamber and Kemerer (1994). This includes coupling between object (CBO), response for a class (RFC), number of children (NOC), depth of inheritance (DIT), lack of cohesion in methods (LCOM), weighted methods per class (WMC), and LOC. The definitions of these metrics can be found in Chapter 3.

The binary dependent variable is fault proneness. Fault proneness is defined as the probability of fault detection in a class (Briand et al. 2000; Pai and Bechta Dugan 2007; Aggarwal et al. 2009). The dependent variable will be predicted based on the faults found during the software development life cycle.

4.6 Terminology Used in Study Types

The choice of the empirical process depends on the type of the study. There are two types of processes that can be followed based on the study type:

- Hypothesis testing without model prediction
- Hypothesis testing after model prediction

For example, if the researcher wants to find whether a UML tool is better than a traditional tool and the effectiveness of the tool is measured in terms of productivity of the persons using the tool, then hypothesis testing can be used directly using the data given in Table 4.3.

Consider another instance where the researcher wants to compare two machine learning techniques to find the effect of software metrics on probability of occurrence of faults. In this problem, first the model is predicted using two machine learning techniques. In the next step, the model is validated and performance is measured in terms of performance evaluation metrics (refer Chapter 7). Finally, hypothesis testing is applied on the results obtained in the previous step for verifying whether the performance of one technique is better than the other technique.

Figure 4.5 shows that the term independent and dependent variables is used in both experimental studies and multivariate analysis. In multivariate analysis, the independent and dependent variables are used in model prediction. The independent variables are used as predictor variables to predict the dependent variable. In experimental studies, factors for a statistical test are also termed as independent variables that may have one or more

TABLE 4.3

Productivity for Tools

UML Tool	Traditional Tool
14	52
67	61
13	14

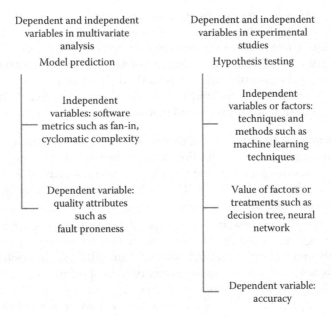

FIGURE 4.5
Terminology used in experimental studies and multivariate analysis studies.

levels called treatments or samples as suitable for a specific statistical test. For example, a researcher may wish to test whether the mean of two samples is equal or not such as in the case when a researcher wants to explore different software attributes like coupling before and after a specific treatment like refactoring. Another scenario could be when a researcher wants to explore the performance of two or more learning algorithms or whether two treatments give uniform results. Thus, the dependent variable in experimental study refers to the behavior measures of a treatment. In software engineering research, in some cases, these may be the performance measures. Similarly, one may refer to performances on different data sets as data instances or subjects, which are exposed to these treatments.

In software engineering research, the performance measures on data instances are termed as the outcome or the dependent variable in case of hypothesis testing in experimental studies. For example, technique A when applied on a data set may give an accuracy (performance measure, defined as percentage of correct predictions) value of 80%. Here, technique A is the treatment and the accuracy value of 80% is the outcome or the dependent variable. However, in multivariate analysis or model prediction, the independent variables are software metrics and the dependent variable may be, for example, a quality attribute.

To avoid confusion, in this book, we use terminology related to multivariate analysis unless and until specifically mentioned.

4.7 Hypothesis Formulation

After the variables have been identified, the next step is to formulate the hypothesis in the research. This is one of the important steps in empirical research.

4.7.1 Experiment Design Types

In this section, we discuss the experimental design types used in experimental studies. The selection of appropriate statistical test for testing hypothesis depends on the type of experimental design. There are four experimental design types that can be used for designing a given case study. Factor is the technique or method used in an empirical study such as machine learning technique or verification method. Treatment is the type of techniques such as DT is a machine learning technique and inspection is a verification technique. The types of experiment design are summarized below.

Case 1: One factor, one treatments—In this case, there is one technique under observation. For example, if the distribution of the data needs to be checked for a given variable, then this design type can be used. Consider a scenario where 25 students had developed the same program. The cyclomatic complexity values of the program can be evaluated using chi-square test.

Case 2: One factor, two treatments—This type of design may be purely randomized or paired design. For example, a researcher wants to compare the performance of two verification techniques such as walkthroughs and inspections. Another instance is when a researcher wants to compare the performance of two machine learning techniques, naïve Bayes and DT, on a given or over multiple data sets. In these two examples, factor is one (verification method or machine learning technique) but treatments are two. Paired *t*-test or Wilcoxon test can be used in these cases. Chapter 6 provides examples for these tests.

TABLE 4.4

Factors and Levels of Example

Factor	Level 1	Level 2
Paradigm type	Structural	OO
Software complexity	Difficult	Simple

Case 3: One factor, more than two treatments—In this case, the technique that is to be analyzed contains multiple values. For example, a researcher wants to compare multiple search-based techniques such as genetic algorithm, particle swarm optimization, genetic programming, and so on. Friedman test can be used to solve this example. Section 6.4.13 provides solution for this example.

Case 4: Multiple factors and multiple treatments—In this case, more than one factor is considered with multiple treatments. For instance, consider an example where a researcher wants to compare paradigm types such as structured paradigm with OO paradigm. In conjunction to the paradigm type, the researcher also wants to check the complexity of the software being difficult or simple. This example is shown in Table 4.4 along with the factors and levels. ANOVA test can be used to solve such examples.

The examples of the above experimental design types are given in Section 6.4. After determining the appropriate experiment design type, the hypothesis needs to be formed in an empirical study.

4.7.2 What Is Hypothesis?

The main objective of an experiment usually is to evaluate a given relationship or hypothesis formed between the cause and the effect. Many authors understand the definition of hypothesis differently:

> A hypothesis may be precisely defined as a tentative proposition suggested as a solution to a problem or as an explanation of some phenomenon. (Ary et al. 1984)
>
> Hypothesis is a formal statement that presents the expected relationship between an independent and dependent variable. (Creswell 1994a)

Hence, hypothesis can be defined as a mechanism to formally establish the relationship between variables in the research. The things that a researcher intends to investigate are formulated in the form of a hypothesis. By formulating a hypothesis, the research objectives or the key concepts involved in the research are defined more specifically. Each hypothesis can be tested for its verifiability or falsifiability. Figure 4.6 shows the process of generation of hypothesis in a research. As shown in the figure research questions can either be generated through problem statement or from well-formed ideas extracted from literature survey. After the development of the research questions, the research hypothesis can be formed.

4.7.3 Purpose and Importance of Hypotheses in an Empirical Research

Aquino (1992) defined the importance of formation of hypothesis in an empirical study. The key advantages of hypothesis formation are given below:

- It provides the researcher with a relational statement that can be directly tested in a research study.

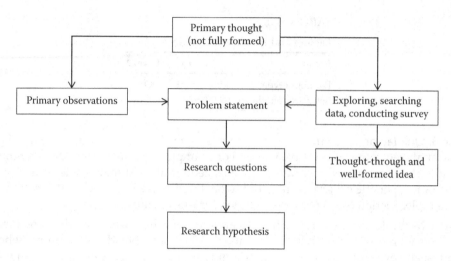

FIGURE 4.6
Generation of hypothesis in a research.

- It helps in formulation of conclusions of the research.
- It helps in forming a tentative or an educated guess about any phenomena in a research.
- It provides direction to the collection of data for validation of hypothesis and thus helps in carrying the research forward.
- Even if the hypothesis is proven to be false, it leads to a specific conclusion.

4.7.4 How to Form a Hypothesis?

Once the RQs are developed or research problem is clearly defined, hypothesis can be derived from the RQs or research problem by identifying key variables and identifying the relationship between the identified variables. The steps that are followed to form the hypothesis are given below:

1. Understand the problem/situation: Clearly understanding the problem is very important. This can be done by breaking down the problem into smaller parts and understanding each part separately. The problem can be restated in words to have a clear understanding of the problem. The meaning of all the words used in stating the problem should be clear and unambiguous. The remaining steps are based on the problem definition. Hence, understanding the problem becomes a crucial step.

2. Identify the key variables required to measure the problem/situation: The key variables used in the hypothesis testing must be selected from the independent and dependent variables identified in Section 4.5. The effect of independent variable on the dependent variable needs to be identified and analyzed.

3. Make an educated guess as to understand the relationship between the variables: An "educated guess" is a statement based on the available RQs or given problem and will be eventually tested. Generally, the relationship established between the independent and dependent variable is stated as an "educated guess."

TABLE 4.5

Transition from RQ to Hypothesis

RQ	Corresponding Hypothesis
Is X related to Y?	If X, then Y.
How are X and Y related to Z?	If X and Y, then Z.
How is X related to Y and Z?	If X, then Y and Z.
How is X related to Y under conditions Z and W?	If X, then Y under conditions Z and W.

4. Write down the hypotheses in a format that is testable through scientific research: There are two types of hypothesis—null and alternative hypotheses. Correct formation of null and alternative hypotheses is the most important step in hypothesis testing. The null hypothesis is also known as hypothesis of no difference and denoted as H_0. The null hypothesis is the proposition that implies that there is no statistically significant relationship within a given set of parameters. It denotes the reverse of what the researcher in his experiment would actually expect or predict. Alternative hypothesis is denoted as H_a. The alternative hypothesis reflects that a statistically significant relationship does exist within a given set of parameters. It is the opposite of null hypothesis and is only reached if H_0 is rejected. The detailed explanation of null and alternative hypothesis is stated in the next Section 4.7.5. Table 4.5 presents corresponding hypothesis to given RQs.

Some of the examples to show the transition from an RQ to a hypothesis are stated below:

RQ: What is the relation of coupling between classes and maintenance effort?

Hypothesis: Coupling between classes and maintenance effort are positively related to each other.

RQ: Are walkthroughs effective in finding faults than inspections?

Hypothesis: Walkthroughs are more effective in finding faults than inspections.

Example 4.1:

There are various factors that may have an impact on the amount of effort required to maintain a software. The programming language in which the software is developed can be one of the factors affecting the maintenance effort. There are various programming languages available such as Java, C++, C#, C, Python, and so on. There is a need to identify whether these languages have a positive, negative, or neutral effect on the maintenance effort. It is believed that programming languages have a positive impact on the maintenance effort. However, this needs to be tested and confirmed scientifically.

Solution:

The problem and hypothesis derived from it is given below:

1. Problem: Need to identify the relationship between the programming language used in a software and the maintenance effort.
2. RQ: Is there a relation between programming language and maintenance effort?
3. Key variables: Programming language and maintenance effort

4. Educated guess: Programming language is related to effort and has a positive impact on the effort.
5. Hypothesis: Programming language and maintenance effort are positively related to each other.

4.7.5 Steps in Hypothesis Testing

The hypothesis testing involves a series of steps. Figure 4.7 depicts the steps in hypothesis testing. Hypothesis testing is based on the assumption that null hypothesis is correct. Thus, we prove that the assumption of no difference (null hypothesis) is not consistent with the research hypothesis. For example, if we strongly believe that technique A is better than technique B, despite our strong belief, we begin by assuming that the belief is not true, and hence we want to fail the test by rejecting null hypothesis.

The various steps involved in hypothesis testing are described below.

4.7.5.1 Step 1: State the Null and Alternative Hypothesis

The null hypothesis is popular because it is expected to be rejected, that is, it can be shown to be false, which then implies that there is a relationship between the observed data. One needs to be specific about what it means if the null hypothesis is not rejected. It only means that there is no sufficient evidence present against null hypothesis (H_0), which is in favor of alternative hypothesis (H_a). There might actually be a difference, but on the basis of the sample result such a difference has not been detected. This is analogous to a legal scenario where if a person is declared "not guilty," it does not mean that he is innocent.

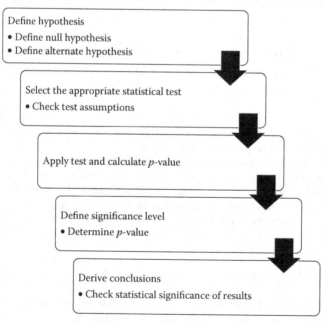

FIGURE 4.7
Steps in hypothesis testing.

The null hypothesis can be written in mathematical form, depending on the particular descriptive statistic using which the hypothesis is made. For example, if the descriptive statistic is used as population mean, then the general form of null hypothesis is,

$$H_o : \mu = X$$

where:
μ is the mean
X is the predefined value

In this example, whether the population mean equals X or not is being tested.

There are two possible scenarios through which the value of X can be derived. This depends on two different types of RQs. In other words, the population parameter (mean in the above example) can be assigned a value in two different ways. First reason is that the predetermined value is selected for practical or proved reasons. For example, a software company decides that 7 is its predetermined quality parameter for mean coupling. Hence, all the departments will be informed that the modules must have a value of <7 for coupling to ensure less complexity and high maintainability. Similarly, the company may decide that it will devote all the testing resources to those faults that have a mean rating above 3. The testers will therefore want to test specifically all those faults that have mean rating >3.

Another situation is where a population under investigation is compared with another population whose parameter value is known. For example, from the past data it is known that average productivity of employees is 30 for project A. We want to see whether the average productivity of employees is 30 or not for project B? Thus, we want to make an inference whether the unknown average productivity for project B is equal to the known average productivity for project A.

The general form of alternative hypothesis when the descriptive parameter is taken as mean (μ) is,

$$H_a : \mu \neq X$$

where:
μ is the mean
X is the predefined value

The above hypothesis represents a nondirectional hypothesis as it just denotes that there will be a difference between the two groups, without discussing how the two groups differ. The example is stated in terms of two popularly used methods to measure the size of software, that is, (1) LOC and (2) function point analysis (FPA). The nondirectional hypothesis can be stated as, "The size of software as measured by the two techniques is different." Whereas, when the hypothesis is used to show the relationship between the two groups rather than simply comparing the groups, then the hypothesis is known as directional hypothesis. The comparison terms such as "greater than," "less than," and so on is used in the formulation of hypothesis. In other words, it specifies how the two groups differ. For example, "The size of software as measured by FPA is more accurate than LOC." Thus, the direction of difference is mentioned. The same concept is represented by one-tailed and two-tailed tests in statistical testing and is explained in Section 6.4.3.

One important point to note is that the potential outcome that a researcher is expecting from his/her experiment is denoted in terms of alternative hypothesis. What is believed to be the theoretical expectation or concept is written in terms of alternative hypothesis.

Thus, sometimes the alternative hypothesis is referred to as the research hypothesis. Now, if the alternative hypothesis represents the theoretical expectation or concept, then what is the reason for performing the hypothesis testing? This is done to check whether the formed or assumed concepts are actually significant or true. Thus, the main aim is check the validity of the alternative hypothesis. If null hypothesis is accepted, it signifies that the idea or concept of research is false.

4.7.5.2 Step 2: Choose the Test of Significance

There are a number of tests available to assess the null hypothesis. The choice among them is to be made to check which test is applicable in a particular situation. The four important factors that are based on assumptions of statistical tests and help in test selection are as follows:

- Type of distribution—Whether data is normally distributed or not?
- Sample size—What is the sample size of the data set?
- Type of variables—What is the measurement scale of variables?
- Number of independent variables—What is the number of factors or variables in the study?

There are various tests available in research for verifying hypothesis and are given as follows:

1. *t*-test for the equality of two means
2. ANOVA for equality of means
3. Paired *t*-test
4. Chi-square test for goodness-of-fit
5. Friedman test
6. Mann–Whitney test
7. Kruskal–Wallis test
8. Wilcoxon signed-rank test

The details of all the tests can be found in Section 6.4.

4.7.5.3 Step 3: Compute the Test Statistic and Associated p-Value

In this step, the descriptive statistic is calculated, which is specified by the null hypothesis. There can be many statistical tests as discussed above that can be applied in practice. But the statistic that is actually calculated depends on the statistic used in hypothesis. For example, if the null hypothesis were defined by the statistic μ, then the statistics computed on the data set would be the mean and the standard deviation. Usually, the calculated statistic does not conform to the value given by null hypothesis. But this is not a cause for concern. What is actually needed to calculate the probability of obtaining the test statistic result that has the value specified in the null hypothesis? This is called as the significance of the test statistic, known as the *p*-value. This *p*-value is compared with the certain significance level determined in the next step. This step is carried out in result execution phase of an empirical study.

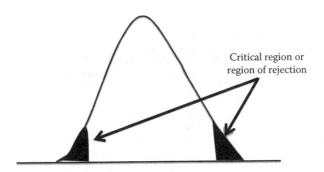

FIGURE 4.8
Critical region.

4.7.5.4 Step 4: Define Significance Level

The critical value or significance value (typically known as α) is determined at this step. The level of significance or α-value is a threshold. When a researcher plans to perform a significance test in an empirical study, a decision has to be made on what is the maximum significance that will be tolerated such that the null hypothesis can be rejected. Figure 4.8 depicts the critical region in a normal curve as shaded portions at two ends.

Generally, this significance value is taken as 0.05 or 0.01. The critical value signifies the critical region or region of rejection. The critical region or region of rejection specifies the range of values, which makes the null hypothesis to be rejected. Using the significance value, the researcher determines the region of rejection and region of acceptance for the null hypothesis.

4.7.5.5 Step 5: Derive Conclusions

The conclusions about the acceptance or rejection of the formed hypothesis are made in this step. Using the decided significance value, the region of rejection is determined. The significance value is used to decide whether or not to reject the null hypothesis. The lower the observed p-value, the more are the chances of rejecting the null hypothesis. If the computed p-value is less than the defined significance threshold then the null hypothesis is rejected and the alternative hypothesis is accepted. In other words, if the p-value lies in the rejection region then the null hypothesis is rejected. Figure 4.9 shows the significance levels of p-value.

The meaning or inference from the results must be determined in this step rather than just repeating the statistics. This step is part of the execution phase of empirical study.

Consider the data set given in Table 4.6. The data consists of six data points. In this example, the coupling aspect for faulty and nonfaulty classes is to be compared. The coupling for faulty classes and coupling of nonfaulty classes for a given software is shown in Table 4.6.

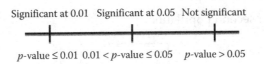

FIGURE 4.9
Significance levels.

TABLE 4.6

A Sample Data Set

S. No.	CBO for Faulty Modules	CBO for Nonfaulty Modules
1	45	9
2	56	9
3	34	9
4	71	7
5	23	10
6	9	15
Mean	39.6	9.83

Step 1: RQ from problem statement

RQ: Is there a difference in coupling values for faulty classes and coupling values for nonfaulty classes?

Step 2: Deriving hypothesis from RQ

In the first step, the hypothesis is derived from the RQ:

H_0: There is no statistical difference between the coupling for faulty classes and coupling for nonfaulty classes.

H_a: The coupling for faulty classes is more than the coupling for nonfaulty classes.

Mathematically,

$$H_0 : \mu\left(CBO_{faulty}\right) = \mu\left(CBO_{nonfaulty}\right)$$

$$H_a : \mu\left(CBO_{faulty}\right) > \mu\left(CBO_{nonfaulty}\right) \text{ or } \mu\left(CBO_{faulty}\right) < \mu\left(CBO_{nonfaulty}\right)$$

Step 3: Determining the appropriate test to apply

As the problem is of comparing means of two dependent samples (collected from same software), the paired t-test is used. In Chapter 6, the conditions for selecting appropriate tests are given.

Step 4: Calculating the value of test statistic

Table 4.7 shows the intermediary calculations of t-test.

The t-statistics is given as:

$$t = \frac{\mu_1 - \mu_2}{\sigma_d / \sqrt{n}}$$

where:

μ_1 is the mean of first population

μ_2 is the mean of second population

$$\sigma_d = \sqrt{\frac{\sum d^2 - \left[\left(\sum d\right)^2 / n\right]}{n - 1}}$$

TABLE 4.7

T-Test Calculations

CBO for Faulty Modules	CBO for Nonfaulty Modules	Difference (*d*)	D^2
45	9	36	1,296
56	9	47	2,209
34	9	25	625
71	7	64	4,096
23	10	13	169
9	15	−6	36

where:

    n represents number of pairs and not total number of samples

    d is the difference between values of two samples

Substituting the values of mean, variance, and sample size in the above formula, the *t*-score is obtained as:

$$\sigma_d = \sqrt{\frac{\sum d^2 - \left[\left(\sum d\right)^2 / n\right]}{n-1}} = \sqrt{\frac{8431 - \left[(179)^2 / 6\right]}{5}} = 24.86$$

$$t = \frac{\mu_1 - \mu_2}{\sigma_d / \sqrt{n}} = \frac{39.66 - 9.83}{24.86 / \sqrt{6}} = 2.93$$

As the alternative hypothesis is of the form, H_1: $\mu > X$ or $\mu < X$, the tail of sampling distribution is nondirectional. Let us take the level of significance (α) for one-tailed test as 0.05.

Step 5: Determine the significance value

The *p*-value at significance level of 0.05 (two-tailed test) is considered and *df* as 5. From the *t*-distribution table, it is observed that the *p*-value is 0.032 (refer to Section 6.4.6 for computation of *p*-value).

Step 6: Deriving conclusions

Now, to decide whether to accept or reject the null hypothesis, this *p*-value is compared with the level of significance. As the *p*-value (0.032) is less than the level of significance (0.05), the H_0 is rejected. In other words, the alternative hypothesis is accepted. Thus, it is concluded that there is statistical difference between the average of coupling metrics for faulty classes and the average of coupling metrics for nonfaulty classes.

4.7.6 Example: Hypothesis Formulation in FPS

There are few RQs that the study intends to answer (stated in Section 4.3.3). Based on these RQs, the study built some hypotheses that are tested. There are two sets of hypothesis, "Hypothesis Set A" and "Hypothesis Set B." Hypothesis set A focuses on the hypothesis related to the relationship between OO metrics and fault proneness; whereas hypothesis set B focuses on the comparison in the performance of machine learning techniques and LR method. Thus, hypothesis in set A deals with the RQs 1, 2, and 3, whereas hypothesis in set B deals with the RQ 4.

4.7.6.1 Hypothesis Set A

There are a number of OO metrics used as independent variables in the study. These are CBO, RFC, LCOM, NOC, DIT, WMC, and source LOC (SLOC). The hypotheses given below are tested to find the individual effect of each OO metric on fault proneness at different severity levels of faults:

CBO hypothesis—H_0: There is no statistical difference between a class having high import or export coupling and a class having less import or export coupling.

H_a: A class with high import or export coupling is more likely to be fault prone than a class with less import or export coupling.

RFC hypothesis—H_0: There is no statistical difference between a class having a high number of methods implemented within a class and the number of methods accessible to an object class because of inheritance, and a class with a low number of methods implemented within a class and the number of methods accessible to an object class because of inheritance.

H_a: A class with a high number of methods implemented within a class and the number of methods accessible to an object class because of inheritance is more likely to be fault prone than a class with a low number of methods implemented within a class and the number of methods accessible to an object class because of inheritance.

LCOM hypothesis—H_0: There is no statistical difference between a class having less cohesion and a class having high cohesion.

H_a: A class with less cohesion is more likely to be fault prone than a class with high cohesion.

NOC hypothesis—H_0: There is no statistical difference between a class having greater number of descendants and a class having fewer descendants.

H_a: A class with a greater number of descendants is more likely to be fault prone than a class with fewer descendants.

DIT hypothesis—H_0: There is no statistical difference between a class having large depth in inheritance tree and a class having small depth in inheritance tree.

H_a: A class with a large depth in inheritance tree is more likely to be fault prone than a class with a small depth in inheritance tree.

WMC hypothesis—H_0: There is no statistical difference between a class having a large number of methods weighted by complexities and a class having a less number of methods weighted by complexities.

H_a: A class with a large number of methods weighted by complexities is more likely to be fault prone than a class with a fewer number of methods weighted by complexities.

4.7.6.2 Hypothesis Set B

The study constructs various fault proneness prediction models using a statistical technique and two machine learning techniques. The statistical technique used is the LR and the machine learning techniques used are DT and ANN. The hypotheses given below are tested to compare the performance of regression and machine learning techniques at different severity levels of faults:

1. H_0: LR models do not outperform models predicted using DT.

   H_a: LR models do outperform models predicted using DT.

2. H_0: LR models do not outperform models predicted using ANN.

   H_a: LR models do outperform models predicted using ANN.

3. H_0: ANN models do not outperform models predicted using DT.

   H_a: ANN models do outperform models predicted using DT.

4.8 Data Collection

Empirical research involves collecting and analyzing data. The data collection needs to be planned and the source (people or repository) from which the data is to be collected needs to be decided.

4.8.1 Data-Collection Strategies

The data collected for research should be accurate and reliable. There are various data-collection techniques that can be used for collection of data. Lethbridge et al. (2005) divides the data-collection techniques into the following three levels:

First degree: The researcher is in direct contact or involvement with the subjects under concern. The researcher or software engineer may collect data in real-time. For example, under this category, the various methods are brainstorming, interviews, questionnaires, think-aloud protocols, and so on. There are various other methods as depicted in Figure 4.10.

Second degree: There is no direct contact of the researcher with the subjects during data collection. The researcher collects the raw data without any interaction with the subjects. For example, observations through video recording and fly on the wall (participants taping their work) are the two methods that come under this category.

Third degree: There is access only to the work artifacts. In this, already available and compiled data is used. For example, analysis of various documents produced from an organization such as the requirement specifications, failure reports, document change logs, and so on come under this category. There are various reports that can be generated using different repositories such as change report, defect report, effort data, and so on. All these reports play an important role while conducting a research. But the accessibility of these reports from the industry or any private organization is not an easy task. This is discussed in the next subsection, and the detailed collection methods are presented in Chapter 5.

The main advantage of the first and second degree methods is that the researcher has control over the data to a large extent. Hence, the researcher needs to formulate and decide

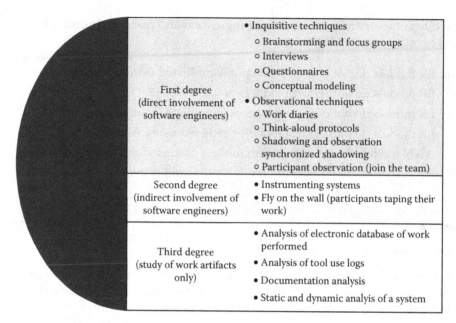

First degree (direct involvement of software engineers)	• Inquisitive techniques ○ Brainstorming and focus groups ○ Interviews ○ Questionnaires ○ Conceptual modeling • Observational techniques ○ Work diaries ○ Think-aloud protocols ○ Shadowing and observation synchronized shadowing ○ Participant observation (join the team)
Second degree (indirect involvement of software engineers)	• Instrumenting systems • Fly on the wall (participants taping their work)
Third degree (study of work artifacts only)	• Analysis of electronic database of work performed • Analysis of tool use logs • Documentation analysis • Static and dynamic analyis of a system

FIGURE 4.10
Various data-collection strategies.

on data-collection methods in the experimental design phase. The methods under these categories require effort from both the researcher and the subject. Because of this reason, first degree methods are most expensive than the second or third degree methods. Third degree methods are least expensive, but the control over data is minimum. This compromises the quality of the data as the correctness of the data is not under the direct control of the researcher.

Under first degree category, the interviews and questionnaires are the most easy and straightforward methods. In interview-based data collection, the researcher prepares a list of questions about the areas of interest. Then, an interview session takes place between the researcher and the subject(s), wherein the researcher can ask various research-related questions. Questions can be either open, inviting multiple and broad range of answers, or closed, offering a limited set of answers. The drawback of collecting data from interviews and questionnaires is that they produce typically an incomplete picture. For example, if one wants to know the number of LOC in a software program. Conducting interviews and questionnaires will only provide us general opinions and evidence, but the accurate information is not provided. Methods such as think-aloud protocols and work diaries can be used for this strategy of data collection. Second degree requires access to the environment in which participants or subject(s) work, but without having direct contact with the participants. Finally, the third degree requires access only to work artifacts, such as source code or bugs database or documentation (Wohlin 2012).

4.8.2 Data Collection from Repositories

The empirical study is based on the data that is often collected from software repositories. In general, it is seen in the literature that data collected is either from academic or

TABLE 4.8

Differences between the Types of Data Sets

S. No.	Academic	Industrial	Open Source
1	Obtained from the projects made by the students of some university	Obtained from the projects developed by experienced and qualified programmers	Obtained from the projects developed by experienced developers located at different geographical locations
2	Easy to obtain	Difficult to obtain	Easy to obtain
3	Obtained from data set that is not necessarily maintained over a long period of time	Obtained from data set maintained over a long period of time	Obtained from data set maintained over a long period of time
4	Results are not reliable and acceptable	Results are highly reliable and acceptable	Results may be reliable and acceptable
5	It is freely available	May or may not be freely available	It is generally freely available
6	Uses ad hoc approach to develop projects	Uses very well planned approach	Uses well planned and mature approach
7	Code may be available	Code is not available	Code is easily available
8	Example: Any software developed in university such as LALO (Briand et al. 2001), UMD (Briand et al. 2000), USIT (Aggarwal et al. 2009)	Example: Performance Management traffic recording (Lindvall 1998), commercial OO system implemented in C++ (Bieman et al. 2003), UIMS (Li and Henry 1993), QUES (Li and Henry 1993)	Example: Android, Apache Tomcat, Eclipse, Firefox, and so on

university systems, industrial or commercial systems, and public or open source software. The academic data is the data that is developed by the students of some university. Industrial data is the proprietary data belonging to some private organization or a company. Public data sets are available freely to everyone for use and does not require any payment from the user. The differences between them are stated in Table 4.8.

It is relatively easy to obtain the academic data as it is free from confidentiality concerns and, hence, gaining access to such data is easier. However, the accuracy and reliability of the academic data is questionable while conducting research. This is because the university software is developed by inexperienced, small number of programmers and is typically not applicable in real-life scenarios. Besides the university data sets, there is public or open source software that is widely used for conducting empirical research in the area of software engineering. The use of open source software allows the researchers to access vast repositories of reasonable quality, large-sized software. The most important type of data is the proprietary/industrial data that is usually owned by a corporation/organization and is not publically available.

The usage of open source software has been on the rise, with products such as Android and Firefox becoming household names. However, majority of the software developed across the world, especially the high-quality software, still remains proprietary software. This is because of the fact that given the voluntary nature of developers for open source software, the attention of the developers might shift elsewhere leading to lack of understanding and poor quality of the end product. For the same reason, there are also challenges with timeliness of the product development, rigor in testing and documentation, as well as characteristic lack of usage support and updates. As opposed to this, the proprietary software is typically developed by an organization with clearly demarcated manpower for design, development, and testing of the software. This allows for committed, structured development of software for a well-defined end use, based on robust requirement gathering. Therefore, it is imperative that the empirical studies in software engineering be validated over data from proprietary systems, because the developers of such proprietary software would be the key users of the research. Additionally, industrial data is better suited for empirical research because the development follows a structured methodology, and each step in the development is monitored and documented along with its performance measurement. This leads to development of code that follows rigorous standards and robustly captures the data sets required by the academia for conducting their empirical research.

At the same time, access to the proprietary software code is not easily obtained. For most of the software development organizations, the software constitutes their key intellectual asset and they undertake multiple steps to guard the privacy of the code. The world's most valuable products, such as Microsoft Windows and Google search, are built around their closely held patented software to guard against competition and safeguard their products developed with an investment of billions of dollars. Even if there is appreciation of the role and need of the academia to access the software, the enterprises typically hesitate to share the data sets, leading to roadblocks in the progress of empirical research.

It is crucial for the industry to appreciate that the needs of the empirical research do not impinge on their considerations of software security. The data sets required by the academia are the metrics data or the data from the development/testing process, and does not compromise on security of the source code, which is the primary concern of the industry. For example, assume an organization uses commercial code management system/test management system such as HP Quality Center or HP Application Lifecycle Management. Behind the scenes, a database would be used to store information about all modules, including all the code and its versions, all development activity in full detail, and the test cases and their results. In such a scenario, the researcher does not need access to the data/code stored in the database, which the organization would certainly be unwilling to share, but rather specific reports corresponding to the problem he wishes to address. As an illustration, for a defect prediction study, only a list of classes with corresponding metrics and defect count would be required, which would not compromise the interests of the organization. Therefore, with mutual dialogue and understanding, appropriate data sets could be shared by the industry, which would create a win-win situation and lead to betterment of the process. The key challenge, which needs to be overcome, is to address the fear of the enterprises regarding the type of data sets required and the potential hazards. A constructive dialogue to identify the right reports would go a long way towards enabling the partnership because access to the wider database with source code would certainly be impossible.

Once the agreement with the industry has been reached and the right data sets have been received, the attention can be shifted to actual conducting of the empirical research with the more appropriate industrial data sets. The benefits of using the industrial database would be apparent in the thoroughness of the data sets available and the consistency of the software system. This would lead to more accurate findings for the empirical research.

4.8.3 Example: Data Collection in FPS

This empirical study given in Section 4.2 makes use of the public domain data set KC1 from the NASA metrics date program (MDP) (NASA 2004; PROMISE 2007). The NASA data repository stores the data, which is collected and validated by the MDP (2006). The data in KC1 is collected from a storage management system for receiving/processing ground data, which is implemented in the C++ programming language. Fault data for KC1 is collected since the beginning of the project (storage management system) but that data can only be associated back to five years (MDP 2006). This system consists of 145 classes that comprise 2,107 methods, with 40K LOC. KC1 provides both class-level and method-level static metrics. At the method level, 21 software product metrics based on product's complexity, size, and vocabulary are given. At the class level, values of ten metrics are computed, including six metrics given by Chidamber and Kemerer (1994). The seven OO metrics are taken in this study for analyses. In KC1, six files provide association between class/method and metric/defect data. In particular, there are four files of interest, the first representing the association between classes and methods, the second representing association between methods and defects, the third representing association between defects and severity of faults, and the fourth representing association between defects and specific reason for closure of the error report.

First, defects are associated with each class according to their severities. The value of severity quantifies the impact of the defect on the overall environment with 1 being most severe to 5 being least severe as decided in data set KC1. The defect data from KC1 is collected from information contained in error reports. An error either could be from the source code, COTS/OS, design, or is actually not a fault. The defects produced from the source code, COTS/OS, and design are taken into account. The data is further processed by removing all the faults that had "not a fault" keyword used as the reason for closure of error report. This reduced the number of faults from 669 to 642. Out of 145 classes, 59 were faulty classes, that is, classes with at least one fault and the rest were nonfaulty.

In this study, the faults are categorized as high, medium, or low severity. Faults with severity rating 1 were classified as high-severity faults. Faults with severity rating 2 were classified as medium-severity faults and faults with severity rating 3, 4, and 5 as low-severity faults, as at severity rating 4 no class is found to be faulty and at severity rating 5 only one class is faulty. Faults at severity rating 1 require immediate correction for the system to continue to operate properly (Zhou and Leung 2006).

Table 4.9 summarizes the distribution of faults and faulty classes at high-, medium-, and low-severity levels in the KC1 NASA data set after preprocessing of faults in the data set. High-severity faults were distributed in 23 classes (15.56%). There were 48 high-severity faults (7.47%), 449 medium-severity faults (69.93%), and 145 low-severity faults (22.59%). As shown in Table 4.9, majority of the classes are faulty at severity rating medium (58 out of 59 faulty classes). Figure 4.11a–c shows the distribution of high-severity faults, medium-severity faults, and low-severity faults. It can be seen from Figure 4.11a that 22.92% of classes with high-severity faults contain one fault, 29.17% of classes contain two faults, and so on. In addition, the maximum number of faults (449 out of 642) is covered at medium severity (see Figure 4.11b).

TABLE 4.9

Distribution of Faults and Faulty Classes at High-, Medium-, and Low-Severity Levels

Level of Severity	Number of Faulty Classes	% of Faulty Classes	Number of Faults	% of Distribution of Faults
High	23	15.56	48	7.47
Medium	58	40.00	449	69.93
Low	39	26.90	145	22.59

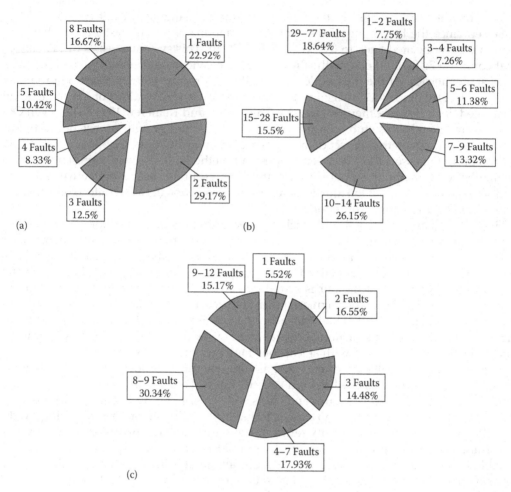

FIGURE 4.11
Distribution of (a) high-, (b) medium-, and (c) low-severity faults.

4.9 Selection of Data Analysis Methods

There are various data analysis methods available in the literature (such as statistical, machine learning) that can be used to analyze different kinds of gathered data. It is very essential to carefully select the methods to be used while conducting a research. But it is very difficult to select appropriate data analysis method for a given research. Among various available data analysis methods, we can select the most appropriate method by comparing different parameters and properties of all the available methods. Besides this, there are very few sources available that provide guidance for selection of data analysis methods.

In this section, guidelines that can be used for the appropriate selection of the data analysis methods are presented. The selection of a data analysis technique can be made based on the following three criteria: (1) the type of dependent variable, (2) the nature of data set, or (3) the important aspects of different methods.

4.9.1 Type of Dependent Variable

The data analysis methods can be selected based on the type of the dependent variable being used. The dependent variable can be either discrete/binary or continuous. A discrete variable is a variable that can only take a finite number of values, whereas a continuous variable can take infinite number of values between any two points. If the dependent variable is binary (e.g., fault proneness, change proneness), then among statistical techniques, the researcher can use the LR and discriminant analysis. The examples of machine learning classifiers that support binary-dependent variable are DT, ANN, support vector machine, random forest, and so on. If the dependent variable is continuous, then the selection of data analysis method depends on whether the variable is a count variable (i.e., used for counting purpose) or not a count variable. The examples of continuous count variable are number of faults, lines of source code, and development effort. ANN is one of the machine learning techniques that can be used in this case. In addition, for noncount continuous-dependent variable, the traditional ordinary least squares (OLS) regression model can be used. The diagrammatic representation of the selection of appropriate data analysis methods based on type of dependent variable is shown in Figure 4.12.

4.9.2 Nature of the Data Set

Other factors to consider when choosing and applying a learning method include the following:

1. Diversity in data: The variables or attributes of the data set may belong to different categories such as discrete, continuous, discrete ordered, counts, and so on. If the attributes are of many different kinds, then some of the algorithms are preferable over others as they are easy to apply. For example, among machine learning techniques, support vector machine, neural networks, and nearest neighbor methods require that the input attributes are numerical and scaled to similar ranges (e.g., to the [–1,1] interval). Among statistical techniques, linear regression and LR require

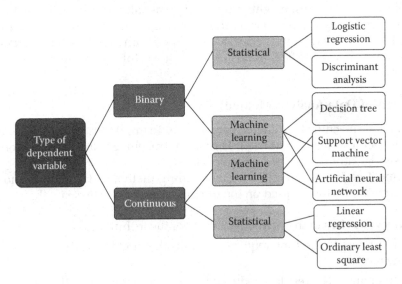

FIGURE 4.12
Selection of data analysis methods based on the type of dependent variable.

the input attributes be numerical. The machine learning technique that can handle heterogeneous data is DT. Thus, if our data is heterogeneous, then one may apply DT instead of other machine learning techniques (such as support vector machine, neural networks, and nearest neighbor methods).

2. Redundancy in the data: There may be some independent variables that are redundant, that is, they are highly correlated with other independent variables. It is advisable to remove such variables to reduce the number of dimensions in the data set. But still, sometimes it is found that the data contains the redundant information. In this case, the researcher should make careful selection of the data analysis methods, as some of the methods will give poor performance than others. For example, linear regression, LR, and distance-based methods, will give poor performance because of numerical instabilities. Thus, these methods should be avoided.

3. Type and existence of interactions among variables: If each attribute makes an independent impact or contribution to the output or dependent variable, then the techniques based on linear functions (e.g., linear regression, LR, support vector machines, naïve Bayes) and distance functions (e.g., nearest neighbor methods, support vector machines with Gaussian kernels) perform well. But, if the interactions among the attributes are complex and huge, then DT and neural network should be used as these techniques are particularly composed to deal with these interactions.

4. Size of the training set: Selection of appropriate method is based on the tradeoff between bias/variance. The main idea is to simultaneously minimize bias and variance. Models with high bias will result in underfitting (do not learn relationship between the dependent and independent variables), whereas models with high variance will result in overfitting (noise in the data). Therefore, a good learning technique automatically adjusts the bias/variance trade-off based on the size of training data set. If the training set is small, high bias/low variance classifiers should be used over low bias/high variance classifiers. For example, naïve Bayes has a high bias/low variance (naïve Bayes is simple and assumes independence of variables) and k-nearest neighbor has a low bias/high variance. But as the size of training set increases, low bias/high variance classifiers show good performance (they have lower asymptotic error) as compared with high bias/low variance classifiers. High bias classifiers (linear) are not powerful enough to provide accurate models.

4.9.3 Aspects of Data Analysis Methods

There are various machine learning tasks available. To implement each task, there are various learning methods that can be used. The various machine learning tasks along with the data analysis algorithms are listed in Table 4.10.

To implement each task, among various learning methods, it is required to select the appropriate method. This is based on the important aspects of these methods.

1. Accuracy: It refers to the predictive power of the technique.
2. Speed: It refers to the time required to train the model and the time required to test the model.
3. Interpretability: The results produced by the technique are easily interpretable.
4. Simplicity: The technique must be simple in its operation and easy to learn.

TABLE 4.10

Data Analysis Methods Corresponding to Machine Learning Tasks

S. No.	Machine Learning Tasks	Data Analysis Methods
1	Multivariate querying	Nearest neighbor, farthest neighbor
2	Classification	Logistic regression, decision tree, nearest neighbor classifier, neural network, support vector machine, random forest
3	Regression	Linear regression, regression tree
4	Dimension reduction	Principal component analysis, nonnegative matrix factorization, independent component analysis
5	Clustering	k-means, hierarchical clustering

Besides the four above-mentioned important aspects, there are some other considerations that help in making a decision to select the appropriate method. These considerations are sensitivity to outliers, ability to handle missing values, ability to handle nonvector data, ability to handle class imbalance, efficacy in high dimensions, and accuracy of class probability estimates. They should also be taken into account while choosing the best data analysis method. The procedure for selection of appropriate learning technique is further described in Section 7.4.3.

The methods are classified into two categories: parametric and nonparametric. This classification is made on the basis of the population under study. Parametric methods are those for which the population is approximately normal, or can be approximated to normal using a normal distribution. Parametric methods are commonly used in statistics to model and analyze ordinal or nominal data with small sample sizes. The methods are generally more interpretable, faster but less accurate, and more complex. Some of the parametric methods include LR, linear regression, support vector machine, principal component analysis, k-means, and so on. Whereas, nonparametric methods are those for which the data has an unknown distribution and is not normal. Nonparametric methods are commonly used in statistics to model and analyze ordinal or nominal data with small sample sizes. The data cannot even be approximated to normal if the sample size is so small that one cannot apply the central limit theorem. Nowadays, the usage of nonparametric methods is increasing for a number of reasons. The main reason is that the researcher is not forced to make any assumptions about the population under study as is done with a parametric method. Thus, many of the nonparametric methods are easy to use and understand. These methods are generally simpler, less interpretable, and slower but more accurate. Some of the nonparametric methods are DT, nearest neighbor, neural network, random forest, and so on.

Exercises

4.1. What are the different steps that should be followed while conducting experimental design?

4.2. What is the difference between null and alternative hypothesis? What is the importance of stating the null hypothesis?

4.3. Consider the claim that the average number of LOC in a large-sized software is at most 1,000 SLOC. Identify the null hypothesis and the alternative hypothesis for this claim.

4.4. Discuss various experiment design types with examples.

4.5. What is the importance of conducting an extensive literature survey?

4.6. How will you decide which studies to include in a literature survey?

4.7. What is the difference between a systematic literature review, and a more general literature review?

4.8. What is a research problem? What is the necessity of defining a research problem?

4.9. What are independent and dependent variables? Is there any relationship between them?

4.10. What are the different data-collection strategies? How do they differ from one another?

4.11. What are the different types of data that can be collected for empirical research? Why the access to industrial data is difficult?

4.12. Based on what criteria can the researcher select the appropriate data analysis method?

Further Readings

The book provides a thorough and comprehensive overview of the literature review process:

A. Fink, *Conducting Research Literature Reviews: From the Internet to Paper.* 2nd edn. Sage Publications, London, 2005.

The book provides an excellent text on mathematical statistics:

E. L. Lehmann, and J.P. Romano, *Testing Statistical Hypothesis,* 3rd edn., Springer, Berlin, Germany, 2008.

A classic paper provides techniques for collecting valid data that can be used for gathering more information on development process and assess software methodologies:

V. R. Basili, and D. M. Weiss, "A methodology for collecting valid software engineering data," *IEEE Transactions on Software Engineering,* vol. 10, no. 6, pp. 728–737, 1984.

The following book is a classic example of concepts on experimentation in software engineering:

V. R. Basili, R. W. Selby, and D. H. Hutchens, "Experimentation in software engineering," *IEEE Transactions on Software Engineering,* vol. 12, no. 7, pp. 733–743, 1986.

A taxonomy of data-collection techniques is given by:

T. C. Lethbridge, S. E. Sim, and J. Singer, "Studying software engineers: Data collection techniques for software field studies," *Empirical Software Engineering*, vol. 10, pp. 311–341, 2005.

The following paper provides an overview of methods in empirical software engineering:

S. Easterbrook, J. Singer, M.-A. Storey, and D. Damian, "Selecting empirical methods for software engineering research," In: F. Shull, J. Singer, and D.I. Sjøberg (eds.), *Guide to Advanced Empirical Software Engineering*, Springer, London, 2008.

5

Mining Data from Software Repositories

One of the problems faced by the software engineering community is scarcity of data for conducting empirical studies. However, the software repositories can be mined to collect and gather the data that can be used for providing empirical results by validating various techniques or methods. The empirical evidence gathered through analyzing the data collected from the software repositories is considered to be the most important support for software engineering community these days. These evidences can allow software researchers to establish well-formed and generalized theories. The data obtained from software repositories can be used to answer a number of questions. Is design A better than design B? Is process/method A better than process/method B? What is the probability of occurrence of a defect or change in a module? Is the effort estimation process accurate? What is the time taken to correct a bug? Is testing technique A better than testing technique B? Hence, the field of extracting data from software repositories is gaining importance in organizations across the globe and has a central and essential role in aiding and improving the software engineering research and development practice.

As already mentioned in Chapter 1 and 4 the data can either be collected from proprietary software, open source software (OSS), or university software. However, obtaining data from proprietary software is extremely difficult as the companies are not usually willing to share the source code and information related to the evolution of the software. Another source for collecting empirical data is academic software developed by universities. However, collecting data from software developed by student programmers is not recommended, as the accuracy and applicability of this data cannot be determined. In addition, the university software is developed by inexperienced, small number of programmers and thus does not have applicability in the real-life scenarios.

The rise in the popularity of the use of OSS has made vast amount of data available for use in empirical research in the area of software engineering. The information from open source repositories can be easily extracted in a well-structured manner. Hence, now researchers have access to vast repositories containing large-sized software maintained over a period of time.

In this chapter, the basic techniques and procedures for extracting data from software repositories is provided. A detailed discussion on how change logs and bug reports are organized and structured is presented. An overview of existing software engineering repositories is also given. In this chapter, we present defect collection and reporting system that can be used for collecting changes and defects from maintenance phase.

5.1 Configuration Management Systems

Configuration management systems are central to almost all software projects developed by the organizations. The aim of a configuration management system is to control and manage changes that occur in all the artifacts produced during the software development life cycle. The artifacts (also known as deliverables) produced during the software development life cycle include software requirement specification, software design document, source code listings, user manuals, and so on (Bersoff et al. 1980; Babich 1986).

A configuration management system also controls any changes incurred in these artifacts. Typically, configuration management consists of three activities: configuration identification, configuration control, and configuration accounting (IEEE/ANSI Std. 1042–1987, IEEE 1987).

5.1.1 Configuration Identification

Each and every software project artifact produced during the software development life cycle is uniquely named. The following terminologies are related to configuration identification:

- Release: The first issue of a software artifact is called a release. This usually provides most of the functionalities of a product, but may contain a large number of bugs and thus is prone to issue fixing and enhancements.
- Versions: Significant changes incurred in the software project's artifacts are called versions. Each version tends to enhance the functionalities of a product, or fix some critical bugs reported in the previous version. New functionalities may or may not be added.
- Editions: Minor changes or revisions incurred in the software artifacts are termed as editions. As opposed to a version, an edition may not introduce significant enhancements or fix some critical issues reported in the previous version. Rather, small fixes and patches are introduced.

5.1.2 Configuration Control

Configuration control is a critical process of versioning or configuration management activities. This process incorporates the approval, control, and implementation of changes to the software project artifact(s), or to the software project itself. Its primary purpose is to ensure that each and every change incurred to any software artifact is carried out with the knowledge and approval of the software project management team. A typical change request procedure is presented in Figure 5.1.

Figure 5.2 presents the general format of a change request form. The request consists of some important fields such as severity (impact of failure on software operation) and priority (speed with which the defect must be addressed).

The change control board (CCB) is responsible for the approval and tracking of changes. The CCB carefully and closely reviews each and every change before approval. After the changes are successfully implemented and documented, they must be notified so that they are tracked and recorded in the software repository hosted at version control systems (VCS). Sometimes, it is also known as the software library, archive, or repository, wherein the entire official artifacts (documents and source code) are maintained during the software development life cycle.

The changes are notified through a software change notice. The general format of a change notice is presented in Figure 5.3.

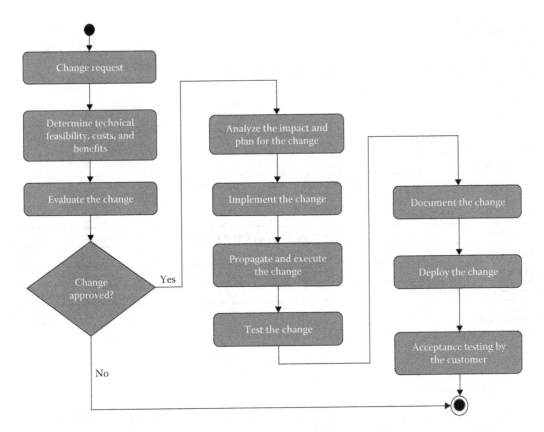

FIGURE 5.1
Change cycle.

Change Request Form				
Change Request ID				
Type of Change Request	☐ Enhancement	☐ Defect Fixing		☐ Other (Specify)
Project				
Requested By	*Project team member name*			
Brief Description of the Change Request	*Description of the change being requested*			
Date Submitted				
Date Required				
Priority	☐ Low	☐ Medium	☐ High	☐ Mandatory
Severity	☐ Trivial	☐ Moderate	☐ Serious	☐ Critical
Reason for Change	*Description of why the change is being requested*			
Estimated Cost of Change	*Estimates for the cost of incurring the change*			
Other Artifacts Impacted	*List other artifacts affected by this change*			
Signature				

FIGURE 5.2
Change request form.

Change Notice Form			
Change Request ID			
Type of Change Request	☐ Enhancement	☐ Defect Fixing	☐ Other (Specify)
Project			
Module in which change is made			
Change Implemented by	*Project team member name*		
Date and time of change implementation			
Change Approved By	*CCB member who approved the change*		
Brief Description of the Change Request	*Description of the change incurred*		
Decision	☐ Approved ☐ Approved with Conditions	☐ Rejected	☐ Other
Decision Date			
Conditions	*Conditions imposed by the CCB*		
Approval Signature			

FIGURE 5.3
Software change notice.

5.1.3 Configuration Accounting

Configuration accounting is the process that is responsible for keeping track of each and every activity, including changes, and any action that affects the configuration of a software product artifact, or the software product itself. Generally, the entire data corresponding to each and every change is maintained in the VCS. Configuration accounting also incorporates recording and reporting of all the information required for versioning or configuration management of a software project. This information includes the status of software artifacts under versioning control, metadata, and other related information for the proposed changes, and the implementation status of the changes that were approved in the configuration control process.

A typical configuration status report includes

- A list of software artifacts under versioning. These comprise a baseline.
- Version-wise date as to when the baseline of a version was established.
- Specifications that describe each artifact under versioning.
- History of changes incurred in the baseline.
- Open change requests for a given artifact.
- Deficiencies discovered by reviews and audits.
- The status of approved changes.

In the next section, we present the importance of mining information from software repositories, that is, information gathered from historical data such as defect and change logs.

5.2 Importance of Mining Software Repositories

Software repositories usually provide a vast array of varied and valuable information regarding software projects. By applying the information mined from these repositories, software engineering researchers and practitioners do not need to depend primarily on their intuition and experience, but more on field and historical data (Kagdi et al. 2007).

However, past experiences, dominant methodologies, and patterns still remain the driving force for significant decision-making processes in software organizations (Hassan 2008). For instance, software engineering practitioners mostly rely on their experience and gut feeling while making essential decisions. Even the managers tend to allocate their organization's development and testing resources on the grounds of their experience in previous software projects, and their intuition regarding the complexity and criticality of the new project when compared with the previous projects. Developers generally employ their experience while adding new features or issue fixing. Testers tend to prioritize the testing of modules, classes, and other artifacts that are discovered to be error prone based on historical data and bug reports.

A major reason behind the ignorance of how valuable is the information provided in software engineering repositories, is perhaps the lack of effective mining techniques that can extract the right kind of information from these repositories in the right form.

Recognizing the need for effective mining techniques, the mining software repositories (MSR) field has been developed by software engineering practitioners. The MSR field analyzes and cross-links the rich and valuable data stored in the software repositories to discover interesting and applicable information about various software systems as well as projects. However, software repositories are generally employed in practice as mere record-keeping data stores and are rarely used to facilitate decision-making processes (Hassan 2008). Therefore, MSR researchers also aim at carrying out a significant transformation of these repositories from static record-keeping repositories into active ones for guiding the decision-making process of modern software projects.

Figure 5.4 depicts that after mining the relevant information from software repositories, data mining techniques can be applied and useful results can be obtained, analyzed, and interpreted. These results will guide the practitioners in decision making. Hence, mining data from software repositories will exhibit the following potential benefits:

- Enhance maintenance of the software system
- Empirical validation of techniques and methods
- Supporting software reuse
- Proper allocation of testing and maintenance resources

5.3 Common Types of Software Repositories

This section describes different types of software repositories and various artifacts provided by them that may be used to extract useful information (Hassan 2008). Figure 5.5 presents the different types of software repositories commonly employed.

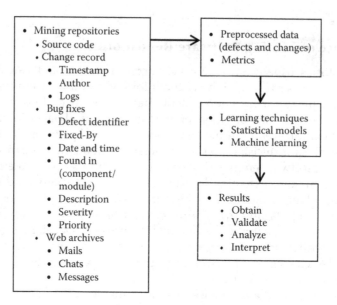

FIGURE 5.4
Data analysis procedure after mining software repositories.

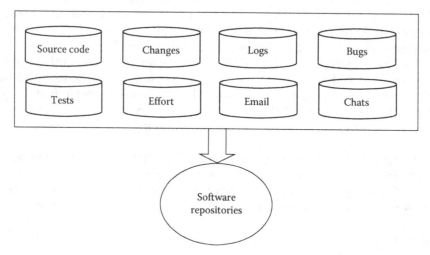

FIGURE 5.5
Commonly used software repositories.

5.3.1 Historical Repositories

Historical repositories record varied information regarding the evolution and progress of a software project. They also capture significant historical dependencies prevalent between various artifacts of a project, such as functions (in the source code), documentation files, or configuration files (Gall et al. 1998). Developers can possibly employ the information extracted from these historical repositories for various purposes. A major area of application is propagating the changes to related artifacts, instead of analyzing only static and/or dynamic code dependencies that may not be able to capture significant dependencies.

For example, consider an application that consists of a module (say module 1) that takes in some input and writes it to a data store, and another module (module 2) that reads the data from that data store. If there is a modification in the source code of the module that saves data to the data store, we may be required to perform changes to module 2 that retrieves data from that data store, although there are no traditional dependencies (such as control flow dependency) between the two modules. Such dependencies can be determined if and only if we analyze the historical data available for the software project. For this example, data extracted from historical repositories will reveal that the two modules, for saving the data to the data store and reading the data from that data store, are co-changing, that is, a change in module 1 has resulted in a change in module 2.

Historical repositories include source control repositories, bug repositories, and archived communications.

- Source control repositories

  Source control repositories record and maintain the development trail of a project. They track each and every change incurred in any of the artifacts of a software system, such as the source code, documentation manuals, and so on. Additionally, they also maintain the metadata regarding each change, for instance, the developer or project member who carried out the change, the time-stamp when the change was performed, and a short description of the change. These are the most readily available repositories, and also the most employed in software projects (Ambriola et al. 1990). Git, CVS, subversion (SVN), Perforce, and ClearCase are some of the popular source control repositories that are used in practice. Source control repositories, also known as VCS, are discussed in detail later in Section 5.5.

- Bug repositories

  These repositories track and maintain the resolution history of defect/bug reports, which provide valuable information regarding the bugs that were reported by the users of a large software project, as well as the developers of that project. Bugzilla and Jira are the commonly used bug repositories.

- Archived communications

  Discussions regarding the various aspects of a software project during its life cycle, such as mailing lists, emails, instant messages, and internet relay chats (IRCs) are recorded in the archived communications.

5.3.2 Run-Time Repositories or Deployment Logs

Run-time repositories, also known as deployment logs, record information regarding the execution of a single deployment, or different deployments of a software system. For example, run-time repositories may record the error messages reported by a software application at varied deployment sites. Deployment logs are now being made available at a rapidly increasing rate, owing to their use for remote defect fixing and issue resolution and because of some legal acts. For example, the Sarbanes-Oxley Act of 2002 states that it is mandatory to log the execution of every commercial, financial, and telecommunication application in these repositories.

Run-time repositories can possibly be employed to determine the execution anomalies by discovering dominant execution or usage patterns across various deployments, and recording the deviations observed from such patterns.

5.3.3 Source Code Repositories

Source code repositories maintain the source code for a large number of OSS projects. Sourceforge.net and Google code are among the most commonly employed code repositories, and host the source code for a large number of OSS systems, such as Android OS, Apache Foundation Projects, and many more. Source code is arguably one of the most important artifacts of any software project, and its application is discussed in detail later in Section 5.8.

5.4 Understanding Systems

Understanding large software systems still remains a challenging process for most of the software organizations. This is probably because of various reasons. Most importantly, documentation manuals and files pertaining to large systems rarely exist and even if such data exists, they are often not updated. In addition, system experts are usually too preoccupied to guide novice developers, or may no longer be a part of the organization (Hassan 2008). Evaluating the system characteristics and tracing its evolution history thus have become important techniques to gain an understanding about the system.

5.4.1 System Characteristics

A software system may be analyzed by the following general characteristics, which may prove helpful in decision-making process on whether data should be collected from a software system and used in research-centric applications or not.

1. Programming language(s): The computer language(s) in which a software system has been written and developed. Java remains the most popular programming language for many OSS systems, such as Apache projects, Android OS, and many more. C, C++, Perl, and Python are also other popular programming languages.

2. Number of source files: This attribute gives the total number of source code files contained in a software system. In some cases, this measure may be used to depict the complexity of a software system. A system with greater number of source files tends to be more complex than those with lesser number of source files.

3. Number of lines of code (LOC): It is an important size metric of any software system that indicates the total number of LOC of the system. Many software systems are classified on the basis of their LOC as small-, medium-, and large scale systems. This attribute also gives an indication of the complexity of a software system. Generally, systems with larger size, that is, LOC, tend to be more complex than those with smaller size.

4. Platform: This attribute indicates the hardware and software environment (predominantly software environment) that is required for a particular software system to function. For example, some software systems are meant to work only on Windows OS.

5. Company: This attribute provides information about the organization that has developed, or contributed to the development of a software system.

6. Versions and editions: A software system is typically released in versions, with each version being rolled out to incorporate some significant changes in the

previous version of that software system. Even for a given version, several editions may be released to incorporate some minor changes in the software system.

7. Application/domain: A software system usually serves a fundamental purpose or application, along with some optional or secondary features. Open source systems typically belong to one of these domains: graphics/media/3D, IDE, SDK, database, diagram/visualization, games, middleware, parsers/generators, programming language, testing, and general purpose tools that combine multiple such domains.

5.4.2 System Evolution

Software evolution primarily aims to incorporate and revalidate the probable significant modifications or changes to a software system without being able to predict in advance how the customer or user requirements will eventually evolve (Gall et al. 1997). The existing, large software system can never be entirely complete and hence continuously evolves. As the software system continues to evolve, its complexity will tend to increase until and unless we turn up with a better solution to solve or mitigate these issues.

Software system evolution also aims to ensure the reliability and flexibility of the system. However, to adapt to the ever-changing real-world environment, a system should evolve once in every few months. Faster evolution is achievable and necessary too, owing to the rapidly increasing resources over the Internet, which makes it easier for the users to extract useful information.

The concept of software evolution has led to the phenomenon of OSS development. Any user can easily obtain the project artifacts and modify them according to his/her requirements. The most significant advantage of this open source movement is that it promotes the evolution of new ideas and methodologies that aim to improve the overall software process and product life cycle. This is the basic principal and agenda of software engineering. However, a negative impact is that it is difficult to keep a close and continuous check on the development and modification of a software project, if it has been published as open source (Livshits and Zimmermann et al. 2005).

It can be stated that the software development is an ongoing process, and it is truly a never-ending cycle. After going through various methodologies and enhancements, evolutionary metrics were consequently proposed in the literature to cater to the matter of efficiency and effectiveness of the programs. A software system may be analyzed by various evolutionary and change metrics (suggested by Moser et al. 2008), which may prove helpful in understanding the evolution and release history of a software system (Moser et al. 2008). The details of the evolution metrics are given in Chapter 3.

5.5 Version Control Systems

VCS, also known as source control systems or simply versioning systems, are systems that track and record changes incurred to a single artifact or a set of artifacts of a software system.

5.5.1 Introduction

In this section we provide classification of VCS. Each and every change, no matter how big or small, is recorded over time so that we may recall specific revisions or versions of the system artifacts later.

The following general terms are associated with a VCS (Ball et al. 1997):

- Revision numbers: VCS typically tend to distinguish between different version numbers of the software artifacts. These version numbers are usually called revision numbers and indicate various versions of an artifact.
- Release numbers: With respect to software products, revision numbers are termed as release numbers and these indicate different releases of the software product.
- Baseline or trunk: A baseline is the approved version or revision of a software artifact from which changes can be made subsequently. It is also called trunk or master.
- Tag: Whenever a new version of a software product is released, a symbolic name, called the tag, is assigned to the revision numbers of current software artifacts. The tag indicates the release number. In the header section of every tagged artifact, the relation tag (symbolic name)—revision number is stored.
- Branch: They are very common in a VCS and a single branch indicates a self-maintained line of development. In other words, a developer may create a copy of some project artifacts for his own use, and give an appropriate identification to the new line of development. This new line of development created from the originally stored software artifacts is referred to as a branch. Hence, multiple copies of a file may be created independent of each other. Each branch is characterized by its branch number or identification.
- Head: It (sometimes also called "tip") refers to the commit that has been made most recently, either to a branch or to the trunk. The trunk and every branch have their individual heads. Head is also sometimes referred to the trunk.

Figure 5.6 depicts the branches that come out of a baseline or trunk.

The major functionalities provided by a VCS include the following:

- Revert project artifacts back to a previously recorded and maintained state
- Revert the entire software project back to a previously recorded state
- Review any change made over time to any of the project artifacts
- Retrieve metadata about any change, such as the developer or project member who last modified any artifact that might be causing a problem, and more

Employing a VCS also means that if we accidentally modify, damage, or even lose some project artifacts, we can generally recover them easily by simply cloning or downloading those artifacts from the VCS. Generally, this can be achieved with insignificant costs and overheads.

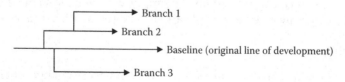

FIGURE 5.6
Trunk and branches.

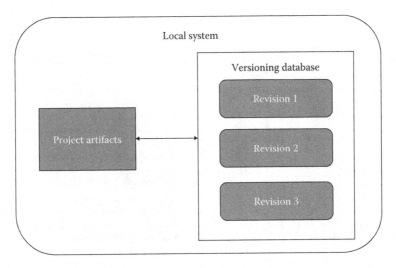

FIGURE 5.7
Local version control.

5.5.2 Classification of VCS

VCS may be categorized as follows (http://git-scm.org).

5.5.2.1 Local VCS

Local VCS employ a simple database that records and maintains all the changes to artifacts of the software project under revision control. Figure 5.7 presents the concept of a local VCS.

A system named revision control system (RCS) was a very popular local versioning system, which is still being used by many organizations as well as the end users. This tool operates by simply recording the patch sets (i.e., the differences between two artifacts) while moving from one revision to the other in a specific format on the user's system. It can then easily recreate the image of a project artifact at any point of time by summing up all the maintained patches.

However, the user cannot collaborate with other users on other systems, as the database is local and not maintained centrally. Each user has his/her own copy of the different revisions of project artifacts, and thus there are consistency and data sharing problems. Moreover, if one user loses the versioning data, recovering it is impossible until and unless a backup is maintained from time to time.

5.5.2.2 Centralized VCS

Owing to the drawbacks of local versioning systems, centralized VCS (CVCS) were developed. The main aim of CVCS is to allow the user to easily collaborate with different users on other systems. These systems, such as CVS, Perforce, and SVN, employ a single centralized server that records and maintains all the versioned artifacts of a software project under revision control, and there are a number of clients or users that check out (obtain) the project artifacts from that central server. For several years, this has been the standard methodology followed in various organizations for version control.

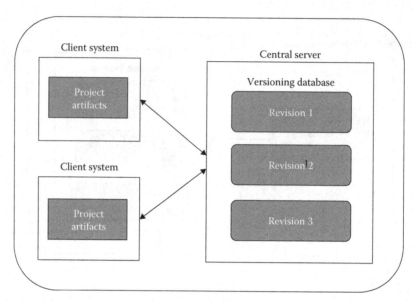

FIGURE 5.8
Centralized version control.

However, if the central server fails or the data stored at central server is corrupted or lost, there are no chances of recovery unless we maintain periodic backups. Figure 5.8 presents the concept of a CVCS.

5.5.2.3 Distributed VCS

To overcome the limitations of CVCS, distributed VCS (DVCS) were introduced. As opposed to CVCS, a DVCS (such as Bazaar, Darcs, Git, and Mercurial) ensures that the clients or users do not just obtain or check out the latest revision or snapshot of the project artifacts, but clone, mirror, or download the entire software project repository to obtain the artifacts.

Thus, if any server of the DVCS fails or its data is corrupted or lost, any of the software project repositories stored at the client machine can be uploaded as back up to the server to restore it. Therefore, every checkout carried out by a client is essentially a complete backup of the entire software project data.

Nowadays, DVCS have earned the attention of various organizations across the globe, and these organizations are relying on them for maintaining their software project repositories. Git is the most popular DVCS employed in practice and hosts a large number of software project repositories. Google and Apache Software Foundation also employ Git to maintain the source code and change control data for their various projects, including Android OS (https://android.googlesource.com), Chromium OS, Chrome browser (https://chromium.googlesource.com), Open Office, log4J, PDFBox, and Apache-Ant, respectively (https://apache.googlesource.com). The concept of a DVCS is presented in Figure 5.9. The figure shows that a copy of entire software project repository is maintained at each client system.

The next section discusses the information maintained by the bug tracking system.

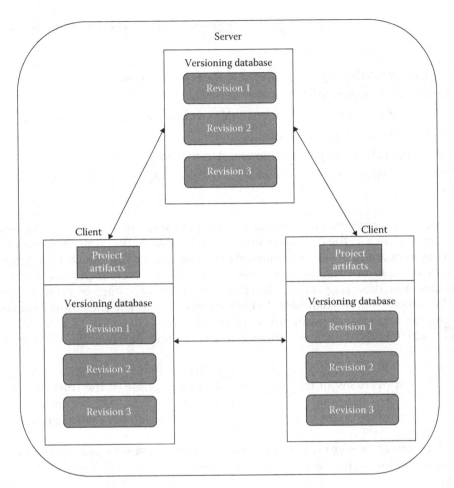

FIGURE 5.9
Distributed version control systems.

5.6 Bug Tracking Systems

A bug tracking system (also known as defect tracking system) is a software system/application that is built with the intent of keeping a track record of various defects, bugs, or issues in software development life cycle. It is a type of issue tracking system. Bug tracking systems are commonly employed by a large number of OSS systems and most of these tracking systems allow the users to generate various types of defect reports directly. Typical bug tracking systems are integrated with other software project management tools and methodologies. Some systems are also used internally by some organizations (http://www.dmoz.org).

A database is a crucial component of a bug tracking system, which stores and maintains information regarding the bugs reported by the users and/or developers. These bugs are generally referred to as known bugs. The information about a bug typically includes the following:

- The time when the bug was reported in the software system
- Severity of the reported bug
- Behavior of the source program/module in which the bug was encountered
- Details on how to reproduce that bug
- Information about the person who reported that bug
- Developers who are possibly working to fix that bug, or will be assigned the job to do so

Many bug tracking systems also support tracking through the status of a bug to determine what is known as the concept of bug life cycle. Ideally, the administrators of a bug tracking system are allowed to manipulate the bug information, such as determining the possible values of bug status, and hence the bug life cycle states, configuring the permissions based on bug status, changing the status of a bug, or even remove the bug information from the database. Many systems also update the administrators and developers associated with a bug through emails or other means, whenever new information is added in the database corresponding to the bug, or when the status of the bug changes.

The primary advantage of a bug tracking system is that it provides a clear, concise, and centralized overview of the bugs reported in any phase of the software development life cycle, and their state. The information provided is valuable for defining the product road map and plan of action, or even planning the next release of a software system (Spolsky 2000).

Bugzilla is one of the most widely used bug tracking systems. Several open source projects, including Mozilla, employ the Bugzilla repository.

5.7 Extracting Data from Software Repositories

The procedure for extracting data from software repositories is depicted in Figure 5.10. The example shows the data-collection process of extracting defect/change reports. The first step in the data-collection procedure is to extract metrics using metrics-collection tools such as understand and chidamber and kemerer java metrics (CKJM). The second step involves collection of bug information to the desired level of detail (file, method, or class) from the defect report and source control repositories. Finally, the report containing the software metrics and the defects extracted from the repositories is generated and can be used by the researchers for further analysis. The data is kept in software repositories in various types such as CVS, Git, SVN, ClearCase, Perforce, Mercurial, Veracity, and Fossil. These repositories are used for management of software content and changes, including documents, programs, user documentation, and other related information. In the next subsections, we discuss the most popular VCS, namely CVS, SVN, and Git. We also describe Bugzilla, the most popular bug tracking system.

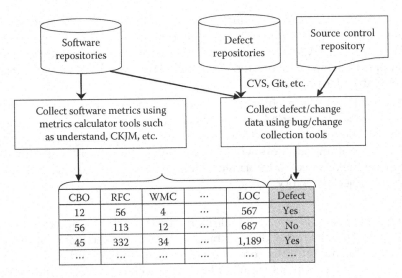

FIGURE 5.10
The procedure for defect/change data collection.

5.7.1 CVS

CVS is a popular CVCS that hosts a large number of OSS systems (Cederqvist et al. 1992). CVS has been developed with the primary goal to handle different revisions of various software project artifacts by storing the changes between two subsequent revisions of these artifacts in the repository. Thus, CVS predominantly stores the change logs rather than the actual artifacts such as binary files. It does not imply that CVS cannot store binary files. It can, but they are not handled efficiently.

The features provided by CVS are discussed below (http://cvs.savannah.gnu.org):

Revision numbers: Each new revision or version of a project artifact stored in the CVS repository is assigned a unique revision number by the VCS itself. For example, the first version of a checked in artifact is assigned the revision number 1.1. After the artifacts are modified (updated) and the changes are committed (permanently recorded) to the CVS repository, the revision number of each modified artifact is incremented by one. Since some artifacts may be more affected by updation or changes than the others, the revision numbers of the artifacts are not unique. Therefore, a release of the software project, which is basically a snapshot of the CVS repository, comprises of all the artifacts under version control where the artifacts can have individual revision numbers.

Branching and merging: CVS supports almost all of the functionalities pertaining to branches in a VCS. The user can create his/her own branch for development, and view, modify, or delete a branch created by the user as well as other users, provided the user is authorized to access those branches in the repository. To create a new branch, CVS chooses the first unused even integer, starting with 2, and appends it to the artifacts' revision number from where the branch is forked off, that is, the user who has created that branch wishes to work on those particular artifacts only. For example, the first branch, which is created at the revision number 1.2 of an artifact, receives the branch number 1.2.2 but CVS internally stores it as 1.2.0.2. However, the main issue with branches is that the detection of branch merges is not supported by CVS. Consequently, CVS does not boast of enough mechanisms that support tracking of evolution of typically large-sized software systems as well as their particular products.

Version control data: For each artifact, which is under the repository's version control, CVS generates detailed version control data and saves it in a change log or simply log files. The recorded log information can be easily retrieved by using the CVS log command. Moreover, we can specify some additional parameters so as to allow the retrieval of information regarding a particular artifact or even the complete project directory.

Figure 5.11 depicts a sample change log file stored by the CVS. It shows the versioning data for the source file "nsCSSFrameConstructor.cpp," which is taken from the Mozilla project. The CVS change log file typically comprises of several sections and each section presents the version history of an artifact (source file in the given example). Different sections are always separated by a single line of "=" characters.

However, a major shortcoming of CVS that haunts most of the developers is the lack of functionality to provide appropriate mechanisms for linking detailed modification reports and classifying changes (Gall et al. 2003).

The following attributes are recorded in the above commit record:

- RCS file: This field contains the path information to identify an artifact in the repository.

- Locks and AccessList: These are file content access and security options set by the developer during the time of committing the file with the CVS. These may be used to prevent unauthorized modification of the file and allow the users to only download certain file, but does not allow them to commit protected or locked files with the CVS repository.

- Symbolic names: This field contains the revision numbers assigned to tag names. The assignment of revision numbers to the tag names is carried out individually for each artifact because the revision numbers might be different.

- Description: This field contains the *modification reports* that describe the change history of the artifact, beginning from the first commit until the current version. Apart from the changes incurred in the head or main trunk, changes in all the branches are also recorded there. The revisions are separated by a few number of "-" characters.

- Revision number: This field is used to identify the revision of source code artifact (main trunk, branch) that has been subject to change(s).

- Date: This field records the date and time of the check in.

- Author: This field provides the information of the person who committed the change.

- State: This field provides information about the state of the committed artifact and generally assumes one of these values: "Exp" (experimental) and "dead" (file has been removed).

```
RCS file:
/cvsroot/mozilla/layout/html/style/src/nsCSSFrameConstructor.cpp,v
Working file: nsCSSFrameConstructor.cpp

head: 1.804
branch:
locks: strict
access list:

symbolic names:
    MOZILLA_1_3a_RELEASE: 1.800
    NETSCAPE_7_01_RTM_RELEASE: 1.727.2.17
    PHOENIX_0_5_RELEASE: 1.800
    ...
    RDF_19990305_BASE: 1.46
    RDF_19990305_BRANCH: 1.46.0.2

keyword substitution: kv
total revisions: 976; selected revisions: 976

description:
----------------------------
revision 1.804
date: 2002/12/13 20:13:16; author: doe@netscape.com; state: Exp; lines: +15 - 47

bug     950151:    crash    in    mozalloc_abort(char     const*    const)     |
mozalloc_handle_oom(unsigned             int)          |          moz_xmalloc         |
mozilla::SharedBuffer::Create(unsigned int)

....
----------------------------
....
========================================================================

RCS file:
/cvsroot/mozilla/layout/html/style/src/nsCSSFrameConstructor.h,v
```

FIGURE 5.11
Example log file from Mozilla project at CVS.

- Lines: This field counts the lines added and/or deleted of the newly checked in revision compared with the previous version of a file. If the current revision is also a branch point, a list of branches derived from this revision is listed in the branches field. In the above example, the *branches* field is blank, indicating that the current revision is not a branch point.
- Free Text: This field provides the comments entered by the *author* while committing the artifact.

5.7.2 SVN

SVN is a commonly employed CVCS provided by the Apache organization that hosts a large number of OSS systems, such as Tomcat and other Apache projects. It is also free and open source VCS.

Being a CVCS, SVN has the capability to operate across various networks, because of which people working on different locations and devices can use SVN. Similar to other VCS, SVN also conceptualizes and implements a version control database or repository in the same manner. However, different from a working copy, a SVN repository can be considered as an abstract entity, which has the ability to be accessed and operated upon almost exclusively by employing the tools and libraries, such as the Tortoise-SVN.

The features provided by SVN are discussed below:

Revision numbers: Each revision of a project artifact stored in the SVN repository is assigned a unique natural number, which is one more than the number assigned to the previous revision. This functionality is similar to that of CVS. The initial revision of a newly created repository is typically assigned the number "0," indicating that it consists of nothing other than an empty trunk or main directory. Unlike most of the VCS (including CVS), the revision numbers assigned by SVN apply to the entire repository tree of a project, not the individual project artifacts. Each revision number represents an entire tree, or a specific state of the repository after a change is committed. In other words, revision "i" means the state of the SVN repository after the "ith" commit. Since some artifacts may be more affected by updation or changes than the others, it implies that the two revisions of a single file may be the same, since even if one file is changed the revision number of each and every artifact is incremented by one. Therefore, every artifact has the same revision number for a given version of the entire project.

Branching and merging: SVN fully provides the developers with various options to maintain parallel branches of their project artifacts and directories. It permits them to create branches by simply replicating or copying their data, and remembers that the copies which are created are related among themselves. It also supports the duplication of changes from a given branch to another. SVN's repository is specially calibrated to support efficient branching. When we duplicate or copy any directory to create a branch, we need not worry that the entire SVN repository will grow in size. Instead, SVN does not copy any data in reality. It simply creates a new directory entry, pointing to an existing tree in the repository. Owing to this mechanism, branches in the SVN exist as normal directories. This is opposed to many of the other VCS, where branches are typically identified by some specific "labels" or identifiers to the concerned artifacts.

SVN also supports the merging of different branches. As an advantage over CVS, SVN 1.5 had incorporated the feature of merge tracking to SVN. In the absence of this feature, a great deal of manual effort and the application of external tools were required to keep track of merges.

Version control data: This functionality is similar to CVS. For each artifact, which is under version control in the repository, SVN also generates detailed version control data and stores it to change log or simply log files. The recorded log information can be easily retrieved by using a SVN client, such as Tortoise-SVN client, and also by the "svn log" command. Moreover, we can also specify some additional parameters so as to allow the retrieval of information regarding a particular artifact or even the complete project directory.

Figure 5.12 depicts a sample change log file stored by the SVN repository. It presents the versioning data for the source file "mbeans-descriptors.dtd" of the Apache's Tomcat project.

Although the SVN classifies changes to the files as modified, added, or deleted, there are no other classification types for the incurred changes that are directly provided by it,

```
=================================================================

Revision: 1561635

Actions: Modified

Author: kkolinko

Date: Monday, January 27, 2014 4:49:44 PM

Bugzilla ID: Nil

Modified:
/tomcat/trunk/java/org/apache/tomcat/util/modeler/mbeans-descriptors.dtd

Added:  Nil

Deleted: Nil

Message:

Followup to r1561083
Remove svn:mime-type property from *.dtd files.
The value was application/xml-dtd.

A mime-type is not needed on these source files, and it is inconvenient:
application/* mime types make SVN to treat the files as binary ones

=================================================================
```

FIGURE 5.12
Example log file from Apache Tomcat project at SVN.

such as classifying changes for enhancement, bug-fixing, and so on. Even though we have a "Bugzilla-ID" field, it is still optional and the developer committing the change is not bound to specify it, even if he has fixed a bug already reported in the Bugzilla database.

The following attributes are recorded in the above commit record:

- Revision number: This field identifies the source code revision (main trunk, branch) that has been modified.
- Actions: This field specifies the type of operation(s) performed with the file(s) being changed in the current commit. Possible values include "Modified" (if a file has been changed), "Deleted" (if a file has been deleted), "Added" (if a file has been added), and a combination of these values is also possible, in case there are multiple files affected in the current commit.
- Author: This field identifies the person who did the check in.
- Date: Date and time of the check in, that is, permanently recording changes with the SVN, are recorded in the date field.
- Bugzilla ID (optional): This field contains the ID of a bug (if the current commit fixes a bug) that has also been reported in the Bugzilla database. If specified, then this field may be used to link the two repositories: SVN and Bugzilla, together.

We may obtain change logs from the SVN (through version control data) and bug details from the Bugzilla.

- **Modified:** This field lists the source code files that were modified in the current commit. In the above log file, the file "mbeans-descriptors.dtd" was modified.
- **Added:** This field lists the source code files that were added to the project in the current commit. In the above log file, this field is not specified, indicating that no files have been added.
- **Message:** The following *message* field contains informal data entered by the author during the check in process.

5.7.3 Git

Git is a popular DVCS, which is being increasingly employed by a large number of software organizations and software repositories throughout the world. For instance, Google hosts maintains the source control data for a large number of its software projects through Git, including the Android operating system, Chrome browser, Chromium OS, and many more (http://git-scm.com).

Git stores and accesses the data as content addressable file systems. It implies that a simple Hash-Map, or a key–value pair is the fundamental concept of Git's data storage and access mechanism. The following terms are specific to Git and are an integral part of Git's data storage mechanism:

- Git object: It is an abstraction of the key–value pair storage mechanism of Git. It is also called a hash–object pair, and each object stores a secure hash value (sha) and a corresponding pointer to the data or files stored by Git for that hash value. The fields of a Git object are as follows:
  - SHA, a unique identifier for each object
  - Type of the Git object (string), namely, tree, tag, commit, or blob
  - Size of the Git object (integer)
  - Content of the Git object, represented in bytes.
- Tree: It eliminates the problem of storing different file names but supports storing different files together. A tree corresponds to branches or file directories. The fields of a Tree object are as follows:
  - Name of the Tree object (string)
  - Type of the Tree object, having the fixed value as "Tree"
- Blob: It corresponds to the Inodes in a Tree object, which store the File's information and content. Different blobs are created for different commits of a single file, and each blob is associated with a unique tag value. The fields of a Blob object are as follows:
  - Name of the Blob object (string)
  - Type of the Blob object, having the fixed value as "Blob"
- Commit: It is one that connects the Tree objects together. The fields of a Commit object are as follows:
  - Message specified while committing the changes (string)
  - Type of the Commit object, having the fixed value as "Commit"

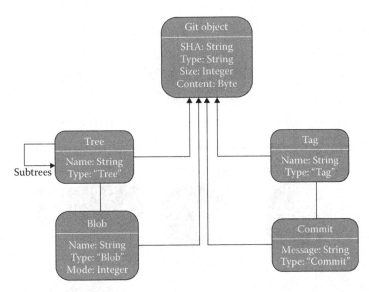

FIGURE 5.13
Git storage structure.

- Tag: It contains a reference to another Git object and may also hold some metadata of another object. The fields of a Tag object are as follows:
  - Name of the Tag object (string)
  - Type of the Tag object, having the fixed value as "Tag"

Figure 5.13 depicts the data structure or data model of Git VCS. Figure 5.14 presents an example of how data is stored by Git.

The tree has three edges, which correspond to different file directories. The first two edges point to blob objects, which store the actual file content. The third edge points to another tree or file directory, which stores the file "simplegit.rb" in the blob object.

However, Git visualizes and stores the information much differently than the other VCS, such as CVS, even though it provides a similar user interface. The important differences between Git and CVS are highlighted below:

Revision numbers: Similar to CVS, each new version of a file stored in the Git repository receives a unique revision number (e.g., 1.1 is assigned to the first version of a committed file) and after the commit operation, the revision number of each modified file is incremented by one. But in contrast to CVS, and many other VCS, that store the change-set (i.e., changes between subsequent versions), Git thinks of its data more like a set of snapshots of a mini file system. Every time a user performs a commit and saves the state of his/her project with Git, Git simply captures a snapshot of what all the files look like at that particular moment of committing, and then reference to that snapshot is stored. For efficiency, Git simply stores the link to the previous file, if the files in current and previous commit are identical.

Local operations: In Git, most of the operations are done using files on client machine, that is, local disk. For example, if we want to know the changes between current version and version created few months back. Git does local calculation by looking up the differences between current version and previous version instead of getting

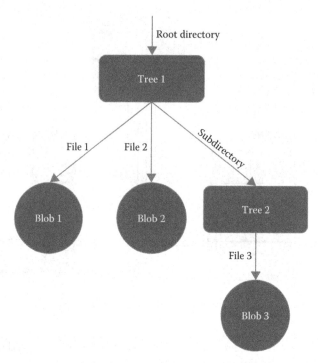

FIGURE 5.14
Example storage structure at Git.

information from remote server or downloading previous version from the remote server. Thus, the user feels the increase in speed as the network latency overhead will be reduced. Further, lots of work can be done offline.

Branching and merging: Git also provides its users to exploit its branching capabilities easily and efficiently. All the basic operations on a branch, such as creation, cloning, modification, and so on, are fully supported by Git. In CVS, the main issue with branches is that CVS does not support detection of branch merges. However, Git determines to use for its merge base, the best common ancestor. This is contrary to CVS, wherein the developer performing the merging has to figure out the best merge base himself. Thus, merging is much easier in Git.

Version control data: Similar to CVS, for each working file, Git generates version control data and saves it in log files. From here, the log file and its metadata can be retrieved by using the "git log" command. The specification of additional parameters allows the retrieval of information regarding a given file or a complete directory. Additionally, the log command can also be used with a "stat" flag to indicate the number of LOC changes incurred in each affected file after a commit is issued. Figure 5.15 presents an example of Git log file for Apache's log4j application (https://apache.googlesource.com/log4j).

In addition to the above differences, Git also maintains integrity (no change can be made without the knowledge of Git) and, generally, Git only adds data. The following attributes are recorded in the above commit record:

- Commit: Indicates the check-sum for this commit. In Git, everything is check-summed prior to being stored and is then referenced by using that checksum.

```
commit 1b0331d1901d63ac65efb200b0e19d7aa4eb2b8b
Author: duho.ro <duho.ro@lge.com>
Date:    Thu Jul 11 09:32:18 2013 + 0900


Bug: 9767739


UICC: fix read EF Image Instance
The EFs(4Fxx) path under DF Graphics are not distinguish with the EFs(4Fxx) path
under DF Phonebook. So, getEFPath(EF_IIDF) is not able to return correct path.
Because getEFPath(EF_IMG)is correct path, DF graphics, getEFPath(EF_IMG) is used
instead of getEFPath(EF_IIDF), EF_IMG is a linear fixed EF. The result of loading
EF_IMG should be processed as a LoadLinearFixedContext. So, it is needed to calculate
the number of EF_IMG records. If those changes are added, the changes are duplicated
with the codes of EVENT_GET_RECORD_SIZE_DONE. The codes of EVENT_GET_RECORD_SIZE_IMG_
DONE are removed and the event is treated by the logic of the EVENT_GET_RECORD_SIZE_
DONE. And then remove incorrect handler events(EVENT_READ_IMG_DONE and
EVENT_READ_ICON_DONE) are moved to the handler events which have the procedure for
loading same type EFs (EVENT_READ_RECORD_DONE and the EVENT_READ_BINARY_DONE).


.../internal/telephony/uicc/IccFileHandler.java   | 140 +++-------
.../internal/telephony/uicc/RuimFileHandler.java  |   8 +-

2 files changed, 38 insertions(+), 110 deletions(-)
```

FIGURE 5.15
Example log file for Android at Git.

It implies that it is impossible to modify the contents of any artifact file or even directory without the knowledge of Git.

- Date: This field records the date and time of the check in.
- Author: The author field provides the information of the person who committed the change.

Free text: This field provides informal data or comments given by the *author* during the commit. This field is of prime importance in extracting information for areas such as defect prediction, wherein bug or issue IDs are required to identify a defect, and these can be obtained after effectively processing this field. Following the free text, we have the list of files that have been changed in this commit. The name of a changed file is followed by a number which indicates the total number of LOC changes incurred in that file, which in turn is followed by the number of LOC insertions (the count of occurrences of "+") and LOC deletions (the count of occurrences of "−"). However, a modified LOC is treated as a line that is first deleted (−) and then inserted (+) after modifying. The last line summarizes the total number of files changed, along with total LOC changes (insertions and deletions).

Table 5.1 compares the following freely available features of the software repositories. These repositories can be mined to obtain useful information for analysis.

- Initial release: The date of initial release is specified.
- Development language: The programming language in which the system is developed.
- Maintained by: The name of the company that is currently responsible of the development and maintenance of the software.

TABLE 5.1

Comparison of Characteristics of CVS, SVN, and Git Repositories

Repository Characteristics	CVS	SVN	GIT
Initial release	July 3, 1986	October 20, 2000	April 3, 2005
Development language	C	C	C, Perl, Shell Script
Maintained by	CVS Team	Apache	Junio Hamano
Repository type	Client server	Client server	Distributed
License type	GNU-GPL (open source)	Apache/BSD style license (open source)	GNU-GPL v2 (open source)
Platforms supported	Windows, Unix, OS X	Windows, Unix, OS X	Windows, Unix, OS X
Revision IDs	Numbers	Numbers	SHA-1 hashes
Speed	Medium	High	Excellent
Ease of deployment	Good	Medium	Good
Repository replication	Indirect	Indirect	Yes (Git clone)
Example	Firefox	Apache, FreeBSD, SourceForge, Google Code	Chrome, Android, Linux, Ruby, Open Office

- Repository type: It describes the type of relationship that the various copies have with each other. In client–server model, the master copy is available on the server and the clients access them, and, in distributed model, users have local repositories available with them.
- License type: The license of the software.
- Platforms supported: The operating system supported by the repository.
- Revision IDs: The unique identifiers to identify releases and versions of the software.
- Speed: The speed of the software.
- Ease of deployment: How easily can the system be deployed?
- Repository replication: How easily the repository can be replicated?
- Example: Names of few popular softwares that use the specified VCS.

5.7.4 Bugzilla

Bugzilla is a popular bug tracking system, which provides access to bug reports for a large number of OSS systems. Bugzilla database can easily be accessed through HTTP and the defect reports can be retrieved from the database in the XML format. Bug reports thus obtained can aid managers and developers in the identification of defect-prone modules or files in the source code, which are candidates for redesign or reimplementation, and hence analyze such files more closely and carefully (http://www.bugzilla.org).

Additionally, contact information, mailing addresses, discussions, and other administrative information are also provided by Bugzilla. Some interesting patterns and information for the evolutionary view of a software system, such as bug severity, affected artifacts, and/or products or component may also be obtained from this bug database. Figure 5.16 depicts the schema diagram of Bugzilla database (referenced from http://bugzilla.org).

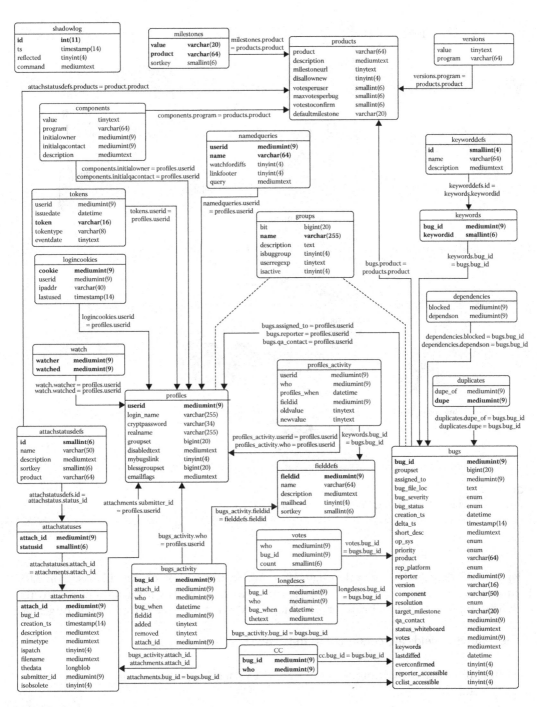

FIGURE 5.16
Bugzilla database schema.

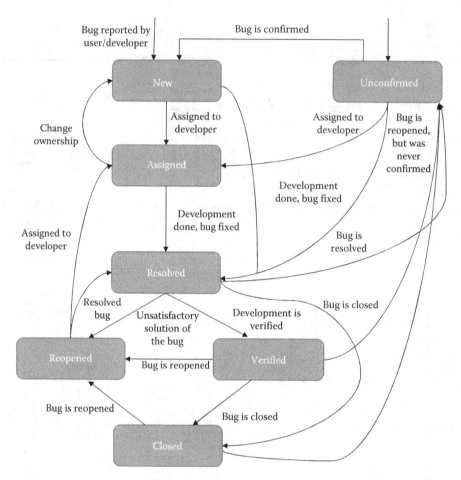

FIGURE 5.17
Life cycle of a bug at Bugzilla (http://www.bugzilla.org).

Figure 5.17 depicts the life cycle stages of a bug, in the context of Bugzilla (version 3.0). Figure 5.18 presents a sample XML report, which provides information regarding a defect, reported in the browser product of Mozilla project (http://www.bugzilla.mozilla.org).

The following fields are contained in the above bug report record:

- Bug ID: This is the unique ID assigned to each bug reported in the software system.

- Bug status: This field contains information about the current status or state of the bug. Some of the possible values include *unconfirmed*, *assigned*, *resolved*, and so on. The status whiteboard can be used to add notes and tags regarding a bug.

- Product: This field implies the product of the software project that is affected by a bug. For Mozilla project, some products are Browser, MailNews, NSPR, Phoenix, Chimera, and so on.

```
<bug_id> 950155 </bug_id>
<bug_status> NEW </bug_status>
<product> Firefox </product>
<priority> -- </priority>
<version> Trunk </version>
<rep_platform> x86 Windows NT </rep_platform>
<assigned_to> Nobody; OK to take and work on it </assigned_to>
<delta_ts> 20020116205154 </delta_ts>
<component>  Untriaged  </component>
<reporter>  alex_mayorga  </reporter>
<target_milestone> --- </target_milestone>
<bug_severity> critical </bug_severity>
<creation_ts> 2013 - 12 - 13 11:18 </creation_ts>
<op_sys> Windows NT </op_sys>
<short_desc> crash in mozalloc_abort(char const* const)    |
mozalloc_handle_oom(unsigned  int)    |       moz_xmalloc    |
mozilla::SharedBuffer::Create(unsigned int) </short_desc>

<keywords> crash </keywords>
<dependson> --- </dependson>
<blocks> --- </blocks>
<long_desc>
    <who> alex_mayorga </who>
    <bug_when> 2013 - 12 - 13 11:18 </bug_when>
    <thetext> --- </thetext>

</long_desc>
```

FIGURE 5.18
Sample bug report from Bugzilla.

- Component: This field determines the component or constituent affected by a reported bug. In Mozilla, some examples of component are Java, JavaScript, Networking, Layout, and so on.
- Depends on: It gives information about the bugs on which the reported bug depends. Those bugs have to be fixed prior to this bug.
- Blocks: Gives details of the bugs that are blocked by the reported bug.
- Bug severity: This field classifies the reported bug into various severity levels. The severity levels include blocker, critical, major, minor, trivial, enhancement, and so on.
- Target milestone: This field stores the possible target version of an artifact, that is, the changes should be merged into the main branch or trunk.

5.7.5 Integrating Bugzilla with Other VCS

We have described in detail the basic concepts and functioning of various VCS (Git, CVS, and SVN), and the most popular bug repository, the Bugzilla.

But, we may wonder that whether both of these two types of repositories (VCS and bug repository) are required, or only one of them is sufficient? Well, we would like to say that although both of these serve different purposes, but their capabilities are such that they complement one another really well.

We know that a VCS maintains its repository, the required information regarding each and every change incurred in the source code of a software project under versioning. Through the change logs maintained by the VCS, we may come across certain changes that had been incurred for bug fixing. So we wonder that if VCS can also provide bug information, then what is the need of a bug-tracking system like Bugzilla? Well, as we have discussed in the previous section, we can obtain detailed information about a bug from Bugzilla, such as the bug life cycle, the current state of a bug, and so on. All this information cannot be obtained by a VCS.

Similarly, a bug repository can neither provide information regarding the changes that were incurred for purposes other than defect-fixing, nor does it store the different versions and details of a project (and its artifacts) that are maintained in the VCS.

Therefore, some organizations typically employ both a VCS and a bug-tracking system to serve the dual purpose of versioning and defect data management. For instance, the Mozilla open source project is subject to versioning under the CVS repository, while the bugs for that project are reported in the Bugzilla database. For such a project, we may obtain the bug IDs from the change logs of the VCS and then link or map these bug IDs to the ones stored in the Bugzilla database. We may then obtain detailed information about both the committed changes, and the bugs that were fixed by these changes. We have also stated in Section 5.7.2 related to SVN that the SVN change logs contain an optional Bugzilla ID field, which may be employed to link the change log information provided by SVN to the bug information in the Bugzilla database.

5.8 Static Source Code Analysis

Source code is one of the most important artifacts available in the software repositories. Generally, various versions of the source code are maintained in the repositories. This allows the researchers and practitioners to examine the changes made between various versions of the software. A number of facts and information can be extracted from the source code.

The source code of a software system may be easily obtained by "cloning" a source code software repository where that particular system is hosted. Cloning simply means to copy and paste the entire software repository from a server (usually a remote server) to the end user's system. Source code is a very crucial artifact of any software system, which can be used to reveal interesting findings for that system through effective analysis.

For example, Git repositories may be easily cloned by applying the git "clone" command. The general syntax of the git clone command is:

```
'git clone [remote repository URL] --branch [exact branch name]
[destination path of end-user machine]'
```

The [remote repository URL] indicates the URL of a git repository and must be specified. The [exact branch name] of the main trunk needs to be specified, if we wish to clone a specific branch/version of the repository. If the branch is not specified, then the trunk (i.e., the latest version) will be cloned. The [destination path of end-user machine] may be specified, if the user wishes to download the repository in a specific location on his machine.

For example, the following clone command will download the repository for Android OS "Contacts" Application, for the branch "android_2.3.2_r1" to the destination "My Documents":

```
'git clone https://android.googlesource.com/platform/packages/apps/
Contacts --branch android_2.3.2_r1 C:/Users/My Documents '
```

5.8.1 Level of Detail

After we have obtained the source code from a software repository, we may perform various types of analysis. The analysis may be carried out at various levels of detail in the source code. The level of detail depends on the kind of information the researchers or practitioners want to extract. The level of detail, in turn, determines the method of source code analysis. A lower or finer detail level usually requires a more robust and complex analysis method(s). The detail levels may be generally classified as follows:

5.8.1.1 Method Level

A method is at the lowest level of granularity or at the highest level of detail in source code. Method-level analysis could report various parameters and measures, such as the number of LOC in a method, the dependencies between different methods in the source code, the complexity measures of a method, and so on. For a differential analysis of the source code of two given versions of a software, method-level analysis could also reveal the addition of new dependencies or removal of previous dependencies between methods in the source code. This level of analysis is more complex and usually more time consuming than the other levels of detail.

5.8.1.2 Class Level

After the methods, we have classes present in the source code at the next level of detail. Class-level analysis of the source code may reveal various measures such as the number of methods in a class, coupling between different classes, cohesion within a class, LOC for a class, and many complexity measures, including the cyclomatic complexity (CC). A differential analysis at the class level may provide information such as the number of lines added, deleted, and modified for a given class, changes in the number of methods in a class, and so on. Class-level analysis remains the most commonly exploited granularity level by most of the researchers and practitioners.

5.8.1.3 File Level

A file in the source code may be considered as the group of one or more classes. File level is therefore at a lower level of detail. File-level analysis reports measures such as the number of classes in a file, LOC in a file, and so on. Additionally, we generally consider file-level detail for a differential analysis of the source code for two versions of a software. This analysis reports measures such as the LOC changes (addition, deletion, and modification) reported for a file, the number of classes changed, and so on.

5.8.1.4 System Level

System level is at the lowest level of detail or at the highest level of granularity. System-level analysis is usually easier than the higher levels of detail and less number of parameters may be reported after analysis, such as the LOC, the total number of files, and so on. For a differential analysis, we may report the system-level measures, including changes in LOC (added, deleted, and modified), changes in the number of files (added, deleted, and modified), and so on.

5.8.2 Metrics

Predominantly, the source code of a software system has been employed in the past to gather software metrics, which are, in turn, employed in various research areas, such as defect prediction, change proneness, evaluating the quality of a software system, and many more (Hassan 2008).

For the validation of impact of OO metrics on defect and change proneness, various studies have been conducted in the past with varied set of OO metrics. These studies show that Chidamber and Kemerer (CK) metric suite remains the most popularly employed metric suite in literature.

Studies carried out by various researchers (Basili et al. 1996; Tang et al. 1999; Briand et al. 2000a; El Emam et al. 2001a; Yu et al. 2002; Gyimothy et al. 2005; Olague et al. 2007; Elish et al. 2011; Malhotra and Jain 2012) show that OO metrics have a significant impact on defect proneness. Several studies have also been carried out to validate the impact of OO metrics on change proneness (Chaumum et al. 1999; Bieman et al. 2003; Han et al. 2010; Ambros et al. 2009; Zhou et al. 2009; Malhotra and Khanna 2013). These also reveal that OO metrics have a significant impact on change proneness.

However, most of these studies relied on metric data that was made publically available, or obtained the data manually, which is a time-consuming and error-prone process.

However, in the study to investigate the relationship between OO metrics and change proneness, Malhotra and Khanna (2013) had effectively analyzed the source code of software repositories (Frinika, FreeMind, and OrDrumbox) to calculate software metrics in an automated and efficient manner. They had gathered the source code for two versions of each of the considered software systems and then, with the help of Understand for Java (http://www.scitools.com/) software, they had collected the metrics for the previous version (Frinika—0.2.0, FreeMind—0.9.0 RC1, OrDrumbox—0.8.2004) of the software systems. Various OO and size metrics were collected and analyzed, including CK metrics. The software Understand for Java gives the metrics at various levels of detail such as files, methods, classes, and so on. Thus, metrics were collected for all the classes in the software systems. They assessed and predicted changes in the classes. The Understand software also provides metrics for the "unknown" classes. These values must be discarded, as such classes cannot be accessed.

Malhotra and Jain (2012) had also carried out a study to propose a defect prediction model for OSS systems, wherein the focus was on the applicability of OO metrics in predicting defect-prone classes. The metrics were collected by using CKJM metrics tool for calculating CK metrics. It is an open source application that calculates metrics for various suites, including CK and quality metrics for OO design (QMOOD). It operates on the Java byte code files (i.e., .class files), which can be obtained from the source code of OSS systems hosted at various repositories.

These studies have proven to be appropriate examples for how source code obtained from software repositories may be analyzed effectively and thus add to the value to software repository mining field. Tools such as Understand and CKJM, to a certain extent, advocate the importance of analyzing the source code obtained from software repositories and its application in popular research areas.

Now, we discuss some of the tools that generate the data for OO metrics for a given software project.

5.8.3 Software Metrics Calculation Tools

There are various metrics tools available in the literature to calculate OO metrics. These tools are listed below:

1. Understand

   It is a proprietary and paid application developed by SciTools (http://www.sci-tools.com). It is a static code analysis software tool and is mainly employed for purposes such as reverse engineering, automatic documentation, and calculation of source code metrics for software projects with large size or code bases. Understand basically functions through an integrated development environment (IDE), which is designed to aid the maintenance of old code and understanding new code by employing detailed cross references and a wide range of graphical views. Understand supports a large number of programming languages, including Ada, C, the style sheet language CSS, ANSI C, and C++, C#, Cobol, JavaScript, PHP, Delphi, Fortran, Java, JOVIAL, Python, HTML, and the hardware description language VHDL. The calculated metrics include complexity metrics (such as McCabe's CC), size and volume metrics (such as LOC), and other OO metrics (such as depth of inheritance tree [DIT] and coupling between object classes [CBO]).

2. CKJM Calculator

   CKJM is an open source application written in the Java programming language (http://gromit.iiar.pwr.wroc.pl/p_inf/ckjm/metric.html). It is intended to calculate a total of 19 OO metrics for systems developed using Java. It supports various OO metrics, including coupling metrics (CBO, RFC, etc.), cohesion metrics (lack of cohesion in methods [LCOM] of a class, cohesion among methods of a class, etc.), inheritance metrics (DIT, number of children [NOC], etc.), size metrics (LOC), complexity metrics (McCabe's CC, average CC, etc.), and data abstraction metrics. The tool operates on the compiled source code of the applications, that is, on the byte code or .class files and then calculates different metrics for the same. However, it simply pipes the output report to the command line. But, it can be easily embedded in another application to generate metric reports in the desired format.

3. Source Monitor

   It is a freeware application written in C++ and can be used to measure metrics for software projects written in C++, C, C#, VB.NET, Java, Delphi, Visual Basic (VB6), or HTML (www.campwoodsw.com/sourcemonitor.html). This tool collects metrics in a fast, single pass through the source files. The user can print the metrics in tables and charts, and even export metrics to XML or CSV files for further processing. Source monitor supports various OO metrics, predominantly code metrics. Some of the code metrics provided by source monitor include: percent branch statements, methods per class, average statements per method, and maximum method or function complexity.

4. NDepend

It is a proprietary application developed using .NET Framework that can perform various tasks for systems written in .NET, including the generation of 82 code and quality metrics, trend monitoring for these metrics, exploring the code structure, detect dependencies, and many more (www.ndepend.com). Currently, it provides 12 metrics on application (such as number of methods, LOC, etc.), 17 metrics on assemblies (LOC and other coupling metrics), 12 metrics on namespaces (such as afferent coupling and efferent coupling at the namespace level), 22 metrics on type (such as NOC and LCOM), 19 metrics on methods (coupling and size metrics), and two metrics on fields (size of instance and afferent coupling at the field level). It can also be integrated with Visual Studio and performs lightweight and fast analysis to generate metrics. It is useful for the real-world applications.

5. Vil

It is a freeware application that provides different functionalities, including metrics, visualization, querying, and analysis of the different components of applications developed in .NET (www.1bot.com). The .NET components supported are assemblies, classes, and methods. It works for all of the .NET languages, including C# and Visual Basic.NET. It provides a large (and growing) suite of metrics pertaining to different entities such as classes, methods, events, parameters, fields, try/catch blocks, and so on, reported at multiple levels. Vil also supports various class cohesion, complexity, inheritance, coupling dependencies, and data abstraction metrics. Few of these metrics include: CC, LCOM, CBO, instability, distance, afferent, and efferent couplings.

6. Eclipse Metrics Plugin

It is an open source Eclipse plugin that calculates various metrics for code written in Java language during build cycles (eclipse-metrics.sourceforge.net). Currently, the supported metrics include McCabe's CC, LCOM, LOC in method, number of fields, number of parameters, number of levels, number of locals in scope, efferent couplings, number of statements, and weighted methods per class. It also warns the user of "range violations" for each of the calculated metric. This enables the developer to stay aware of the quality of the code he has written. The developer may also export the metrics to HTML web page or to CSV or XML file formats for further analysis.

7. SonarQube

It is an open source application written in the Java programming language (www.sonarqube.org). Mostly employed for Java-based applications, however, it also supports various other languages such as C, C++, PHP, COBOL, and many more. It also offers the ability to add our own rules to these languages. SonarQube provides various OO metrics, including complexity metrics (class complexity, method complexity, etc.), design metrics (RFC, package cycles, etc.), documentation metrics, (comment lines, comments in procedure divisions, etc.), duplication metrics (duplicated lines, duplicated files, etc.), issues metrics (total issues, open issues, etc.), size metrics (LOC, number of classes, etc.), and test metrics (branch coverage, total coverage, etc.).

8. Code Analyzer

   It is an OSS written in the Java programming language, which is intended for applications developed in C, C++, Java, Assembly, and HTML languages (source-forge.net/projects/codeanalyze-gpl). It calculates the OO metrics across multiple source trees of a software project. It offers flexible report capabilities and a nice tree-like view of software projects being analyzed. The metrics calculated include Total Files (for multiple file metrics), Code Lines/File (for multiple file metrics), Comment Lines/File (for multiple file metrics), Comment Lines, Whitespace Lines, Total Lines, LOC, Average Line Length, Code/Comments ratio, Code/Whitespace ratio, and Code/(Comments + Whitespace) ratio. In addition to the predefined metrics, it also supports user-defined software source metrics.

9. Pylint

   It is an open source application developed using Python programming language, which is intended for analyzing the source code of applications written in the Python language, and looks for defects or bugs and reveals possibly signs of poor quality (www.pylint.org). Pylint displays to the user a number of messages as it analyzes the Python source code, as well as some statistics regarding the warnings and errors found in different files. The messages displayed are generally classified into different categories such as errors and warnings. Different metrics are generated on the basis of these statistics. The metrics report displays summaries gathered from the source code analysis. The details include: a list of external dependencies found in the code; where they appear; number of processed modules; the total number of errors and warnings for each module; the percentage of errors and warnings; percentage of classes, functions, and modules with docstrings; and so on.

10. phpUnderControl and PHPUnit

    phpUnderControl is an add-on tool for the well-known continuous integration tool named the CruiseControl (http://phpUnderControl.org). It is an open source application written in Java and PHP languages, which aims at integrating some of the best PHP development tools available, including testing and software metrics calculator tools. PHPUnit is a tool that provides a framework for automated software tests and generation of various software metrics (http://phpunit.de). The software predominantly generates a list of various code, coverage, and test metrics, such as unit coverage, LOC, test to code ratio, and so on. The reports are generated in XML format, but phpUnderControl comes with a set of XSL style sheets that can format the output for further analysis.

Table 5.2 summarizes the features of the above-stated source code analysis tools.

5.9 Software Historical Analysis

As described in Section 5.3.1, software historical repositories record several kinds of information regarding the evolution and progress of a software project. Historical repositories may be mined to extract useful information for future trend analysis and other research areas. Here, we discuss in detail various approaches and applications of historical

TABLE 5.2

Summary of Source Code Analysis Tools

Tool	Availability	Source Language	Programming Language(s) Supported	Metrics Provided
Understand	Proprietary (paid)	IDE	C, C++, Ada, Java, C#, etc.	Complexity, size, volume, and a few other OO metrics
CKJM calculator	Open source	Java	Java byte code	OO design metrics
Source monitor	Proprietary (freeware)	C++	C, C++, C#, .NET, Java, etc.	Code metrics
NDepend	Proprietary (paid)	.NET	.NET languages	Various quality and code metrics
Vil	Proprietary (freeware)	.NET	.NET languages	OO design metrics
Eclipse metrics plugin	Open source	Java	Java (Eclipse IDE)	Code metrics
SonarQube	Open source	Java	Java, C, C++, PHP, Cobol, etc.	OO design, documentation size and test metrics
Code analyzer	Open source	Java	C, C++, Java, Assembly, HTML	Code metrics and allows user defined metrics
Pylint	Open source	Python	Python	Code and dependency metrics
phpUnderControl	Open source	PHP, Java	PHP	Code, coverage, and test metrics

repository analysis for any software system. Figure 5.19 gives an overview of the various applications of software historical analysis. The applications are explained in the subsections presented below.

5.9.1 Understanding Dependencies in a System

As stated earlier, information recorded in historical software repositories, such as archived communications and bug reports, is very valuable for the team members of any software project. This information can be used effectively for understanding various dependencies in a software system, which the traditional methods may not be able to do (Hassan 2008).

Existing methods for understanding and resolving dependencies in a software system include dependency graphs and also the source code documentation. However, these offer merely a static view of a software system and generally fail to reveal any kind of information regarding the history of a system or even the rationale and agenda behind the system's current state or design.

Hassan and Holt (2004b) have proposed mining source control repositories, a type of historical repositories, and attaching historical sticky notes corresponding to each code dependency in a software system. These notes record various information pertaining to a dependency such as the timestamp when it was introduced in the system, the developer who introduced it, and the motive behind adding it. For instance, by employing the historical sticky notes on a large open source operating system, the NetBSD system, many unexpected dependencies could be easily explained and analyzed.

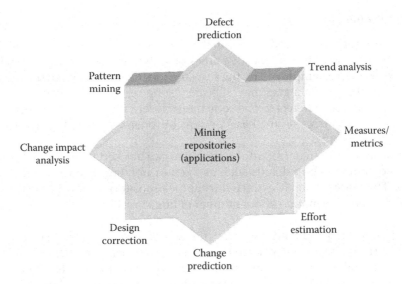

FIGURE 5.19
Applications of data mined from software repositories.

Thus, using the historical data, many unexpected dependencies can be easily revealed, explained, and rationalized.

5.9.2 Change Impact Analysis

In software configuration management, one of the important issues is to predict the impact of the requested change on the system, so that the decision related to implementation of change can be made. It can also increase program understanding and estimate the cost of implementing the change.

The software repositories maintain the information about short and long descriptions of bugs, the associated components with the bug, the details of the person who submitted the bug, and the details of the person who was assigned the bug. These details can be used to predict the modules/classes that will be affected by the change, and the developers that can handle these modules/classes (Canfora and Cerulo 2005).

5.9.3 Change Propagation

Change propagation is defined as the process of ensuring that the changes incurred in one module of a software system are propagated to the other modules to ensure that the assumptions and dependencies in the system are consistent, after the particular module has changed. As in the case of understanding dependencies, instead of employing traditional dependency graphs for propagating changes, we may use historical co-changes. Here, the intuition is that entities or modules that were co-changing frequently in the past are also likely to be co-changing in the future (Zimmermann et al. 2005). Change propagation is applied after change impact analysis is completed.

Hassan and Holt (2004a) also depicted that historical dependencies tend to outperform the traditional information when we carry out change propagation for various open source projects. Kagdi et al. (2007) presented the applicability of historical data in change propagation from code to documentation entities, where no structural code dependencies are present.

5.9.4 Defect Proneness

Defect proneness (or defect prediction), that is, predicting the occurrence of defects or bugs truly remains one of the most active areas of software engineering research. Defect-prediction results may be employed by the software practitioners of large software projects effectively. For instance, managers may allocate testing resources appropriately, developers may review and verify defect-prone code more closely, and testers may prioritize their testing efforts and resources on the basis of defect-proneness data (Aggarwal et al. 2005; Malhotra and Singh 2012).

The repositories maintain defect information that can be used for analysis. The available information contains the bug details such as fixed or not, bug fixed by, and the affected components. This information is extracted from the repositories for creation of defect prediction models. The information about number of bugs must be related to either a module or a class.

The defects may be predicted based on the severity levels. The testing professionals can select from the list of prioritized defects, according to the available testing resources using models predicted at various severity levels of defects. Zhou and Leung (2006) validated the NASA data set to predict defect-proneness models with respect to two categories of defects: high and low. They categorized defects with severity rating 1 as high-severity defects and defects with other severity levels as low-severity defects. Singh et al. (2010) categorized defects into three severity levels: high, medium, and low severity of defects. They also analyzed two different ML methods (artificial neural network and decision tree) for predicting defect-proneness models and evaluated the performance of these models using receiver operating characteristic (ROC) analysis.

5.9.5 User and Team Dynamics Understanding

Archived communications, such as mailing lists, IRC channels, and instant messaging, are commonly employed by the users as well as team members of many large projects for communications and various discussions. Various significant and intricate details such as project policies, design details, future plans, source code, and patch reviews are covered under these discussions (Rigby and Hassan 2007). Consequently, these discussions maintain a rich source of historical information about the internal operations and workings of large software projects. Mining these discussions and communications can aid us to better analyze and understand large software development teams' dynamics.

In the case of bug repositories, the users and developers are continually recording bugs for large software projects. However, each bug report must be prioritized to determine whether the report should be considered for addressing, and which developer should be assigned that report. Prioritizing defects is a time-consuming process and generally requires extensive knowledge regarding a project and the expertise of its developer team. Therefore, past bug reports may be employed to aid the bug-prioritization process. In a study carried out by Anvik et al. (2006), they were able to speed up the bug-prioritization process by employing prior or past bug reports so as to determine the most suitable developers to whom a bug should be assigned.

5.9.6 Change Prediction

Change proneness may be defined as the probability that a given component of the software will change. Along with defect prediction or defect proneness, change proneness is also very crucial and needs to be evaluated accurately. Prediction of change-prone classes may aid in maintenance and testing. A class that is highly probable to change in the later releases of a software needs to be tested rigorously, and proper tracking is required for that class, while modifying and maintaining the software (Malhotra and Khanna 2013).

Therefore, various studies have also been carried out in the past for predicting effective change-proneness models, and to validate the impact of OO metrics on change proneness. Historical analysis may be effectively employed for change-proneness studies. Similar to defect prediction studies, researchers have also adopted various novel techniques that analyze vast and varied data regarding a software system, which is available through historical repositories to discover probable changes in a software system.

5.9.7 Mining Textual Descriptions

The defect-tracking systems of OSS keep track of day-to-day defect reports that can be used for making strategic decisions. An example could be like "Which technique is more effective than the other for finding defects?" These repositories can be used to maintain unstructured data on defects that are encountered during a project's life cycle. While extensive testing can minimize these defects, it is not possible to completely remove these defects. After extracting textual descriptions from defect reports in software repositories, we can apply text mining techniques to extract relevant attributes from each report. Text mining involves processing of thousands of words extracted from the textual descriptions.

The bug-tracking systems contain information about defect description (in short and long form) along with bug severity. Menzies and Marcus (2008) mined defect description using text mining and used rule-based learning to establish relationship between information extracted from defect descriptions and severity. The results in the study were validated using defect reports obtained from NASA projects.

Harman et al. (2012b) mined textual descriptions from mobile apps to extract relationships from technical (features offered), business (price), and customer (ratings and downloads) perspective. The study used 32,108 apps that were priced non-zero from the Blackberry app store.

5.9.8 Social Network Analysis

In social and behavioral sciences, social network analysis (Wasserman et al. 1994) is a widely used technique that can be used to derive and measure unseen relationships between people, that is, the social entities. In the context of MSR, social network analysis may be effectively applied for discovering information related to software development, such as developer roles, contributions, and associations.

Huang and Liu (2005) had proposed an approach for social network analysis that was based on the analysis of CVS logs (deltas) to group the developers. The developers' contributions at a directory (module) level were also determined by the analysis of these logs.

Ohira et al. (2005) devised a visualization tool, named Graphmania, whose goal was to provide support for cross-project or cross-company knowledge sharing as well as collaboration. Over 90,000 projects hosted on SourceForge were analyzed by the authors. It was observed that small projects typically consisted of relatively less number of developers. The tool was targeted for supporting the developers involved in a small project, so as to perform tasks by utilization of both the knowledge of the developers assigned to other different projects and also the relevant information extracted from other projects.

5.9.9 Change Smells and Refactoring

Over the years, researchers and practitioners have employed software historical data for refactoring source code based on change smells. Ratzinger et al. (2005) devised a graph visualization technique to identify change smells, wherein the nodes represent the classes and edges represent logical couplings.

Fowler (1999) introduced another notion of change smells that was based on the strength of logical couplings present between different entities. This notion was presented with an analogy to bad smell. Change smells may be treated as the indicators of structural deficiencies. These are the candidates for reengineering or refactoring based on the software's change history. Fowler discussed two change smells: man-in-the-middle and data containers, and refactorings based on these two. To remove the man-in-the-middle difficulty, he suggested standard refactoring methods such as move method and move field. For improving the code that exhibits data-container smell, move method and extract method refactoring techniques were suggested.

5.9.10 Effort Estimation

Accurate estimation of resources, costs, manpower, and so on is critical to be able to monitor and control the project completion within the time schedule. Software effort estimation can provide a key input to planning processes. The overestimation of software development effort may lead to the risk of too many resources being allocated to the project. On the other hand, the underestimation of the software development effort may lead to tight schedules. The importance of software development effort estimation has motivated the construction of models to predict software development effort in recent years. In Malhotra and Jain (2011), various ML techniques have been used for predicting effort using date collected from 499 projects.

5.10 Software Engineering Repositories and Open Research Data Sets

Table 5.3 summarizes the characteristics of various software repositories and data sets that are discussed in detail in this section.

5.10.1 FLOSSmole

Its former name was OSS mole. FLOSSmole is a project that has been collaboratively designed to gather, share, and store comparable data and is used for the analysis of free and OSS development for the purpose of academic research (http://flossmole.org). FLOSSmole maintains data and results about FLOSS projects that have not been developed in a centralized manner.

The FLOSSmole repository provides data that includes source code, project metadata and characteristics (e.g., programming languages, platform, target audience, etc.), developer-oriented information, and issue-tracking data for various software systems.

The purpose of FLOSSmole is to provide widely used data sets of high quality, and sharing of standard analyses for validation, replication, and extension. The project contains the results of collective efforts of many research groups.

Table 5.3

Summary of Software Repositories and Data Sets

Repository	Web Link	Source	Data Format	Sources	Public
FLOSSmole	http://flossmole.org	OSS	DB dumps, text, DB access	Multiple	Yes
FLOSSMetrics	http://flossmetrics.org	OSS	DB dumps, web service, web	Multiple	Yes
PROMISE	http://promisedata.org	Mostly proprietary	Mostly ARFF	Multiple	Yes
Qualitas Corpus	http://qualitascorpus.com	OSS	CSV, source code, JAR	Multiple	Yes
Sourcerer project	http://sourcerer.ics.uci.edu	OSS	DB dumps	Multiple	Yes
UDB	http://udd.debian.org	OSS	DB dump	Single (Debian)	Yes
Bug prediction data set	http://bug.inf.usi.ch	OSS	CSV	Multiple	Yes
ISBSG	http://www.isbsg.org	Proprietary	Spreadsheet	Multiple	No
Eclipse bug data	http://www.st.cs.uni-saarland.de/softevo/bug-data/eclipse	OSS	ARFF, CSV	Single (Eclipse)	Yes
SIR	http://sir.unl.edu	OSS	C/Java/C#	Multiple	Needs registration
Ohloh	http://www.ohloh.net	OSS	Web service (limited)	Multiple	Yes
SRDA	http://zerlot.cse.nd.edu	OSS	BD dumps	Multiple	Needs registration
Helix data set	http://www.ict.swin.edu.au/research/projects/helix	OSS	CSV	Multiple	Yes
Tukutuku	http://www.metriq.biz/tukutuku	OSS	N/A	Multiple	No
SECOLD	http://www.secold.org	OSS	Web service, dumps	Multiple	Yes

5.10.2 FLOSSMetrics

FLOSSMetrics is a research project funded by the European Commission Sixth Framework Program. The primary goal of FLOSSMetrics is to build, publish, and analyze a large scale of projects, and also to retrieve information and metrics regarding the libre software development using pre-existing methodologies and tools that have already been developed. The project also provides its users with a public platform for validating and industrially exploiting the obtained results.

As of now, four types of repository metrics are offered: source code management information, code metrics (only for files written in C), mailing lists (archived communications), and bug-tracking system details. The project, while focusing on the software project development itself, also provides valuable information regarding the actors or developers, the project artifacts and source code, and the software processes.

The FLOSSMetrics is currently in its final stage, and some of the results and databases/data sets are already available. To be specific, the FLOSS community has proven to be a great opportunity for enhancing and stimulating empirical research in the scope of software engineering. Additionally, there are thousands of projects available in that community, and a majority of them provide their source code and artifact repositories to everyone for any application.

5.10.3 PRedictOr Models In Software Engineering

PRedictOr Models In Software Engineering (PROMISE) software management decisions are recommended to be based on well-understood and well-supported prediction models. It is problematic to collect data from real-world software projects. Because such data is not easy to attain, we must appropriately employ whatever data is available. This is the main goal of PROMISE repository (http://promisedata.org).

PROMISE repository hosts abundant data for a large number of applications and software systems, including Eclipse, Mozilla, and some popular Apache Software Foundation's Projects such as log-4j, ivy, ant, and many more.

Typical data provided by the PROMISE repository includes issue-tracking data and effort-related data for estimation models, including COnstructive COst MOdel (COCOMO), project metadata, general characteristics, and much more.

5.10.4 Qualitas Corpus

The qualitas corpus (QC) repository is a collection of software systems and projects developed using the Java programming language (http://qualitascorpus.com). The repository is intended to be employed for empirical studies of source code artifacts of Java-based applications. The primary goal of this project is to provide researchers and practitioners with resources that enable them to conduct reproducible or replicated studies of software systems and their properties. The repository data reduces the overall cost of conducting large empirical studies of source code and also supports the comparison of various measures of the same artifacts.

The collection of systems hosted at the repository is such that each of the systems consists of a set of versions. Each version comprises of the original application distribution (in compressed form) and two unpacked or uncompressed forms, bin (Java byte code or .class files, usually in .jar format) and src (Java source code or .java files).

The current release is version 20130901. It has 112 systems, 15 systems with 10 or more versions, and 754 versions total. Various domains represented in the corpus include 3D/media/graphics, IDE, SDK, database, tool, testing, games, middleware, programming languages, database, and so on.

5.10.5 Sourcerer Project

Sourcerer is an ongoing research project at the University of California, Irvine (http://sourcerer.ics.uci.edu). It is primarily meant for exploring the open source benefits and features through the application of code analysis. As we all know, the open source movement that has garnered tremendous support over the years has resulted in the generation of an extremely large body of code. This provides a tremendous opportunity to software engineering practitioners and researchers.

Sourcerer's managed repository stores and maintains the local copies of software projects that have been garnered from numerous open source repositories. As of now, the repository hosts as many as 18,000 Java-based projects obtained from Apache, Java.net, Google Code, and SourceForge.

Additionally, the project provides Sourcerer DB, which is a relational database whose structure and reference information are extracted from the projects' source code. Moreover, a code search engine has also been developed using the Sourcerer infrastructure.

5.10.6 Ultimate Debian Database

Ultimate Debian Database (UDD) project gathers a lot of data about various aspects of the Debian project (a free, Linux kernel-based OS) in the same SQL database where the project's data is stored (http://udd.debian.org). It allows the users to easily access and combine all these data.

Data that is currently subject to import includes bugs from the Debian bug-tracking system, packages, and sources files from Debian and Ubuntu, popularity contest, history of migrations to testing, history of uploads, and so on. UDD-based services include:

- Bugs search, that is, multicriteria search engine for information related to bugs.
- Debian maintainer dashboard that provides information regarding the Debian project and its development.
- Bugs usertags, which allow the users to search for user-specified tag on bugs.
- Sponsors stats, which provides some statistics regarding who is sponsoring uploads to Debian.
- Bapase, which allows the users to look for different packages using various criteria.

5.10.7 Bug Prediction Data Set

The bug prediction repository data set is a collection of metrics and models of software projects and as well as their historical data (http://bug.inf.usi.ch). The goal of such kind of data set is to allow practitioners and researchers to compare various defect prediction methodologies and to evaluate whether a new prediction technique is better than the existing ones. In particular, the repository contains the data required to:

- Apply a prediction or modeling technique based on the source code metrics, historical measures, or software process information (obtained from the CVS log data, etc.)
- Evaluate the performance of the prediction technique by comparing the results obtained with the actual number of postrelease bugs reported in a bug-tracking system

The repository has been designed to perform bug prediction in a given software system at the class level. However, we can also derive the package or subsystem information by aggregating the data for each class, because with each class, the package that contains it is specified. For each system hosted at the repository, the data set includes the following information:

- Biweekly versions of the software projects that are converted to OO models
- Historical information obtained from the CVS change log
- Fifteen metric values calculated from the CVS change log data, for each class or source code file of the systems
- Seventeen source code metrics (CKJM suite and eleven more OO metrics), for each version or revision of each class file in the source code
- Postrelease defects for each class file, categorized by severity and priority

5.10.8 International Software Benchmarking Standards Group

The International Software Benchmarking Standards Group (ISBSG) keeps a record of a repository of data from numerous organizations' software projects that have been completed (http://www.isbsg.org). The ISBSG data repository has many uses, including best-practice networking, project benchmarking, and summary analyses.

The goal of ISBSG is to improve IT performance through estimation, benchmarking, project planning and management, and IT infrastructure planning, and enhance management performance of software portfolio maintenance and support.

ISBSG repository data is truly a valuable asset for the software industry, practitioners and researchers, and for all the organizations that develop and produce software. The repository, which is open to the general public, has provided different research data sets on various topics, including project duration, function points structure, software cost estimation, and so on.

5.10.9 Eclipse Bug Data

Eclipse bug data (EBD) repository has been populated by mining the Eclipse's bug and versioning or version control databases (http://www.st.cs.uni-saarland.de/softevo/bug-data/eclipse). The primary goal of this repository is to map failures to Eclipse components. The data set obtained as a result provides information regarding the defect density for all the Eclipse components. The bug data set can be easily used for mapping and relating the source code, software processes, and developers to reported defects.

As of now, the repository has been populated from the analysis of three versions of Eclipse, namely 2.0, 2.1, and 3.0. All the three versions are open source and can be downloaded readily. The defect data reports are provided in the XML file format for each version, and the source code files and packages are structured hierarchically.

The data set populated in EBD is publicly available for download and use by anyone. A typical application of this data set is in defect prediction research and to validate various hypotheses on the nature and cause of defects as they occur in a software system during its life cycle.

5.10.10 Software-Artifact Infrastructure Repository

Software-artifact infrastructure repository (SIR) is a repository of software projects' artifacts that are meant to aid practitioners and researchers in performing rigorous and controlled experimentation with the project's source code analysis and software testing techniques (http://sir.unl.edu).

The repository contains data sets for many Java, C, C++, and C#-based software systems, in multiple versions, together with their supporting artifacts such as user manuals, test suites, defect data, and so on. The repository also maintains documentation on how to employ these artifacts for experimentation purposes, supporting tools, and methodologies that facilitate experimentation and gathering of useful information regarding the processes used to maintain and enhance the artifacts, and supporting tools that aid these processes.

The SIR repository data is freely made available to the users after they register with the project community and agree to the terms specified in SIR license.

5.10.11 Ohloh

Ohloh is a free, public directory of FOSS (Free and/or OSS), and also contains the information about the members and contributors who develop and maintain it (http://www .ohloh.net). Ohloh source code and repository is publicly available. Basically, it provides a free code search location or site that keeps an indexing for most of the projects hosted at Ohloh. Ohloh can be edited or modified by everyone, just like a wiki. Anyone can join and add new projects, and even make modifications to the existing project pages. Such public reviews have helped to make Ohloh one of the biggest, most accurate, and up-to-date FOSS directories available.

Ohloh does not host software projects and source code. Instead, it is a community, a directory, and analytics and search service. Ohloh can generate reports regarding the composition and activity of software project source code by connecting to the corresponding source code repositories, analyzing the source code's history updates being made currently, and attributing the updates to their respective contributors. It also aggregates this data to track the changing nature of the FOSS world.

Additionally, Ohloh provides various tools and methodologies for comparing projects, languages, repositories, and analyzing language statistics. Popular projects accessible from Ohloh include Google Chrome, Mozilla, WebKit, MySQL, Python, OpenGL, and many more. Ohloh is owned and operated by Black Duck software.

5.10.12 SourceForge Research Data Archive

SourceForge research data archive (SRDA) is a repository of FLOSS research project's data set. The data that is made available from the FLOSS research repository has been derived from the NSF funded project: "Understanding Open Source Software Development" (http://zerlot.cse.nd.edu). The SRDA research project aims to spread an awareness regarding the understanding of FOSS phenomenon and for predicting the pattern of growth that has been exhibited by FOSS projects over time.

The FOSS project community over the years has been able to develop a substantial amount of the Internet infrastructure, and has the support of several organizations and communities that develop OSS, including Apache, Perl, and Linux.

The following are various types of data that can be extracted from the SourceForge.net research data archive:

- Variation in project development team size over time, that is, the number of developers as a function of time.
- Participation of developers on projects, that is, number of projects in which individual developers participate.

- The above two measures are used to form what is known as a "collaboration social-network." This is used to obtain scale-free distributions among project activity and developer activity.
- The extended-community size for each project, which includes the number of project developers along with the registered members who have participated in any way in the development life cycle of a project, such as discussions on forums, bug reporting, patch submission, and so on.
- Date of creation at SourceForge.net for each software project.
- Date of the first release for each software project.
- Ranking of the projects as per SourceForge.net.
- Category-wise distribution of projects, for example, databases, games, communications, security, and so on.

5.10.13 Helix Data Set

The Helix data set is a collection of historical releases of many nontrivial Java OSS systems (http://www.ict.swin.edu.au/research/projects/helix). To aid researchers in empirical software engineering, Helix data set has been developed with a focus on software evolution.

Currently, there are over forty open source Java software systems with approximately 1,000 releases with more than 65,000 classes. Each and every system has at least 100 classes, with majority of them being far larger, that is, nontrivial. Also, each system has at least fifteen releases with more than 18 months of release history.

The data set (available as a ZIP file) provides an evolution history, with consistent metadata that also includes license information and a classification of software type. Also, more than fifty different metrics have been extracted for each release and have been made available in a simple CSV file format.

5.10.14 Tukutuku

The objectives of the Tukutuku benchmarking project are: first, data gathering on web projects that will be used to build company-specific or generic cost estimation models that will enable a web company to enhance its current cost estimation practices; and second, to enable a web company to benchmark its productivity within and across web companies (http://www.metriq.biz/tukutuku).

Till date the Tukutuku benchmarking project has gathered data on 169 web projects worldwide, and this data has been used to help several web companies.

5.10.15 Source Code ECO System Linked Data

The first ever online Linked Data repository containing source code facts is the Source code ECO system Linked Data (SECOLD). SECOLD V. 001 was the first version released, which was published on January 20, 2011. This version was an Ambient Software Evolution Group's (Concordia University) research project (http://www.secold.org).

SECOLD provides both implicit and explicit fact of any type that can be found in software repositories, for example, source code file, tokens, ast nodes, authors, licenses, bugs, commits, code clones, and so on. The SECOLD is independent of programming language.

Factual information from any source code or version control can be published by it. Nevertheless, the first release only contains Java code and SVN. The second release is expected to cover C#, C++, CVS, and Git.

The information contained in it has been extracted from source code, versioning, and bug/issue systems. The information pieces have been interconnected explicitly. The data has been extracted from approximately 18,000 open source projects with as much as 1,500,000 files and nearly 400,000,000 LOC. It is a multipurpose project. Its applications include mostly software research, software documentation/traceability, and enhancing the future of software development.

5.11 Case Study: Defect Collection and Reporting System for Git Repository

5.11.1 Introduction

Defect collection and reporting system (DCRS) is a software developed by undergraduate students of Delhi Technological University, Delhi, India (see Figure 5.20). The tool is intended for mining or extracting useful data from a certain subset of open source repositories (Malhotra et al. 2014).

OSS repositories provide rich data that can be extracted and employed in various empirical studies. Primary focus is on defect and change data that can be obtained from version control repositories of OSS systems. An important question here is that what

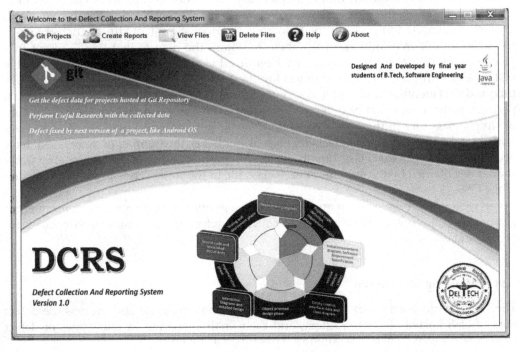

FIGURE 5.20
Defect collection and reporting system.

kind of software repositories are suitable for extracting such kind of data. After a rigorous study and analysis, we have chosen those software repositories that employ "Git" as the VCS (http://git-scm.com). The reasons for selecting Git-based software systems are as follows:

1. Git is the most popular VCS and the defect and change data can be extracted in relatively easier ways through the change logs maintained by Git.
2. A large number of OSS systems are maintained through Git, including Google's Android OS.

Thus, by employing DCRS, we can easily obtain defect and change data for a large number of OSS repositories that are based on Git VCS.

5.11.2 Motivation

Previous studies have shown that bug or defect data collected from open source projects may be employed in research areas such as defect prediction (defect proneness) (Malhotra and Singh 2012). For instance, some commonly traversed topics in defect prediction include analysis and validation of the effect of a given metric suite (such as CKJM and QMOOD) on defect-proneness in a system (Aggarwal et al. 2009); and evaluating the performance of pre-existing defect proneness models, such as machine learning methods (Gondra 2008). But, unfortunately, there exists no mechanism that can collect the defect data for Git-based OSS such as Android, and provide useful information for the above-stated areas.

Thus, a system is required that can efficiently collect defect data for a Git-based OSS, which in turn, to say the least, might be used in the above mentioned research areas. Such a system is expected to perform the following operations: First, obtain the defect logs of the software's source code and filter them to obtain the defects that were present in a given version of that software and have been fixed in the subsequent version. Then, the system should process the filtered defect logs to extract useful defect information such as unique defect identifier and defect description, if any. The next task that the system should perform is the association of defects to their corresponding source files (Java code files, or simply class files in the source code). In the next step, it should perform the computation of total number of fixed defects for each class, that is, the number of defects that have been associated with that class. Finally, the corresponding values of different metric suites should be obtained by the system for each class file in the source code of previous version of the OSS.

The DCRS incorporates each and every functionality stated above and, consequently, generates various reports that contain the collected defect data in a more processed, meaningful, and useful form.

5.11.3 Working Mechanism

The DCRS performs a series of well-defined operations to collect and process the defect data of OSS hosted at Git repository, and then finally generate various types of defect reports of the same. These operations are described as follows:

First, the DCRS processes an OSS's source code (for two of the predetermined consecutive versions) to retrieve what are known as change logs. A change log provides information regarding the modifications that have been made from time to time in the source code. These modifications could be for various purposes, such as defect fixing, refactoring, enhancements, and so on. Each and every change incurred in the source code, no matter how big or small, is recorded in the change log and thus forms an individual change or modification record. An individual change record provides information such as:

- The Timestamp of commit (i.e., recording the changes with Git repository)
- Unique change identifier
- Unique defect identifier, if the change has fixed any bug(s)
- An optional change description
- List of the modified source code files, along with the changes in LOC for each changed file

Figure 5.21 presents the generation of change logs by the DCRS for the Android application of "Mms."

Processing of both the version's source code is necessary, because a change log contains change information from the beginning of time (i.e., when the software was released for the first time), but we are interested only in the changes that have been incurred during the transition from previous version to the next one (e.g., from Android v4.0 to v4.2, for our demonstration of DCRS on Android OS).

Figure 5.22 depicts a change log record for an Android application package named "Mms":

In the next operation, these change logs are further processed, one record at a time, to get defect records (i.e., changes that have been made for defect fixes, not for other reasons like refactoring, enhancement, etc.). Defect IDs and the defect description, if any, are retrieved from the defect logs. Thus, a defect record differs from a change record only in one aspect that a defect record has at least one defect identifier. In other words, a defect log must contain the defect identifier(s) of the defect(s) that was/were fixed in the corresponding change. A description of the defect(s) may or may not be provided. In the latter case, a predefined value of description is stored for such defect records.

These defect IDs are finally mapped to classes in the source code. Only Java source code or class files are considered and other file formats are ignored. The defect data collected is thus used to accordingly generate various reports in .csv format, which are described later in this chapter.

Figure 5.23 presents the screen that generates various defect reports for the Android "Mms" Application.

The change logs required for the above process can only be obtained through the usage of appropriate Git Bash commands. Hence, Git Bash becomes a dependency for the DCRS, and if not installed and configured correctly in the system, DCRS will not work at all.

The entire DCRS system has been implemented in Java programming language (Java SE 1.7), using Eclipse RCP-Juno IDE. The data for required metrics for the class files of previous version of an OSS has been obtained using CKJM tool that covers a wide range of metrics (McCabe 1976; Chidamber and Kemerer 1994; Henderson 1996; Martin 2002). The procedure for defect collection and reporting is presented in Figure 5.24.

FIGURE 5.21
Generation of change logs by DCRS.

5.11.4 Data Source and Dependencies

For our demonstration, we've selected Android OS, with Git as its VCS. The two versions of Android OS, namely, Ice Cream Sandwich and Jelly Bean, have been considered for defect collection. The source code for both these versions has been obtained from Google's own Git repository, which is hosted at the URL: https://android.googlesource.com.

It was observed that the source code for android was not available as a single package. Instead, the code was distributed in as many as 379 application packages of Android OS. These include the packages for kernel, build libraries, compilation headers, and the native applications that are included in the OS, such as gallery, email, contacts, photo viewer, calendar, telephony, music player, and so on.

```
TIMESTAMP 1386010429

BEGINNING OF THE COMMIT RECORD
ClassZeroActivity: Queue messages instead of displaying them all at once

Making every AlertDialog immediately visible can lead to exhaustion
of graphics-related resources, typically memory, resulting in a
broken bufferqueue/hw renderer, and subsequent system crash.

Make ClassZeroActivity a singleTop activity, and queue incoming
messages if one is already being displayed.

Change-Id: Id7a2faa6997036507acad38f43fe17bf1f6a42cd

ENDING OF THE COMMIT RECORD BODY

AndroidManifest.xml                            |  1 +
src/com/android/mms/ui/ClassZeroActivity.java | 77 +++++++++++++++++++++++++------

2 files changed, 63 insertions(+), 15 deletions(-)
```

FIGURE 5.22
Change record from the change log file of Android "Mms" application.

As we are interested only in the Java source code or class files, we have analyzed a few of the available Android application packages to determine the fraction of Java source code files in each of these packages. It was noted that there are significantly fewer number of Java source code files as compared to other files types, such as layout files, media files, string value files, and so on, in every application package that has been analyzed.

The Android application packages were "cloned" (downloaded) by the DCRS itself. This functionality is discussed in detail in Section 5.8.

5.11.5 Defect Reports

From the defect data collected as stated in the above section, the tool generates the following primary reports.

5.11.5.1 Defect Details Report

This report provides useful information about each defect: The defect ID, description (if any), and the source file(s) (specifically Java source code files) that had been changed for fixing that defect.

The report also provides LOC changes source file-wise, corresponding to the defect that was fixed by modifying that particular file.

The fields that are contained in this report, corresponding to each defect, are the following:

- Java source file name or simply the class name
- Unique defect identifier
- Defect description, if any
- LOC inserted to fix the defect

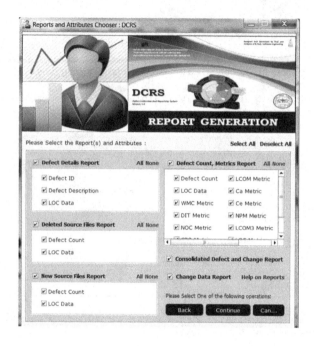

FIGURE 5.23
Generation of defect reports by DCRS.

- LOC deleted to fix the defect
- Total LOC changes to fix the defect (modification data was not available in the change logs)

Figure 5.25 presents the partial defect details report for Android "Mms" application.

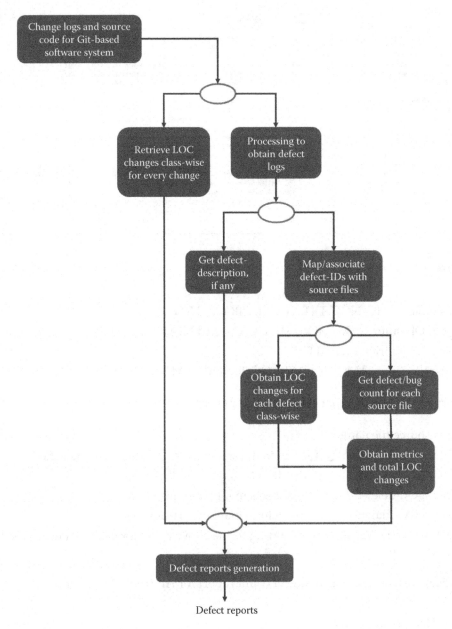

FIGURE 5.24
Flowchart for defect collection and reporting process.

5.11.5.2 *Defect Count and Metrics Report*

This report provides the total number of defects fixed source file-wise (i.e., total number of times a source file was modified to fix defects), along with total LOC changes, with CKJM and some additional metrics data. The metrics studied and incorporated are as follows (McCabe 1976; Chidamber and Kemerer 1994; Henderson 1996; Martin 2002):

1	SOURCE FILES CHANGED	DEFECT-ID	DESCRIPTION	INSERTION	DELETION	TOTAL CHANGES
2	src/com/android/mms/transaction/MessagingNotification.java	10495015	Turn off the DEBUG flag in MessagingNotifications before shi|	1	1	2
3	src/com/android/mms/transaction/TransactionService.java	10955896	Mms app should stop sending messages if it is not the default	3	1	4
4	src/com/android/mms/transaction/SmsReceiverService.java	10955896	Mms app should stop sending messages if it is not the default	2	1	3
5	src/com/android/mms/ui/SlideshowActivity.java	11084519	Messaging app crashed after playing a slide show Bug 110845!	6	4	10
6	src/com/android/mms/ui/ConversationList.java	10819173	Improve Messaging app behavior when not default SMS app. [	68	31	99
7	src/com/android/mms/ui/ConversationList.java	11006233	Improve Messaging app behavior when not default SMS app. [	68	31	99
8	src/com/android/mms/ui/ComposeMessageActivity.java	10819173	Improve Messaging app behavior when not default SMS app. [	19	23	42
9	src/com/android/mms/ui/ComposeMessageActivity.java	11006233	Improve Messaging app behavior when not default SMS app. [	19	23	42
10	src/com/android/mms/ui/MessageListItem.java	10641371	Remove smiley support Bug 10641371 - Remove ascii --> emo	7	6	13
11	src/com/android/mms/util/SmileyParser.java	10641371	Remove smiley support Bug 10641371 - Remove ascii --> emo	0	200	200
12	src/com/android/mms/ui/ConversationListItem.java	10641371	Remove smiley support Bug 10641371 - Remove ascii --> emo	2	2	4
13	tests/src/com/android/mms/util/SmileyParserUnitTests.java	10641371	Remove smiley support Bug 10641371 - Remove ascii --> emo	0	99	99
14	src/com/android/mms/ui/ComposeMessageActivity.java	10641371	Remove smiley support Bug 10641371 - Remove ascii --> emo	0	91	91
15	src/com/android/mms/MmsApp.java	10641371	Remove smiley support Bug 10641371 - Remove ascii --> emo	0	2	2
16	src/com/android/mms/widget/MmsWidgetService.java	10641371	Remove smiley support Bug 10641371 - Remove ascii --> emo	3	3	6
17	src/com/android/mms/data/WorkingMessage.java	8633269	When message is converted from SMS to MMS, there is no po	4	7	11
18	src/com/android/mms/ui/MessagingPreferenceActivity.java	141	Fix NPE trying to enable/disable removed preference. We wer	24	28	52
19	src/com/android/mms/ui/MessagingPreferenceActivity.java	141	Fix NPE trying to enable/disable removed preference. We wer	24	28	52
20	src/com/android/mms/MmsConfig.java	10870624	Update API used to determine default SMS app. In order to di	6	1	7
21	src/com/android/mms/ui/ConversationList.java	10819173	Update Mms app to better handle not being the default app A	60	4	64
22	src/com/android/mms/MmsConfig.java	10819173	Update Mms app to better handle not being the default app A	33	0	33
23	src/com/android/mms/ui/ComposeMessageActivity.java	10819173	Update Mms app to better handle not being the default app A	33	11	44

Defect Report ⊕

FIGURE 5.25
Example of defect details report records.

- CK suite: WMC, NOC, DIT, LCOM, CBO, and RFC
- QMOOD suite: DAM, MOA, MFA, CAM, and NPM
- Martin's metrics: Ca and Ce
- Miscellaneous: AMC, LCOM3, LOC, IC, and CBM

To summarize, the fields that are contained in this report are as follows:

- Java source file name
- Total number of defects for which the above source file has been modified
- Total LOC inserted for all the defects mapped to this class
- Total LOC deleted for all the defects mapped to this class
- Total LOC changes for all the defects mapped to this class
- List of metric values, corresponding to the above class, for each and every metric listed above

Figure 5.26 presents the partial defect count report for Android "Mms" application.

5.11.5.3 LOC Changes Report

This report gives the total number of LOC changes source file-wise, corresponding to each change (irrespective of the purpose of that change, i.e., defect fixing, refactoring, etc.).
To summarize, the fields that are contained in this report are as follows:

- Java source file name
- Total LOC inserted for all the changes (including defect fixes, enhancement, etc.) mapped to this class
- Total LOC deleted for all the changes mapped to this class
- Total LOC changes for all the changes mapped to this class

1	CLASS NAME	DEFEC	DEFE	INSERTION	DELETION	TOTAL-CH	WMC	DIT	NOC
2	src/com/android/mms/ui/SlideshowActivity.java	yes	2	11	4	15	9	0	0
3	src/com/android/mms/ui/ConversationListItem.java	yes	1	2	2	4	15	0	0
4	src/com/android/mms/TempFileProvider.java	yes	2	13	7	20	14	0	0
5	src/com/android/mms/ui/SlideshowAttachmentView.java	yes	1	2	9	11	22	0	0
6	src/com/android/mms/ui/ComposeMessageActivity.java	yes	7	96	142	238	128	0	1
7	src/com/android/mms/widget/MmsWidgetService.java	yes	1	3	3	6	2	0	0
8	src/com/android/mms/MmsConfig.java	yes	4	46	8	54	38	1	0
9	src/com/android/mms/transaction/SmsReceiverService.java	yes	3	33	4	37	22	0	0
10	src/com/android/mms/transaction/MmsSystemEventReceiver.java	yes	3	20	41	61	5	0	0
11	src/com/android/mms/transaction/SendTransaction.java	yes	1	1	1	2	4	0	0
12	src/com/android/mms/transaction/PushReceiver.java	yes	1	2	2	4	4	0	0
13	src/com/android/mms/data/Contact.java	yes	1	12	3	15	41	1	0
14	src/com/android/mms/data/WorkingMessage.java	yes	1	4	7	11	73	1	0
15	src/com/android/mms/MmsApp.java	yes	2	21	2	23	12	0	0
16	src/com/android/mms/transaction/SmsReceiver.java	yes	1	1	1	2	6	0	1
17	src/com/android/mms/ui/SlideListItemView.java	yes	1	2	9	11	22	0	0
18	src/com/android/mms/ui/ConversationList.java	yes	3	128	35	163	27	0	0
19	src/com/android/mms/transaction/TransactionService.java	yes	4	114	58	172	19	0	0
20	src/com/android/mms/util/SmileyParser.java	yes	1	0	200	200	6	1	0
21	src/com/android/mms/transaction/NotificationTransaction.java	yes	1	2	2	4	7	0	0
22	src/com/android/mms/transaction/Transaction.java	yes	2	17	42	59	16	0	4
23	src/com/android/mms/util/ThumbnailManager.java	yes	1	11	11	22	9	0	0

Defect Count ⊕

CBO	RFC	LCOM	Ca	Ce	NPM	LCOM3	LOC	DAM	MOA	MFA	CAM	IC	CBM	AMC
8	10	36	0	8	5	1.125	63	1	3	0	0.2222	0	0	0
13	16	105	0	13	10	1.0714	102	0.9167	1	0	0.219	0	0	0
8	41	85	0	8	12	0.9615	194	0.75	0	0	0.3846	0	0	0
8	23	231	0	8	21	1.0476	135	1	0	0	0.2121	0	0	0
46	129	8128	2	45	23	1.0079	879	0.8829	9	0	0.0656	0	0	0
3	3	1	0	3	2	2	14	1	0	0	0.75	0	0	0
4	39	703	1	3	37	1.027	269	1	0	0	0.0395	0	0	5
10	23	231	0	10	7	1.0476	150	0.7222	0	0	0.2614	0	0	0
3	6	10	0	3	4	1.25	32	1	1	0	0.4667	0	0	0
4	5	6	0	4	4	1.3333	27	1	0	0	0.4	0	0	0
3	5	6	0	3	2	1.3333	27	1	0	0	0.375	0	0	0
13	42	820	5	8	34	1.025	276	0.8	0	0	0.1328	0	0	5
11	74	2628	2	10	38	1.0139	484	0.6957	2	0	0.1041	0	0	5
9	13	66	2	7	12	1.0909	82	0.9	3	0	0.5	0	0	0
6	7	15	1	5	5	1.2	39	0.3333	1	0	0.3333	0	0	0
8	23	231	0	8	21	1.0476	137	1	0	0	0.2121	0	0	0
22	28	351	0	22	15	1.0385	193	0.7742	1	0	0.1068	0	0	0
12	20	171	0	12	7	1.0556	137	0.7391	0	0	0.2719	0	0	0
1	7	15	0	1	3	1.2	44	0.625	1	0	0.3889	0	0	5
4	8	21	0	4	6	1.1667	48	1	0	0	0.3571	0	0	0
8	17	120	4	4	10	1.0667	95	0.5556	2	0	0.25	0	0	0
9	10	36	1	8	7	1.125	68	0.7857	2	0	0.4	0	0	0

Defect Count ⊕

FIGURE 5.26
Example of defect count report records.

Figure 5.27 presents the partial LOC changes report for Android "Mms" application.

In addition to these, the following auxiliary reports are also generated by the DCRS, which might be useful for a statistical comparison of the two versions of Android OS application we have considered:

5.11.5.4 Newly Added Source Files

This report gives the list of source files that were not present in the previous version of Android OS, but have been added in the subsequent version. The total number of defects for each class file and LOC changes are also included in the report.

	FILE NAME	INSERTION	DELETION	TOTAL CHANGES
1				
2	src/com/android/mms/ui/ConversationListItem.java	2	2	4
3	tests/src/com/android/mms/util/SmileyParserUnitTests.java	0	99	99
4	src/com/android/mms/ui/SlideshowAttachmentView.java	2	9	11
5	src/com/android/mms/ui/SmsStorageMonitor.java	59	0	59
6	src/com/android/mms/exif/ExifInvalidFormatException.java	23	0	23
7	src/com/android/mms/transaction/ReadRecTransaction.java	6	2	8
8	src/com/android/mms/widget/MmsWidgetService.java	3	3	6
9	src/com/android/mms/MmsConfig.java	46	8	54
10	src/com/android/mms/exif/OrderedDataOutputStream.java	56	0	56
11	src/com/android/mms/exif/ByteBufferInputStream.java	48	0	48
12	src/com/android/mms/transaction/SmsReceiverService.java	39	4	43
13	src/com/android/mms/model/TextModel.java	1	1	2
14	src/com/android/mms/transaction/SendTransaction.java	1	1	2
15	src/com/android/mms/transaction/PushReceiver.java	2	2	4
16	src/com/android/mms/data/WorkingMessage.java	15	14	29
17	src/com/android/mms/MmsApp.java	21	2	23
18	src/com/android/mms/data/Contact.java	12	3	15
19	src/com/android/mms/transaction/SmsReceiver.java	1	1	2
20	src/com/android/mms/ui/MessageItem.java	1	1	2
21	src/com/android/mms/util/SmileyParser.java	0	200	200
22	src/com/android/mms/exif/ExifOutputStream.java	518	0	518
23	apptests/src/com/android/mms/tests/SmsSendIntentTestActivity.java	5	1	6

      Change Report   ⊕

FIGURE 5.27
Example of LOC changes report records.

To summarize, the fields that are contained in this report are as follows:

- Java source file name
- Total number of defects for which the above source file has been modified
- Total LOC inserted for all the defects mapped to this class file
- Total LOC deleted for all the defects mapped to this class file
- Total LOC changes for all the defects mapped to this class file

Figure 5.28 presents the newly added source files report for Android "Mms" application.

5.11.5.5 Deleted Source Files

This report gives the list of source files that were present in the previous version of Android OS, but have been removed from the subsequent version. The total number of defects and LOC changes are also included in the report (considering the probability of removing a source file in the latest version after fixing some defects).

To summarize, the fields that are contained in this report are as follows:

- Java source file name
- Total number of defects for which the above source file has been modified
- Total LOC inserted for all the defects mapped to this class
- Total LOC deleted for all the defects mapped to this class
- Total LOC changes for all the defects mapped to this class

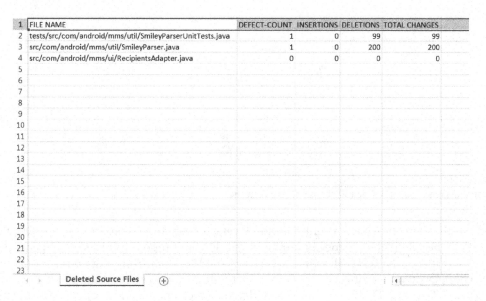

FIGURE 5.28
Example of newly added files report records.

FIGURE 5.29
Example of deleted files report records.

Figure 5.29 presents the deleted source files report for Android "Mms" application.

5.11.5.6 Consolidated Defect and Change Report

This report is similar to the defect or bug count report described in Section 5.11.5.2, but the only difference here is that instead of reporting the LOC changes corresponding to all the defects mapped to a particular class, we report total LOC changes that are incurred in that class. These changes may not only be incurred for issue, defect, or bug fixing, but

1	CLASS NAME	DEFECT	DEFECT-C(	CHANGE	INSERTIO	DELETION	TOTAL-CH	WMC	DIT	NOC
2	src/com/android/mms/ui/SlideshowActivity.java	yes	2	yes	11	4	15	9	0	0
3	src/com/android/mms/ui/ConversationListItem.java	yes	1	yes	2	2	4	15	0	0
4	src/com/android/mms/TempFileProvider.java	yes	2	yes	13	7	20	14	0	0
5	src/com/android/mms/ui/SlideshowAttachmentView.java	yes	1	yes	2	9	11	22	0	0
6	src/com/android/mms/ui/ComposeMessageActivity.java	yes	7	yes	107	150	257	128	0	1
7	src/com/android/mms/widget/MmsWidgetService.java	yes	1	yes	3	3	6	2	0	0
8	src/com/android/mms/MmsConfig.java	yes	4	yes	46	8	54	38	1	0
9	src/com/android/mms/transaction/SmsReceiverService.java	yes	3	yes	39	4	43	22	0	0
10	src/com/android/mms/transaction/MmsSystemEventReceiver.java	yes	3	yes	37	44	81	5	0	0
11	src/com/android/mms/transaction/SendTransaction.java	yes	1	yes	1	1	2	4	0	0
12	src/com/android/mms/transaction/PushReceiver.java	yes	1	yes	2	2	4	4	0	0
13	src/com/android/mms/data/Contact.java	yes	1	yes	12	3	15	41	1	0
14	src/com/android/mms/data/WorkingMessage.java	yes	1	yes	15	14	29	73	1	0
15	src/com/android/mms/MmsApp.java	yes	2	yes	21	2	23	12	0	0
16	src/com/android/mms/transaction/SmsReceiver.java	yes	1	yes	1	1	2	6	0	1
17	src/com/android/mms/ui/SlideListItemView.java	yes	1	yes	2	9	11	22	0	0
18	src/com/android/mms/ui/ConversationList.java	yes	3	yes	129	36	165	22	0	0
19	src/com/android/mms/transaction/TransactionService.java	yes	4	yes	137	73	210	19	0	0
20	src/com/android/mms/util/SmileyParser.java	yes	1	yes	0	200	200	6	1	0
21	src/com/android/mms/transaction/NotificationTransaction.java	yes	1	yes	2	2	4	7	0	0
22	src/com/android/mms/transaction/Transaction.java	yes	2	yes	17	42	59	16	0	4
23	src/com/android/mms/util/ThumbnailManager.java	yes	1	yes	13	11	24	9	0	0

    Consolidated Report   ⊕

CBO	RFC	LCOM	Ca	Ce	NPM	LCOM3	LOC	DAM	MOA	MFA	CAM	IC	CBM	AMC
8	10	36	0	8	5	1.125	63	1	3	0	0.2222	0	0	0
13	16	105	0	13	10	1.0714	102	0.9167	1	0	0.219	0	0	0
8	41	85	0	8	12	0.9615	194	0.75	0	0	0.3846	0	0	0
8	23	231	0	8	21	1.0476	135	1	0	0	0.2121	0	0	0
46	129	8128	2	45	23	1.0079	879	0.8829	9	0	0.0656	0	0	0
3	3	1	0	3	2	2	14	1	0	0	0.75	0	0	0
4	39	703	1	3	37	1.027	269	1	0	0	0.0395	0	0	5
10	23	231	0	10	7	1.0476	150	0.7222	0	0	0.2614	0	0	0
3	6	10	0	3	4	1.25	32	1	1	0	0.4667	0	0	0
4	5	6	0	4	4	1.3333	27	1	0	0	0.4	0	0	0
3	5	6	0	3	2	1.3333	27	1	0	0	0.375	0	0	0
13	42	820	5	8	34	1.025	276	0.8	0	0	0.1328	0	0	5
11	74	2628	2	10	38	1.0139	484	0.6957	2	0	0.1041	0	0	5
9	13	66	2	7	12	1.0909	82	0.9	3	0	0.5	0	0	0
6	7	15	1	5	5	1.2	39	0.3333	1	0	0.3333	0	0	0
8	23	231	0	8	21	1.0476	137	1	0	0	0.2121	0	0	0
22	28	351	0	22	15	1.0385	193	0.7742	1	0	0.1068	0	0	0
12	20	171	0	12	7	1.0556	137	0.7391	0	0	0.2719	0	0	0
1	7	15	0	1	3	1.2	44	0.625	1	0	0.3889	0	0	5
4	8	21	0	4	6	1.1667	48	1	0	0	0.3571	0	0	0
8	17	120	4	4	10	1.0667	95	0.5556	2	0	0.25	0	0	0
9	10	36	1	8	7	1.125	68	0.7857	2	0	0.4	0	0	0

FIGURE 5.30
Example of consolidated defect and change report records.

for any other purpose, such as enhancement, refactoring, and so on. In other words, this report can be considered as the combination of bug count report and LOC changes report (described in Section 5.11.5.3), where we report the bug data from bug count report, but LOC changes incurred for each class are reported from the LOC changes report.

Figure 5.30 presents the partial consolidated defect and change report for Android "Mms" application.

5.11.5.7 Descriptive Statistics Report for the Incorporated Metrics

This report gives various statistical measures for each and every metric incorporated in the tool. The various statistical measures for a metric include the following:

- The minimum value
- The maximum value
- The arithmetic mean

- Standard deviation
- Median (or 50 percentile)
- 25 and 75 percentile

Figure 5.31 presents the descriptive statistics report for Android "Mms" application.

5.11.6 Additional Features

Apart from the core functionalities of defect collection and report generation, the DCRS also provides some additional features or functionalities that might be helpful for the end user. We have placed these functionalities under two broad categories given below.

5.11.6.1 Cloning of Git-Based Software Repositories

The system provides full support for "cloning" software repositories that employ Git as the VCS. "Cloning" simply means copying and pasting the entire software repository from a local or remote hosting server to the end user machine.

Different types of artifacts may be obtained from a software repository depending on the type of that repository. For example, source control repositories provide change logs. Source code may be obtained through code repositories. In our case, the DCRS supports cloning of code repositories that employ Git as the source control repository or VCS. The user can thus obtain the source code as well as Git change logs, which were described earlier. These two artifacts may be then employed for various purposes. For instance, the DCRS employs these artifacts for generating various types of defect reports we have stated earlier. Figure 5.32 presents the cloning operation of DCRS.

	METRIC	MINIMUM	MAXIMUM	MEAN	MEDIAN	STANDARD DEVIATION	25 PERCENTILE	75 PERCENTILE
1								
2	WMC	0	128	11.46829	6	14.623242	3	16
3	DIT	0	4	0.595122	1	0.6529301	0	1
4	NOC	0	5	0.209756	0	0.7456562	0	0
5	CBO	0	46	5.414634	4	5.705692	2	8
6	RFC	0	129	16.45854	10	19.365913	4	22
7	LCOM	0	8128	152.0585	10	630.1567	1	88
8	Ca	0	13	1.253659	0	2.2006752	0	2
9	Ce	0	45	4.307317	3	5.385458	0	6
10	NPM	0	53	8.229268	4	9.900044	2	10
11	LCOM3	0	2	1.362886	1.1667	0.48655528	1.0401	2
12	LOC	0	1450	105.7707	49	183.39568	13	130
13	DAM	0	1	0.567686	0.7667	0.4491532	0	1
14	MOA	0	9	0.44878	0	1.0970286	0	0
15	MFA	0	1	0.017675	0	0.12603785	0	0
16	CAM	0	1	0.433859	0.4	0.2547986	0.2432	0.5556
17	IC	0	3	0.019512	0	0.21999906	0	0
18	CBM	0	5	0.029268	0	0.35492596	0	0
19	AMC	0	88.5	3.94186	0	9.894282	0	5
20								
21								
22								
23								

Descriptive Statisitcs ⊕

FIGURE 5.31
Example of descriptive statistics report records.

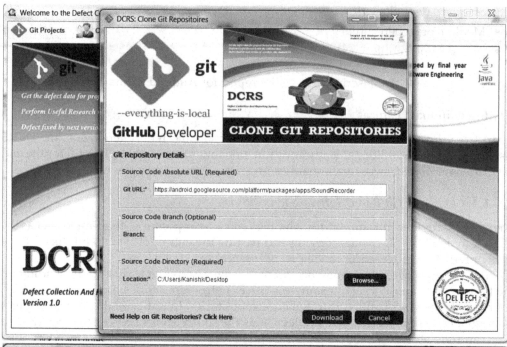

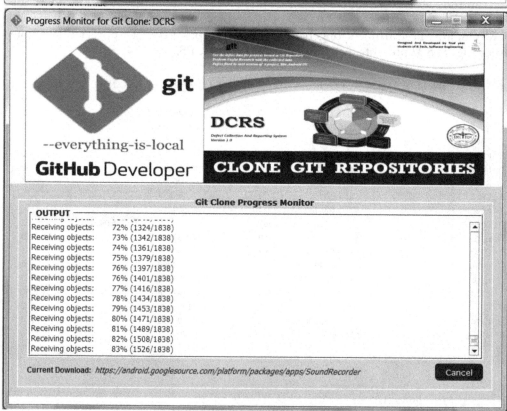

FIGURE 5.32
Cloning operation of DCRS.

FIGURE 5.32 (Continued)

5.11.6.2 Self-Logging

Self-logging or simply logging may be defined as a process of automatically recording events, data, and/or data structures about a tool's execution to provide an audit trail. The recorded information can be employed by developers, testers, and support personal for identifying software problems, monitoring live systems, and for other purposes such as auditing and postdeployment debugging. Logging process generally involves the transfer of recorded data to monitoring applications and/or writing the recorded information and appropriate messages, if any, to files.

The tool also provided the user with the functionality to view the operational logs of the tool. These self-logs are stored as text file(s), indicating the different events, and/or operations that have occurred during the tool's working along with their timestamp. These are ordered by the sequence and hence, the timestamp. Figure 5.33 presents an example self-log file for DCRS.

The self-log file follows a daily rolling append policy, that is, the logs for a given day are appended to the same file, and a new file is created after every 24 hours. The previous day's file is stored with a name that indicates the time of creation. Java Libraries of LogBack and SLF4J have been employed to implement self-logging in the DCRS. They can be downloaded from http://www.slf4j.org/download.html.

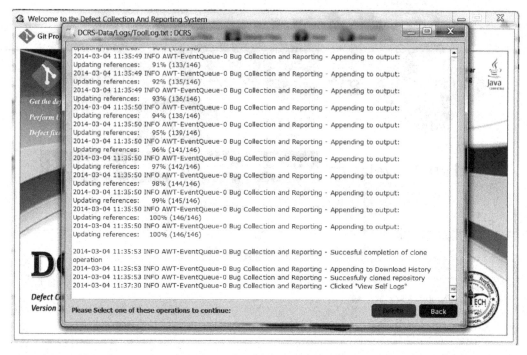

FIGURE 5.33
Example self-log file of DCRS.

5.11.7 Potential Applications of DCRS

As stated in Section 5.11.5, we have demonstrated the working of DCRS on a sample Android OS application (Mms), and analyzed the data and results obtained from the various generated reports. On the basis of the reports we obtained, we can state that the data collected using the DCRS can be potentially employed in the following applications.

5.11.7.1 Defect Prediction Studies

Defect proneness or defect prediction is a useful technique for predicting the defect-prone classes in a given software. Simply, it can be stated as the method of predicting the occurrence and/or number of defects in the software under consideration. For the past many years, defect prediction has been recognized as an important research field so as to organize a software project's testing resources and facilities. For example, consider a scenario wherein we have limited time and/or resources available for software testing procedure and activities. In such a situation, appropriate defect prediction models can aid the testing personnel to focus more attentively on those classes that are highly probable to be defective in the later releases of the software. Various studies have been carried out in the past for predicting effective defect-proneness models, and also for validating the effect of OO metrics on defect proneness (Aggarwal et al. 2009; Gondra 2008; Malhotra and Jain 2012).

We can thus state that the defect reports (defect count and metrics report, and consolidated defect and change report) generated by the DCRS can be effectively employed for such studies.

5.11.7.2 Change-Proneness Studies

Change proneness may be defined as the likelihood that a given component of a software would change in future. Along with defect prediction or defect proneness, change proneness is also a crucial area and needs to be evaluated accurately. Prediction of change-prone classes may aid in maintenance and testing. A class that is highly probable to change in the later releases of a software needs to be tested rigorously, and proper tracking is required for that class while modifying and maintaining the software. Therefore, various studies have also been carried out in the past for predicting effective change-proneness models, and to validate the impact of OO metrics on change proneness (Liang and Li 2007; Malhotra and Khanna 2013s).

The consolidated defect and change report generated by the DCRS can also be used for change analysis, and therefore for change-proneness studies as well.

5.11.7.3 Statistical Comparison

From the newly added and deleted files reports of the DCRS, a statistical comparison of the two versions of the Git-based OSS being considered (in this case, the Android OS) can be performed. Such kind of comparison can also be extended or generalized for a large number of Git-based OSS systems and applications. We may also identify some additional parameters for the comparison through defect and change data analysis, such as the total number of defects reported and total number of changes incurred in the previous version of the considered software.

5.11.8 Concluding Remarks

DCRS can be potentially employed in collection of defect data pertaining to OSS (which employs Git as the VCS) and generating useful reports for the same.

The gathered information can be effectively used for various purposes, including the following two applications:

- Defect prediction and related research work or studies, including analysis and validation of the effect of a given metric suite on defect proneness and the evaluation and comparison of various techniques in developing defect-proneness models, such as statistical and machine learning methods.
- Statistical comparison of two given versions of an OSS (which is based on Git VCS), in terms of the source files that have been added in the newer version, the source files that were present in the previous version but have been deleted in the newer version, and the defects that have been fixed by the newer version.

Exercises

5.1 Briefly describe the importance of mining software repositories.

5.2 How will you integrate repositories and bug tracking systems?

5.3 What are VCS? Compare and contrast different VCS.

5.4 Differentiate between CVS, SVN, and Git repositories.

5.5 Consider an open source software. Select bug analysis and describe how to collect the project data using extraction techniques.

5.6 What is configuration management system? Explain various categories in configuration management.

5.7 Clearly explain the applications of mining software repositories.

5.8 What are the various levels at which a researcher can collect data?

5.9 Explain with the help of the diagram the life cycle of a bug.

5.10 Describe various attributes of Git repository. Explain the procedure for collecting defects from open source repositories.

5.11 List out the attributes of a defect. Give the importance of each attribute.

5.12 What is the importance of mining email servers and chat boards?

5.13 Why is data mining on a Git repository faster than on a CVS repository?

5.14 Define the following terms:

    a. Baseline

    b. Tag

    c. Revision

    d. Release

    e. Version

    f. Edition

    g. Branch

    h. Head

5.15 What are the shortcomings of a CVCS? How does a DVCS overcome these shortcomings?

5.16 What is a commit record? Explain any five attributes present in a commit record of a CVS repository.

5.17 Illustrate the concept of branching and merging in the SVN repository.

5.18 How can the Bugzilla system be integrated with software repositories such as CVS and SVN?

5.19 What is a Git object? Explain all the fields of a Git object.

5.20 Explain the working of DCRS in detail.

Further Readings

An in-depth description the "Git" VCS may be obtained from:

S. Charon, and B. Straut, *Pro Git*, Apress, 2nd edition, 2014, https://git-scm.com/book.

The documentation (user guide, mailing lists, etc.) of the "CVS" client—"TortoiseCVS"—covers the basics and working of the CVS:

https://www.tortoisecvs.org/support.shtml.

Malhotra and Agrawal present a unique defect and change data-collection mechanism by mining CVS repositories:

R. Malhotra, and A. Agrawal, "CMS tool: Calculating defect and change data from software project repositories," *ACM Software Engineering Notes*, vol. 39, no. 1, pp. 1–5, 2014.

The following book documents and describes the detailed of Apache Subversion™ VCS:

B. Collins-Sussman, Brian W. Fitzpatrick, and C. Michael Pilato, *Version Control with Subversion for Subversion 1.7*, TBA, California, 2007, http://svnbook.red-bean.com/.

The following is an excellent tutorial at SVN repository:

SVN Tutorial, http://www.tutorialspoint.com/svn/svn_pdf_version.htm.

A detailed analysis software development history for change propagation in the source code has been carried out by Hassan and Holt:

A.E. Hassan, and R.C. Holt, "Predicting change propagation in software systems," *Proceedings of the 20th IEEE International Conference on Software Maintenance*, IEEE Computer Society Press, Los Alamitos, CA, pp. 284–293, 2004.

Ohira et al. present a case study of FLOSS projects at SourceForge for supporting cross-project knowledge collaboration:

M. Ohira, N. Ohsugi, T. Ohoka, and K. Matsumoto, "Accelerating cross-project knowledge collaboration using collaborative filtering and social networks," *Proceedings of the 2nd International Workshop on Mining Software Repositories*. ACM Press, New York, pp. 111–115, 2005.

An extensive comparison on software repositories can be obtained from:

http://en.wikipedia.org/wiki/Comparison_of_revision_control_software.
Version Control System Comparison. http://better-scm.shlomifish.org/comparison/comparison.html.
D.J. Worth, and C. Greenough, "Comparison of CVS and Subversion," RAL-TR-2006-001.

The details on Mercurial and Perforce repositories can be found in:

B. O'Sullivan, "Distributed revision control with Mercurial," *Mercurial Project*, 2007.
L. Wingerd, *Practical Perforce*. O'Reilly, Sebastopol, CA, 2005.

6

Data Analysis and Statistical Testing

The research data can be analyzed using various statistical measures and inferring conclusions from these measures. Figure 6.1 presents the steps involved in analyzing and interpreting the research data. The research data should be reduced in a suitable form before it can be used for further analysis. The statistical techniques can be used to preprocess the attributes (software metrics) so that they can be analyzed and meaningful conclusions can be drawn out of them. After preprocessing of the data, the attributes need to be reduced so that dimensionality can be reduced and better results can be obtained. Then, the model is predicted and validated using statistical and/or machine learning techniques. The results obtained are analyzed and interpreted from each and every aspect. Finally, hypotheses are tested and decision about the accuracy of model is made.

This chapter provides a description of data preprocessing techniques, feature reduction methods, and tests for statistical testing. As discussed in Chapter 4, hypothesis testing can be done either without model prediction or can be used for model comparison after the models have been developed. In this chapter, we present the various statistical tests that can be applied for testing a given hypothesis. The techniques for model development, methods for model validation, and ways of interpreting the results are presented in Chapter 7. We explain these tests with software engineering-related examples so that the reader gets an idea about the practical use of the statistical tests. The examples of model comparison tests are given in Chapter 7.

6.1 Analyzing the Metric Data

After data collection, descriptive statistics can be used to summarize and analyze the nature of the data. The descriptive statistics are used to describe the data, for example, extracting attributes with very few data points or determining the spread of the data. In this section, we present various statistical measures for summarizing data and graphical techniques for identifying outliers. We also present correlation analysis used to find the relation between attributes.

6.1.1 Measures of Central Tendency

Measures of central tendency are used to summarize the average values of the attributes. These measures include mean, median, and mode. They are known as measures of central tendency as they provide idea about the central values of the data around which all the other values tend to gather.

6.1.1.1 Mean

Mean can be computed by taking the average values of the data set. Mean is defined as the ratio of sum of values of the data points to the total number of data points and is given as,

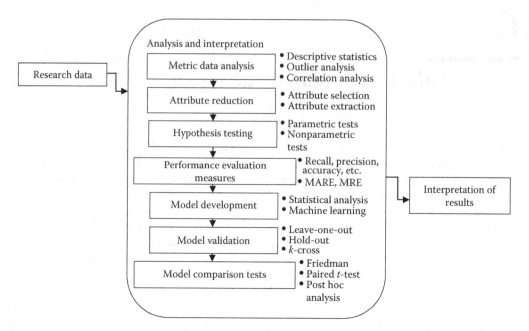

FIGURE 6.1
Steps for analyzing and interpreting data.

$$\text{Mean}(\mu) = \sum_{i=1}^{N} \frac{x_i}{N}$$

where:
  x_i ($i = 1, \ldots N$) are the data points
  N is the number of data points

For example, consider 28, 29, 30, 14, and 67 as values of data points.
  The mean is $(28 + 29 + 30 + 14 + 67)/5 = 33.6$.

6.1.1.2 Median

The median is that value which divides the data into two halves. Half of the number of data points are below the median values and half number of the data points are above the median values. For odd number of data points, median is the central value, and for even number of data points, median is the mean of the two central values. Hence, exactly 50% of the data points lie above the median values and 50% of data points lie below the median values. Consider the following data points:

$$8, 15, 5, 20, 6, 35, 10$$

First, we need to arrange data in ascending order,

$$5, 6, 8, 10, 15, 20, 35$$

The median is at 4th value, that is, 10. If one more additional data point 40 is added to the above distribution then,

$$5, 6, 8, 10, 15, 20, 35, 40$$

$$\text{Median} = \frac{10+15}{2} = 12.5$$

Median is not useful, if number of categories in the ordinal type of scale are very low. In such cases, mode is the preferred measure of central tendency.

6.1.1.3 Mode

Mode gives the value that has the highest frequency in the distribution. For example, consider Table 6.1, the second category of fault severity has the highest frequency of 50. Hence, 2 can be reported as the mode for Table 6.1 as it has the highest frequency.

Unlike the mean and median, the same distribution may have multiple values of mode. Consider Table 6.2, there are two categories of maintenance effort with same frequency: very high and medium. This is known as bimodal distribution.

The major disadvantage of mode is that it does not produce useful results when applied to interval/ratio scales having many values. For example, the following data points represent the number of failures occurred per second, while testing a given software and are arranged in ascending order:

$$15, 17, 18, 18, 45, 63, 64, 65, 71, 75, 79$$

It can be seen that the data is centered around 60–80 number of failures. But the mode of the distribution is 18, since it occurs twice in the distribution whereas the rest of the values only occur once. Clearly, the mode does not represent the central values in this case. Hence, either other measures of central tendency will be useful in this case or the data should be organized in suitable class intervals before mode is computed.

6.1.1.4 Choice of Measures of Central Tendency

The choice of selecting a measure of central tendency depends on

1. The scale type of data at which it is measured.
2. The distribution of data (left skewed, symmetrical, right skewed).

TABLE 6.1

Faults at Severity Levels

Fault Severity	Frequency
0	23
1	19
2	50
3	17

TABLE 6.2

Maintenance Effort

Maintenance Effort	Frequency
Very high	15
High	10
Medium	15

TABLE 6.3

Statistical Measures with Corresponding Relevant Scale Types

Measures	Relevant Scale Type
Mean	Interval and ratio data that are not skewed.
Median	Ordinal, interval, and ratio, but not useful for ordinal scales having few values.
Mode	All scale types, but not useful for scales having multiple values.

Table 6.3 depicts the relevant scale type of data for each statistical measure.
   Consider the following data set:

$$18, 23, 23, 25, 35, 40, 42$$

The mean, median, and mode are shown in Table 6.4, as each measure has different ways for computing "average" values. In fact, if the data is symmetrical, all the three measures (mean, median, and mode) have the same values. But, if the data is skewed, there will always be difference between these measures. Figure 6.2 shows the symmetrical and skewed distributions. The symmetrical curve is a bell-shaped curve, where all the data points are equally distributed.

   Usually, when the data is skewed, the mean is a misleading measure for determining central values. For example, if we calculate average lines of code (LOC) of 10 modules given in Table 6.5, it can be seen that most of the values of the LOC are between 200 and 400, but one module has 3,000 LOC. In this case, the mean will be 531. Only one value has influenced the mean and caused the distribution to skew to the right. However, the median will be 265, since the median is based on the midpoint and is not affected by the extreme values

TABLE 6.4

Descriptive Statistics

Measure	Value
Mean	29.43
Median	25
Mode	23

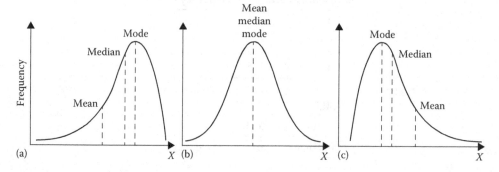

FIGURE 6.2
Graphs representing skewed and symmetrical distributions: (a) left skewed, (b) normal (no skew), and (c) right skewed.

TABLE 6.5

Sample Data of LOC for 10 Modules

Module#	LOC	Module#	LOC
1	200	6	270
2	202	7	290
3	240	8	300
4	250	9	301
5	260	10	3,000

in the data distribution. Hence, the median better reflects the average LOC in modules as compared to the mean and is the best measure when the data is skewed.

6.1.2 Measures of Dispersion

The measures of dispersion indicate the spread or the range of the distributions in the data set. Measures of dispersion include range, standard deviation, variance, and quartiles. The range is defined as the difference between the highest value and the lowest value in the distribution. It is the easiest measure that can be quickly computed. Thus, for the distribution of faults given in Table 6.5, the range of LOC will be

$$\text{Range} = 3000 - 200 = 2800$$

The range of the two distributions may be different even if they have the same mean. The advantage of using range measure is that it is a simple to compute, and the disadvantage is that it only takes into account the extreme values in the distribution and, hence, does not represent actual spread in the distribution. The interquartile range (IQR) can be used to overcome the disadvantage of the simple range measure.

The quartiles are used to compute the IQR of the distribution. The quartile divides the metric data into four equal parts. Figure 6.3 depicts the division of the data set into four equal parts. For the purpose of calculation of quartiles, the data is first required to be arranged in ascending order. The 25% of the metric data is below the lower quartile (25 percentile), 50% of the metric data is below the median value, and 75% of the metric data is below the upper quartile (75 percentile).

The lower quartile (Q_1) is computed by the following methods:

1. Computing the median of the data set
2. Computing the median of the lower half of the data set

The upper quartile (Q_3) is computed by the following methods:

1. Computing the median of the data set
2. Computing the median of the upper half of the data set

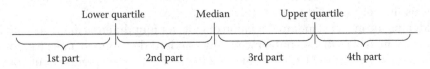

FIGURE 6.3
Quartiles.

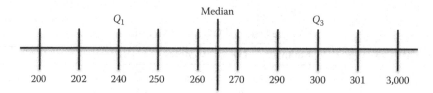

FIGURE 6.4
Example of quartile.

The IQR is defined as the difference between upper quartile and lower quartile and is given as,

$$IQR = Q_3 - Q_1$$

For example, for Table 6.5, the quartiles are shown in Figure 6.4.

$$IQR = Q_3 - Q_1 = 300 - 240 = 60$$

The standard deviation is used to measure the average distance a data point has from the mean. The standard deviation assesses the spread by calculating the distance of the data point from the mean. The standard deviation is large, if most of the data points are near to the mean. The standard deviation (σ_x) for the population is given as:

$$\sigma_x = \sqrt{\frac{\sum (x-\mu)^2}{N}}$$

where:
  x is the given value
  N is the number of values
  μ is the mean of all the values

Variance is a measure of variability and is the square of standard deviation.

6.1.3 Data Distributions

The shape of the distribution of the data is used to describe and understand the metrics data. Shape exhibits the patterns of distribution of data points in a given data set. A distribution can either be symmetrical (half of the data points lie to the left of the median and other half of the data points lie to the right of the median) or skewed (low and/or high data values are imbalanced). A bell-shaped curve is known as normal curve and is defined as, "The normal curve is a smooth, unimodel curve that is perfectly symmetrical. It has 68.3 percent of the area under the curve within one standard deviation of the mean" (Argyrous 2011). For example, for variable LOC the mean is 250 and standard deviation is 50 for the given 500 samples. For LOC to be normally distributed, 342 data points must be between 200 (250 − 50) and 300 (250 + 50). For normal curve to be symmetrical, 171 data points must lie between 200 and 250, and the same number of data points must lie between 250 and 300 (Figure 6.5).

Consider the mean and standard deviation of LOC for four different data sets with 500 data points shown in Table 6.6. Given the mean and standard deviation in Table 6.6, for data set 1 to be normal, the range of LOC consisting 342 (68.3% of 500) data points should be between 200 and 300. Similarly, in data set 2, 342 data points should have LOC ranges between 160 and 280, and in data set 3, 342 data points should have ranges between 170 and 230.

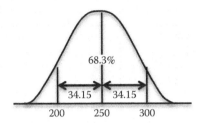

FIGURE 6.5
Normal curve.

TABLE 6.6

Range of Distribution for Normal Data Sets

S. No.	Mean	Standard Deviation	Ranges
1	250	50	200–300
2	220	60	160–280
3	200	30	170–230
4	200	10	190–210

TABLE 6.7

Sample Fault Count Data

Fault Count	Data1	35, 45, 45, 55, 55, 55, 65, 65, 65, 65, 75, 75, 75, 75, 75, 85, 85, 85, 85, 95, 95, 95, 105, 105, 115
	Data2	0, 2, 72, 75, 78, 80, 80, 85, 85, 87, 87, 87, 87, 88, 89, 90, 90, 92, 92, 95, 95, 98, 98, 99, 102
	Data3	20, 37, 40, 43, 45, 52, 55, 57, 63, 65, 74, 75, 77, 82, 86, 86, 87, 89, 89, 90, 95, 107, 165, 700, 705

6.1.4 Histogram Analysis

The normal curves can be used to understand data descriptions. There are a number of methods that can be applied to analyze the normality of the data set. One of the methods is histogram analysis. Histogram is a graphical representation that depicts frequency of occurrence of range of values. For example, consider fault count given for three software systems in Table 6.7. The histograms for all the three data sets are shown in Figure 6.6. The normal curve is superimposed on the histogram to check the normality of the data. Figure 6.6 shows that the data set Data1 is normal. Figure 6.6 also shows that the data set Data2 is left skewed and data set Data3 is right skewed.

6.1.5 Outlier Analysis

Data points that lie away from the rest of the data values are known as outliers. These values are located in an empty space and are extreme or unusual values. The presence of these outliers may adversely affect the results in data analysis. This is because of the following three reasons:

1. The mean no longer remains a true representative to capture central tendency.
2. In regression analysis, the values are squared hence the outliers may overinfluence the results.
3. The outlier may affect the data analysis.

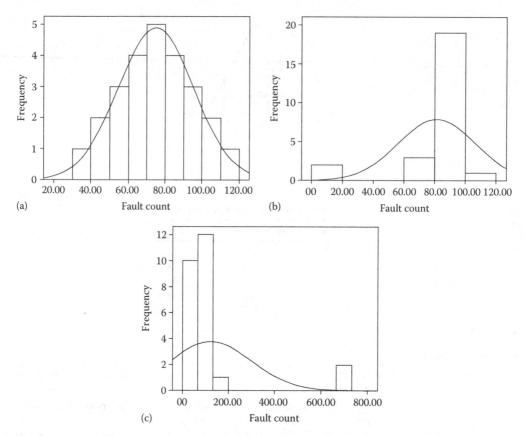

FIGURE 6.6
Histogram analysis for fault count data given in Table 6.7: (a) Data1, (b) Data2, and (c) Data3.

For example, suppose that one calculates the average of LOC, where most values are between 1,000 and 2,000, but the LOC for one module is 15,000. Thus, the data point with the value 15,000 is located far away from the other values in the data set and is an outlier. Outlier analysis is carried out to detect the data points that are overinfluential and must be considered for removal from the data sets.

The outliers can be divided into three types: univariate, bivariate, and multivariate. Univariate outliers are influential data points that occur within a single variable. Bivariate outliers occur when two variables are considered in combination, whereas multivariate outliers occur when more than two variables are considered in combination. Once the outliers are detected, the researcher must make the decision of inclusion or exclusion of the identified outlier. The outliers generally signal the presence of anomalies, but they may sometimes provide interesting patterns to the researchers. The decision is based on the reason of the occurrence of the outlier.

Box plots, z-scores, and scatter plots can be used for detecting univariate and bivariate outliers.

6.1.5.1 Box Plots

Box plots are based on median and quartiles. Box plots are constructed using upper and lower quartiles. An example box plot is shown in Figure 6.7. The two boundary lines

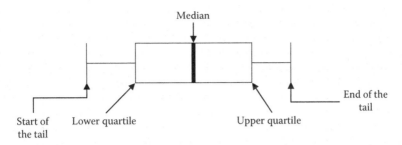

FIGURE 6.7
Example box plot.

signify the start and end of the tail. These two boundary lines correspond to ±1.5 IQR. Thus, once the value of IQR is known, it is multiplied by 1.5. The values shown inside of the box plots are known to be within the boundaries, and hence are not considered to be extreme. The data points beyond the start and end of the boundaries or tail are considered to be outliers. The distance between the lower and the upper quartile is often known as box length.

The start of the tail is calculated as $Q_3 - 1.5 \times$ IQR and end of the tail is calculated as $Q_3 + 1.5 \times$ IQR. To avoid negative values, the values are truncated to the nearest values of the actual data points. Thus, actual start of the tail is the lowest value in the variable above ($Q_3 - 1.5 \times$ IQR), and actual end of the tail is the highest value below ($Q_3 - 1.5 \times$ IQR).

The box plots also provide information on the skewness of the data. The median lies in the middle of the box if the data is not skewed. The median lies away from the middle if the data is left or right skewed. For example, consider the LOC values given below for a software:

$$200, 202, 240, 250, 260, 270, 290, 300, 301, 3000$$

The median of the data set is 265, lower quartile is 240, and upper quartile is 300. The IQR is 60. The start of the tail is $240 - 1.5 \times 60 = 150$ and end of the tail is $300 + 1.5 \times 60 = 390$. The actual start of the tail is the lowest value above 150, that is, 200, and actual end of the tail is the highest value below 390, that is, 301. Thus, the case number 10 with value 30,000 is above the end of the tail and, hence, is an outlier. The box plot for the given data set is shown in Figure 6.8 with one outlier 3,000.

A decision regarding inclusion or exclusion of the outliers must be made by the researchers during data analysis considering the following reasons:

1. Data entry errors
2. Extraordinary or unusual events
3. Unexplained reasons

Outlier values may be present because of combination of data values present across more than one variable. These outliers are called multivariate outliers. Scatter plot is another visualization method to detect outliers. In scatter plots, we simply represent all the data points graphically. The scatter plot allows us to examine more than one metric variable at a given time.

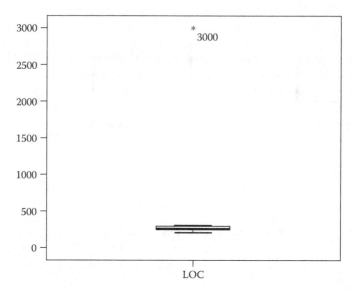

FIGURE 6.8
Box plot for LOC values.

6.1.5.2 Z-Score

Z-score is another method to identify outliers and is used to depict the relationship of a value to its mean, and is given as follows:

$$z\text{-score} = \frac{x - \mu}{\sigma}$$

where:
   x is the score or value
   μ is the mean
   σ is the standard deviation

The z-score gives the information about the value as to whether it is above or below the mean, and by how many standard deviations. It may be positive or negative. The z-score values of data samples exceeding the threshold of ± 2.5 are considered to be outliers.

Example 6.1:

Consider the data set given in Table 6.7. Calculate univariate outliers for each variable using box plots and z-scores.

Solution:

The box plots for Data1, Data2, and Data3 are shown in Figure 6.9. The z-scores for data sets given in Table 6.7 are shown in Table 6.8.

To identify multivariate outliers, for each data point, the Mahalanobis Jackknife distance D measure can be calculated. Mahalanobis Jackknife is a measure of the distance in multidimensional space of each observation from the multivariate mean center of the observations (Hair et al. 2006). Each data point is evaluated using chi-square distribution with 0.001 significance value.

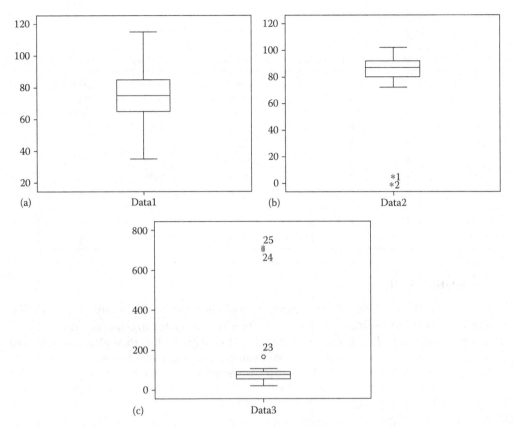

FIGURE 6.9
(a)–(c) Box plots for data given in Table 6.7.

TABLE 6.8

Z-Score for Data Sets

Case No.	Data1	Data2	Data3	Z-scoredata1	Z-scoredata2	Z-scoredata3
1	35	0	20	−1.959	−3.214	−0.585
2	45	2	37	−1.469	−3.135	−0.488
3	45	72	40	−0.979	−0.052	−0.404
4	55	75	43	−0.489	−0.052	−0.387
5	55	78	45	−0.489	0.145	−0.375
6	55	80	52	−0.489	0.145	−0.341
7	65	80	55	−0.489	0.224	−0.330
8	65	85	57	0	0.224	−0.279
9	65	85	63	0	0.224	−0.273
10	65	87	65	0	0.224	−0.262
11	75	87	74	0	0.264	−0.234
12	75	87	75	0	0.303	−0.211
13	75	87	77	0.489	0.343	−0.211
15	75	89	86	0.489	0.422	−0.194
16	85	90	86	0.489	0.422	−0.194

(Continued)

TABLE 6.8 (Continued)

Z-Score for Data Sets

Case No.	Data1	Data2	Data3	Z-scoredata1	Z-scoredata2	Z-scoredata3
17	85	90	87	0.489	0.343	−0.205
18	85	92	89	0.489	0.422	−0.194
19	85	92	89	−1.463	−1.530	−0.652
20	95	95	90	0.979	0.540	−0.188
21	95	95	95	−0.956	−1.354	−0.813
22	95	98	107	0.979	0.659	−0.092
23	105	98	165	1.469	0.659	0.235
24	105	99	700	1.469	0.698	3.264
25	115	102	705	1.959	0.817	3.292
Mean	75	81.35	123.26			
SD	20.41	25.29	176.63			

6.1.6 Correlation Analysis

This is an optional step followed in empirical studies. Correlation analysis studies the variation of two or more independent variables for determining the amount of correlation between them. For example, if the relationship of design metrics to the size of the class is to be analyzed. This is to determine empirically whether the coupling, cohesion, or inheritance metric is essentially measuring size such as LOC. The model that predicts larger classes as more fault prone is not much useful such as these classes cover large part of the system, and thus testing cannot be done very well (Briand et al. 2000; Aggarwal et al. 2009). A nonparametric technique (Spearman's Rho) for measuring relationship between object-oriented (OO) metrics and size can be used, if skewed distribution of the design measures is observed. Hopkins calls a correlation coefficient value between 0.5 and 0.7 as large, 0.7 and 0.9 as very large, and 0.9 and 1.0 as almost perfect (Hopkins 2003).

6.1.7 Example—Descriptive Statistics of Fault Prediction System

Univariate and multivariate outliers are found in FPS study. To identify multivariate outliers, for each data point, the Mahalanobis Jackknife distance is calculated. The input metrics were normalized using min–max normalization. Min–max normalization performs a linear transformation on the original data (Han and Kamber 2001). Suppose that min A and max A are the minimum and maximum values of an attribute A. It maps the value v of A to v' in the range 0–1 using the formula:

$$v' = \frac{v - \min A}{\max A - \min A}$$

Table 6.9 shows "min," "max," "mean," "median," "standard deviation," "25% quartile," and "75% quartile" for all metrics considered in FPS study. The following observations are made from Table 6.9:

- The size of a class measured in terms of lines of source code ranges from 0 to 2,313.
- The values of depth of inheritance tree (DIT) and number of children (NOC) are low in the system, which shows that inheritance is not much used in all the

TABLE 6.9

Descriptive Statistics for Metrics

Metric	Min.	Max.	Mean	Median	Std. Dev.	Percentile (25%)	Percentile (75%)
CBO	0	24	8.32	8	6.38	3	14
LCOM	0	100	68.72	84	36.89	56.5	96
NOC	0	5	0.21	0	0.7	0	0
RFC	0	222	34.38	28	36.2	10	44.5
WMC	0	100	17.42	12	17.45	8	22
LOC	0	2313	211.25	108	345.55	8	235.5
DIT	0	6	1	1	1.26	0	1.5

TABLE 6.10

Correlation Analysis Results

Metric	CBO	LCOM	NOC	RFC	WMC	LOC	DIT
CBO	1						
LCOM	0.256	1					
NOC	−0.03	−0.028	1				
RFC	0.386	0.334	−0.049	1			
WMC	0.245	0.318	0.035	**0.628**	1		
LOC	**0.572**	0.238	−0.039	**0.508**	**0.624**	1	
DIT	0.4692	0.256	−0.031	**0.654**	0.136	0.345	1

systems; similar results have also been shown by others (Chidamber et al. 1998; Cartwright and Shepperd 2000; Briand et al. 2000a).

• The lack of cohesion in methods (LCOM) measure, which counts the number of classes with no attribute usage in common, has high values (upto 100) in KC1 data set.

The correlation among metrics is calculated, which is an important static quantity. As shown in Table 6.10, Gyimothy et al. (2005) and Basili et al. (1996) also calculated the correlation among metrics. The values of correlation coefficient are interpreted using the threshold given by Hopkins (2003). Thus, in Table 6.10, the correlated values with correlation coefficient >0.5 are shown in bold. The correlation coefficients shown in bold are significant at 0.01 level. In this data set, weighted methods per class (WMC), LOC, and DIT metrics are correlated with response for a class (RFC) metric. Similarly, the WMC and coupling between object (CBO) metrics are correlated with LOC metric. Therefore, it shows that these metrics are not totally independent and represent redundant information.

6.2 Attribute Reduction Methods

Sometimes the presence of a large number of attributes in an empirical study reduces the efficiency of the prediction results produced by the statistical and machine learning techniques. Reducing the dimensionality of the data reduces the size of the hypothesis space and allows the methods to operate faster and more effectively. The attribute reduction methods

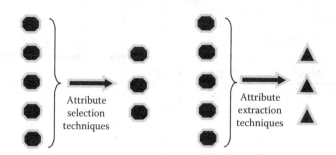

FIGURE 6.10
Attribute reduction procedure.

involve either selection of subset of attributes (independent variables) by eliminating the attributes that have little or no predictive information (known as attribute selection), or combining the relevant attributes into a new set of attributes (known as attribute extraction). Figure 6.10 graphically depicts the procedures of attribute selection and extraction methods.

For example, a researcher may collect a large amount of data that captures various constructs of the design to predict the probability of occurrence of fault in a module. However, much of the collected information may not have any relation or impact on the occurrence of faults. It is also possible that more than one attribute captures the same concept and hence is redundant. The irrelevant and redundant attributes only add noise to the data, increase computational time and may reduce the accuracy of the predicted models. To remove the noise and correlation in the attributes, it is desirable to reduce data dimensionality as a preprocessing step of data analysis. The advantages of applying attribute reduction methods are as follows:

1. Improved model interpretability
2. Faster training time
3. Reduction in overfitting of the models
4. Reduced noise

Hence, attribute reduction leads to improved computational efficiency, lower cost, increased problem understanding, and improved accuracy. Figure 6.11 shows the categories of attribute reduction methods.

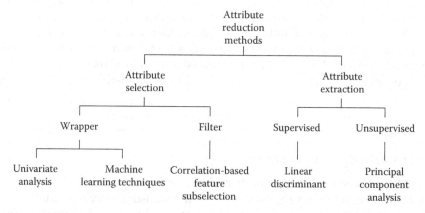

FIGURE 6.11
Classification of attribute reduction methods.

6.2.1 Attribute Selection

Attribute selection involves selecting a subset of attributes from a given set of attributes. For example, univariate analysis and correlation-based feature selection (CFS) techniques can be used for attribute subselection. Different methods, as discussed below, are available for metric selection. These methods can be categorized as wrapper and filter. Wrapper methods use learning techniques to find subset of attributes whereas filter methods are independent of the learning technique. Wrapper methods use learning algorithm for selecting subsets of attributes, hence they are slower in execution as compared to filter methods that compute attribute ranking on the basis of correlation-based and information-centric measures. But at the same time filter methods may produce a subset that does not work very well with the learning technique as attributes are not tuned to specific prediction model. Figure 6.12 depicts the procedure of filter methods, and Figure 6.13 shows the procedure of wrapper methods. Examples of learning techniques used in Wrapper methods include Hill climbing, genetic algorithms, simulated annealing , *Tabu* search. Examples of techniques used in filter methods include correlation coefficient, mutual information, information gain. Two widely used methods for feature selection are explained in sections below.

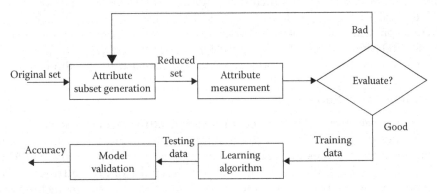

FIGURE 6.12
Procedure of filter method.

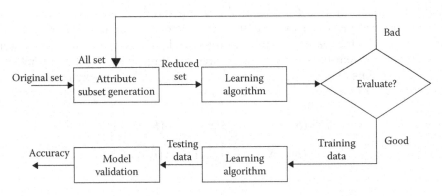

FIGURE 6.13
Procedure of wrapper method.

6.2.1.1 Univariate Analysis

The univariate analysis is done to find the individual effect of each independent variable on the dependent variable. One of the purposes of univariate analysis is to screen out the independent variables that are not significantly related to the dependent variables. For example, in regression analysis, only the independent variables that are significant at 0.05 significance level may be considered in subsequent model prediction using multivariate analysis. The primary goal is to preselect the independent variables for multivariate analysis that seems to be useful predictors. The choice of methods in the univariate analysis depends on the type of dependent variables being used.

In the univariate regression analysis, the independent variables are chosen based on the results of the significance value (see Section 6.3.2), whereas, in the case of other methods, the independent variables are ranked based on the values of the performance measures (see Chapter 7).

6.2.1.2 Correlation-Based Feature Selection

This is a commonly used method for preselecting attributes in machine learning methods. To incorporate the correlation of independent variables, a CFS method is applied to select the best predictors out of the independent variables in the data sets (Hall 2000). The best combinations of independent variables are searched through all possible combinations of variables. CFS evaluates the best of a subset of independent variables, such as software metrics, by considering the individual predictive ability of each attribute along with the degree of redundancy between them. Hall (2000) showed that CFS can be used in drastically reducing the dimensionality of data sets, while maintaining the performance of the machine learning methods.

6.2.2 Attribute Extraction

Unlike attribute selection, which selects the existing attributes with respect to their significance values or importance, attribute extraction transforms the existing attributes and produces new attributes by combining or aggregating the original attributes so that useful information for model building can be extracted from the attributes. Principal component analysis is the most widely used attribute extraction technique in the literature.

6.2.2.1 Principal Component Method

Principal component method (or P.C. method) is a standard technique used to find the interdependence among a set of variables. The factors summarize the commonality of the variables, and factor loadings represent the correlation between the variables and the factor. P.C. method maximizes the sum of squared loadings of each factor extracted in turn. The P.C. method aims at constructing new variable (P_i), called principal component (P.C.) out of a given set of variables X'_js ($j = 1, 2, ..., k$).

$$P_1 = (b_{11} \times X_1) + (b_{12} \times X_2) + \cdots + (b_{1k} \times X_k)$$

$$P_2 = (b_{21} \times X_1) + (b_{22} \times X_2) + \cdots + (b_{2k} \times X_k)$$

$$\vdots$$

$$P_k = (b_{k1} \times X_1) + (b_{k2} \times X_2) + \cdots + (b_{kk} \times X_k)$$

All b_{ij}'s called loadings are worked out in such a way that the extracted P.C. satisfies the following two conditions:

1. P.C.s are uncorrelated (orthogonal).
2. The first P.C. (P_1) has the highest variance, the second P.C. has the next highest variance, and so on.

The variables with high loadings help identify the dimension the P.C. is capturing, but this usually requires some degree of interpretation. To identify these variables, and interpret the P.C.s, the rotated components are used. As the dimensions are independent, orthogonal rotation is used, in which the axes are maintained at 90 degrees. There are various strategies to perform such rotation. This includes quartimax, varimax, and equimax orthogonal rotation. For detailed description refer Hair et al. (2006) and Kothari (2004).

Varimax method maximizes the sum of variances of required loadings of the factor matrix (a table displaying the factor loadings of all variables on each factor). Varimax rotation is the most frequently used strategy in literature. Eigenvalue (or latent root) is associated with each P.C. It refers to the sum of squared values of loadings relating to a dimension. Eigenvalue indicates the relative importance of each dimension for the particular set of variables being analyzed. The P.C. with eigenvalue >1 is taken for interpretation (Kothari 2004).

6.2.3 Discussion

It is useful to interpret the results of regression analysis in the light of results obtained from P.C. analysis. P.C. analysis shows the main dimensions, including independent variables as the main drivers for predicting the dependent variable. It would also be interesting to observe the metrics included in dimensions across various replicated studies; this will help in finding differences across various studies. From such observations, the recommendations regarding which independent variable appears to be redundant and need not be collected can be derived, without losing a significant amount of design information (Briand and Wust 2002). P.C. analysis is a widely used method for removing redundant variables in neural networks.

The univariate analysis is used in preselecting the metrics with respect to their significance, whereas CFS is the widely used method for preselecting independent variables in machine learning methods (Hall 2000). In Hall (2003), the results showed that CFS chooses few attributes, is faster, and overall good performer.

6.3 Hypothesis Testing

As discussed in Section 4.7, hypothesis testing is an important part of empirical research. Hypothesis testing allows a researcher to reach to a conclusion on the basis of the statistical tests. Generally, a hypothesis is an assumption that the researcher wants to accept or reject. For example, an experimenter observes that birds can fly and wants to show that an animal is not a bird. In this example, the null hypothesis can be "the observed animal is a bird." A critical area c is given to test a particular unit x. The test can be formulated as given below:

1. If $x \in c$, then null hypothesis is rejected
2. If $x \notin c$, then null hypothesis is accepted

In the given example, the x is attributes of animals with critical area c = run, walk, sit, and so on. These are the values that will cause null hypothesis to be rejected. The test is "whether $x \neq$ fly"; if yes, reject null hypothesis, otherwise accept it. Hence, if x=fly that means that null hypothesis is accepted.

In real-life, a software practitioner may want to prove that the decision tree algorithms are better than the logistic regression (LR) technique. This is known as assumption of the researcher. Hence, the null hypothesis can be formulated as "there is no difference between the performance of the decision tree technique and the LR technique." The assumption needs to be evaluated using statistical tests on the basis of data to reach to a conclusion. In empirical research, hypothesis formulation and evaluation are the bottom line of research.

This section will highlight the concept of hypothesis testing, and the steps followed in hypothesis testing.

6.3.1 Introduction

Consider a setup where the researcher is interested in whether some learning technique "Technique X" performs better than "Technique Y" in predicting the change proneness of a class. To reach a conclusion, both technique X and technique Y are used to build change prediction models. These prediction models are then used to predict the change proneness of a sample data set (for details on training and testing of models refer Chapter 7) and based on the outcome observed over the sample data set, it is determined which technique is the better predictor out of the two. However, concluding which technique is better is a challenging task because of the following issues:

1. The number of data points in the sample could be very large, making data analysis and synthesis difficult.
2. The researcher might be biased towards one of the techniques and could overlook minute differences that have the potential of impacting the final result greatly.
3. The conclusions drawn can be assumed to happen by chance because of bias in the sample data itself.

To neutralize the impact of researcher bias and ensure that all the data points contribute to the results, it is essential that a standard procedure be adopted for the analysis and synthesis of sample data. Statistical tests allow the researcher to test the research questions (hypotheses) in a generalized manner. There are various statistical tests like the student t-test, chi-squared test, and so on. Each of these tests is applicable to a specific type of data and allows for comparison in such a way that using the data collected from a small sample, conclusions can be drawn for the entire population.

6.3.2 Steps in Hypothesis Testing

In hypothesis testing, a series of steps are followed to verify a given hypothesis. Section 4.7.5 summarizes the following steps; however, we restate them as these steps are followed in each statistical test described in coming sections. The first two steps, however, are part of experimental design process and carried out while the design phase progresses.

Step 1: Define hypothesis—In the first step, the hypothesis is defined corresponding to the outcomes. The statistical tests are used to verify the hypothesis formed in the experimental design phase.

Step 2: Select the appropriate statistical test—The appropriate statistical test is determined in experiment design on the basis of assumptions of a given statistical test.

Step 3: Apply test and calculate p-value—The next step involves applying the appropriate statistical test and calculating the significance value, also known as p-value. There are a series of parametric and nonparametric tests available. These tests are illustrated with example in the coming sections.

Step 4: Define significance level—The threshold level or critical value (also known as α-value) that is used to check the significance of the test statistic is defined.

Step 5: Derive conclusions—Finally, the conclusions on the hypothesis are derived using the results of the statistical test carried out in step 3.

6.4 Statistical Testing

The hypothesis formed in an empirical study is verified using statistical tests. In the following subsections, the overview of statistical tests, the difference between one-tailed and two-tailed tests, and the interpretation of statistical tests are discussed.

6.4.1 Overview of Statistical Tests

The validity of the hypothesis is evaluated using the test statistic obtained by statistical tests. The rejection region is the region within which if a test value falls, then the null hypothesis is rejected. The statistical tests are applied on independent and dependent variables and test value is computed using test statistic. After applying the statistical tests, the actual or test value is compared with the predetermined critical or p-value. Finally, a decision on acceptance or rejection of hypothesis is made (Figure 6.14).

6.4.2 Categories of Statistical Tests

Statistical tests can be classified according to the relationship between the samples, that is, whether they are independent or dependent (Figure 6.15). The decision on the statistical tests can be made based on the number of data samples to be compared. Some tests work on two data samples, such as t-test or Wilcoxon signed-rank, whereas others work on multiple data sets, such as Friedman or Kruskal–Wallis. Further, the tests can be categorized as parametric and nonparametric. Parametric tests are statistical tests that can be applied to a given data set, if it satisfies the underlying assumptions of the test. Nonparametric tests are used when certain assumptions are not satisfied by the data sample. The categorization is depicted in Figure 6.15. Univariate LR can also be applied

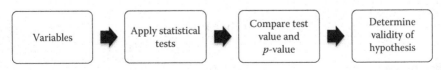

FIGURE 6.14
Steps in statistical tests.

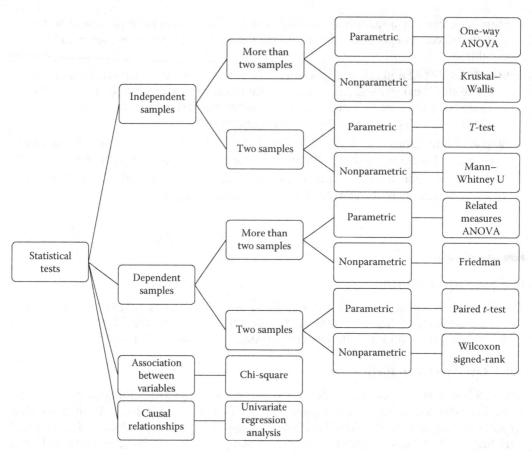

FIGURE 6.15
Categories of statistical tests

for testing the hypothesis for binary dependent variable. Table 6.11 depicts the summary of assumptions, data scale, and normality requirement for each statistical test discussed in this chapter.

6.4.3 One-Tailed and Two-Tailed Tests

In two-tailed test, the deviation of the parameter in each direction from the specified value is considered. When the hypothesis is specified in one direction, then one-tailed test is used. For example, consider the following null and alternative hypotheses for one-tailed test:

$$H_0 : \mu = \mu_0$$

$$H_a : \mu > \mu_0$$

where:
 μ is the population mean
 μ_0 is the sample mean

TABLE 6.11

Summary of Statistical Tests

Test	Assumptions	Data Scale	Normality
One sample *t*-test	The data should not have any significant outliers.	Interval or ratio.	Required
	The observations should be independent.		
Two sample *t*-test	Standard deviations of the two populations must be equal.	Interval or ratio.	Required
	Samples must be independent of each other.	Interval or ratio.	
	The samples are randomly drawn from respective populations.	Interval or ratio.	
Paired *t*-test	Samples must be related with each other.	Interval or ratio.	Required
	The data should not have any significant outliers.		
Chi-squared test	Samples must be independent of each other.	Nominal or ordinal.	Not required
	The samples are randomly drawn from respective populations.		
F-test	All the observations should be independent.	Interval or ratio.	Required
	The samples are randomly drawn from respective populations and there is no measurement error.		
One-way ANOVA	One-way ANOVA should be used when you have three or more independent samples.	Interval or ratio.	Required
	The data should not have any significant outliers.		
	The data should have homogeneity of variances.		
Two-way ANOVA	The data should not have any significant outliers.	Interval or ratio.	Required
	The data should have homogeneity of variances.		
Wilcoxon signed test	The data should consist of two "related groups" or "matched pairs."	Ordinal or continuous.	Not required
Wilcoxon–Mann–Whitney test	The samples must be independent.	Ordinal or continuous.	Not required
Kruskal–Wallis test	The test should validate three or more independent sample distributions.	Ordinal or continuous.	Not required
	The samples are drawn randomly from respective populations.		
Friedman test	The samples should be drawn randomly from respective populations.	Ordinal or continuous.	Not required

Here, the alternative hypothesis specifies that the population mean is strictly "greater than" sample mean. The below hypothesis is an example of two-tailed test:

$$H_0 : \mu = \mu_0$$

$$H_a : \mu < \mu_0 \text{ or } \mu > \mu_0$$

Figure 6.16 shows the probability curve for a two-tailed test with rejection (or critical region) on both sides of the curve. Thus, the null hypothesis is rejected if sample mean lies in either of the rejection region. Two-tailed test is also called nondirectional test.

Figure 6.17 shows the probability curve for one-tailed test with rejection region on one side of the curve. One-tailed test is also referred as directional test.

6.4.4 Type I and Type II Errors

There can be two types of errors that occur in hypothesis testing. They are distinguished as type I and type II errors. Type I or type II error depends directly on the null hypothesis. The goal of the test is to reject the null hypothesis. A statistical test can either reject (prove false) or fail to reject (fail to prove false) a null hypothesis, but can never prove it to be true.

Type I error is the probability of wrongly rejecting the null hypothesis when the null hypothesis is true. In other words, a type I error occurs when the null hypothesis of no difference is rejected, even when there is no difference. A type I error can also be called as "false positive"; a result when an actual "hit" is erroneously seen as a "miss." Type I error is denoted by the Greek letter alpha (α). This means that it usually equals the significance

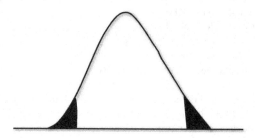

FIGURE 6.16
Probability curve for two-tailed test.

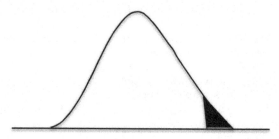

FIGURE 6.17
Probability curve for one-tailed test.

TABLE 6.12

Types of Errors

	H_0 True	H_0 False
Reject H_0	Type I error (false positive)	Correct result (true positive)
Fail to reject H_0	Correct result (true negative)	Type II error (false negative)

level of a test. Type II error is defined as the probability of wrongly not rejecting the null hypothesis when the null hypothesis is false. In other words, a type II error occurs when the null hypothesis is actually false, but somehow, it fails to get rejected. It is also known as "false negative"; a result when an actual "miss" is erroneously seen as a "hit." The rate of the type II error is denoted by the Greek letter beta (β) and related to the power of a test (which equals $1 - \beta$). The definitions of these errors can also be tabularized as shown in Table 6.12.

6.4.5 Interpreting Significance Results

If the calculated value of a test statistic is greater than the critical value for the test then the alternative hypothesis is accepted, else the null hypothesis is accepted and the alternative hypothesis is rejected.

The test results provide calculated p-value. This p-value is the exact level of significance for the outcome. For example, if the p-value reported by the test is 0.01, then the confidence level of the test is $(1 - 0.01) \times 100 = 99\%$ confidence. The obtained p-value is compared with the significance value or critical value, and decision about acceptance or rejection of the hypothesis is made. If the p-value is less than or equal to the significance value, the null hypothesis is rejected. The various tables for obtaining p-values and various test statistic values are presented in Appendix I. The appendix lists t-table test values, chi-square test values, Wilcoxon–Mann–Whitney test values, area under the normal distribution table, F-test table at 0.05 significance level, critical values for two-tailed Nemenyi test at 0.05 significance level, and critical values for two-tailed Bonferroni test at 0.05 significance level.

6.4.6 t-Test

W. Gossett designed the student t-test (Student 1908). The purpose of the t-test is to determine whether two data sets are different from each other or not. It is based on the assumption that both the data sets are normally distributed. There are three variants of t-tests:

1. One sample t-test, which is used to compare mean with a given value.
2. Independent sample t-test, which is used to compare means of two independent samples.
3. Paired t-test, which is used to compare means of two dependent samples.

6.4.6.1 One Sample t-Test

This is the simplest type of t-test that determines the difference between the mean of a data set from a hypothesized value. In this test, the mean from a single sample is collected and is compared with a given value of interest. The aim of one sample *t*-test is to find whether there is sufficient evidence to conclude that there is difference between mean of a given sample from a specified value. For example, one sample *t*-test can be used to determine whether the average increase in number of comment lines per method is more than five after improving the readability of the source code.

The assumption in the one sample *t*-test is that the population from which the sample is derived must have normal distribution. The following null and alternative hypotheses are formed for applying one sample *t*-test on a given problem:

H_0: $\mu = \mu_0$ (Mean of the sample is equal to the hypothesized value.)

H_a: $\mu \neq \mu_0$ (Mean of the sample is not equal to the hypothesized value.)

The *t* statistic is given below:

$$t = \frac{\mu - \mu_0}{\sigma / \sqrt{n}}$$

where:
  μ represents mean of a given sample
  σ represents standard deviation
  n represents sample size

The above hypothesis is based on two tailed *t*-test. The degrees of freedom (DOFs) is $n - 1$ as *t*-test is based on the assumption that the standard deviation of the population is equal to the standard deviation of the sample. The next step is to obtain significance values (*p*-value) and compare it with the established threshold value (α). To obtain *p*-value for the given *t*-statistic, the *t*-distribution table needs to be referred. The table can only be used given the DOF.

Example 6.2:

Consider Table 6.13 where the number of modules for 15 software systems are shown. We want to conclude that whether the population from which sample is derived is on average different than the 12 modules.

TABLE 6.13

Number of Modules

Module No.	Module#	Module No.	Module#	Module No.	Module#
S1	10	S6	35	S11	24
S2	15	S7	26	S12	23
S3	24	S8	29	S13	14
S4	29	S9	19	S14	12
S5	16	S10	18	S15	5

Solution:
The following steps are carried out to solve the example:

Step 1: Formation of hypothesis.
In this step, null (H_0) and alternative (H_a) hypotheses are formed. The hypotheses for the example are given below:
H_0: $\mu = 12$ (Mean of the sample is equal to 12.)
H_a: $\mu \neq 12$ (Mean of the sample is not equal to 12.)
Step 2: Select the appropriate statistical test.
The sample that belongs to a normally distributed population does not contain any significant outliers and has independent observations. The mean of the number of modules can be tested using one sample t-test, as standard deviation for the population is not known.
Step 3: Apply test and calculate p-value.
It can be seen that the mean is 19.933 and standard deviation is 8.172. For one sample t-test, the value of t is,

$$t = \frac{\mu - \mu_0}{\sigma/\sqrt{n}} = \frac{19.93 - 12}{8.17/\sqrt{15}} = 3.76$$

The DOF is 14 (15 − 1) in this example.
To obtain the p-value for a specific t-statistic, we perform the following steps, referring to Table 6.14:
1. For corresponding DOF, named df, identify the row with the desired DOF. In this example, the desired DOF is 14.
2. Now, in the desired row, mark out the t-score values between which the computed t-score falls. In this example, the calculated t-statistic is 3.76. This t-statistic falls beyond the t-score of 2.977.
3. Now, move upward to find the corresponding p-value for the selected t-score for either one-tail or two-tail significance test. In this example, the significance value for one-tail test would be <0.005, and for two-tail test it would be <0.01.
Given 14 DOF and referring the t-distribution table, the obtained p-value is 0.002.

TABLE 6.14

Critical Values of t-Distributions

Level of significance for one-tailed test				
0.10	0.05	0.02	0.01	**0.005**

df	\multicolumn{5}{c}{Level of significance for two-tailed test}
	0.20	0.10	0.05	0.02	**0.01**
1	3.078	6.314	12.706	31.821	**63.657**
2	1.886	2.920	4.303	6.965	**9.925**
3	1.638	2.353	3.182	4.541	**5.841**
...	...	...	...	...	...
14	**1.345**	**1.761**	**2.145**	**2.624**	**2.977**
15	1.341	1.753	2.131	2.602	2.947
...	...	...	...	...	...
120	1.289	1.658	1.980	2.358	2.617
∞	1.282	1.645	1.960	2.326	2.576

Step 4: Define significance level.

After obtaining p-value, we need to decide the threshold or α value. Hence, it can be seen that the results are statistically significant at 0.01 significance value for two-tailed test and 0.005 for one-tailed test.

It is important to note that we will apply two-tail significance in all other examples of the chapter.

Step 5: Derive conclusions.

As computed in Step 4, the results are statistically significant at 0.01 α value. Hence, we reject the null hypothesis and conclude that the average modules in a given software are statistically significantly different than 12 ($t = 3.76$, $p = 0.002$).

6.4.6.2 Two Sample t-Test

The two sample (independent sample) t-test determines the difference between the unknown means of two populations based on the independent samples drawn from the two populations. If the means of two samples are different from each other, then we conclude that the population are different from each other. The samples are either derived from two different populations or the population is divided into two random subgroups and the samples are derived from these subgroups, where each group is subjected to a different treatment (or technique). In both the cases, it is necessary that the two samples are independent to each other. The hypothesis for the application of this variant of t-test can be formulated as given below:

H_0: $\mu_1 = \mu_2$ (There is no difference in the mean values of both the samples.)

H_a: $\mu_1 \neq \mu_2$ (There is difference in the mean values of both the samples.)

The t-statistic for two sample t-test is given as,

$$t = \frac{\mu_1 - \mu_2}{\sqrt{\left(\sigma_1^2/n_1\right) + \left(\sigma_2^2/n_2\right)}}$$

where:

μ_1 and μ_2 are the means of both the samples, respectively

σ_1 and σ_2 are the standard deviations of both the samples, respectively

The DOF is $n_1 + n_2 - 1$, where n_1 and n_2 are the sample sizes of both the samples. Now, obtain the significance value (p-value) and compare it with the established threshold value (α) for the computed t-statistic using the t-distribution.

Example 6.3:

Consider an example for comparing the properties of industrial and open source software in terms of the average amount of coupling between modules (the other modules to which a module is coupled). The purpose of both the software is to serve as text editors developed in Java language. In this example, we believe that the type of software affects the amount of coupling between modules.

Industrial: 150, 140, 172, 192, 186, 180, 144, 160, 188, 145, 150, 141
Open source: 138, 111, 155, 169, 100, 151, 158, 130, 160, 156, 167, 132

Solution:

Step 1: Formation of hypothesis.

In this step, null (H_0) and alternative (H_a) hypotheses are formed. The hypotheses for the example are given below:

TABLE 6.15

Descriptive Statistics

Descriptive Statistic	Industrial Software	Open Source Software
No. of observations	12	12
Mean	162.33	143.92
Standard deviation	20.01	21.99

H_0: $\mu_1 = \mu_2$ (There is no difference in the mean amount of coupling between modules depicted by industrial and open source data sets.)

H_a: $\mu_1 \neq \mu_2$ (There is difference in the mean amount of coupling between modules depicted by industrial and open source data sets.)

Step 2: Select the appropriate statistical test.

As we are using two samples derived from different populations: one sample from industrial software and other from open source software, the samples are independent. Also the test variable, amount of coupling between modules, is measured at the continuous/interval measurement level. Hence, we need to use two sample t-test comparing the difference between average amount of coupling between modules derived from two independent samples.

Step 3: Apply test and calculate p-value.

The summary of descriptive statistic of each sample is given in Table 6.15. The t-statistic is given below:

$$t = \frac{\mu_1 - \mu_2}{\sqrt{\left(\sigma_1^2/n_1\right) + \left(\sigma_2^2/n_2\right)}} = \frac{162.33 - 143.92}{\sqrt{\left(20.01^2/12\right) + \left(21.99^2/12\right)}} = 2.146$$

The DOF is 22 (12 + 12 − 2) in this example. Given 22 DOF and referring the t-distribution table, the obtained p-value is 0.043.

Step 4: Define significance level.

As computed in Step 3, the p-value is 0.043. It can be seen that the results are statistically significant at 0.05 significance value.

Step 5: Derive conclusions.

The results are significant at 0.05 significance level. Hence, we reject the null hypothesis, and the results show that the mean amount of coupling between modules depicted by the industrial software is statistically significantly different than the mean amount of coupling between modules depicted by the open source software ($t = 2.146$, $p = 0.043$).

6.4.6.3 Paired t-Test

The paired t-test can be used if the two samples are related or paired in some manner. The samples are same but subject to different treatments (or technique), or the pairs must consist of before and after measurements on the same sample. We can formulate the following null and alternative hypotheses for application of paired t-test on a given problem:

H_0: $\mu_1 - \mu_2 = 0$ (There is no difference between the mean values of the two samples.)

H_a: $\mu_1 - \mu_2 \neq 0$ (There exists difference between the mean values of the two samples.)

The measure of paired t-test is given as,

$$t = \frac{\mu_1 - \mu_2}{\sigma_d/\sqrt{n}}$$

$$\sigma_d = \sqrt{\frac{\sum d^2 - \left[(\sum d)^2 / n\right]}{n-1}}$$

where:

n represents number of pairs and not total number of samples

d is difference between values of two samples

The DOF is $n - 1$. The p-value is obtained and compared with the established threshold value (α) for the computed t-statistic using the t-distribution.

Example 6.4:

Consider an example where values of the CBO (number of other classes to which a class is coupled to) metric is given before and after applying refactoring technique to improve the quality of the source code. The data is given in Table 6.16.

Solution:

Step 1: Formation of hypothesis.

In this step, null (H_0) and alternative (H_a) hypotheses are formed. The hypotheses for the example are given below:

H_0: $\mu_{CBO1} = \mu_{CBO2}$ (Mean of CBO metric before and after applying refactoring are equal.)

H_a: $\mu_{CBO1} \neq \mu_{CBO2}$ (Mean of CBO metric before and after applying refactoring are not equal.)

Step 2: Select the appropriate statistical test.

The samples are extracted from populations with normal distribution. As we are using samples derived from the same populations and analyzing the before and after effect of refactoring on CBO, these are related samples. We need to use paired t-test for comparing the difference between values of CBO derived from two dependent samples.

Step 3: Apply test and calculate p-value.

We first calculate the mean values of both the samples and also calculate the difference (d) among the paired values of both the samples as shown in Table 6.17. The t-statistic is given below:

$$\sigma_d = \sqrt{\frac{\sum d^2 - \left[(\sum d)^2 / n\right]}{n-1}} = \sqrt{\frac{12 - \left[(8)^2 / 15\right]}{14}} = 0.743$$

$$t = \frac{\mu_1 - \mu_2}{\sigma_d / \sqrt{n}} = \frac{67.6 - 67.07}{0.743 / \sqrt{15}} = 2.779$$

The DOF is 14 (15 − 1) in this example. Given 14 DOF and referring the t-distribution table, the obtained p-value is 0.015.

TABLE 6.16

CBO Values

| CBO before refactoring | 45 | 48 | 49 | 52 | 56 | 58 | 66 | 67 | 74 | 75 | 81 | 82 | 83 | 88 | 90 |
| CBO after refactoring | 43 | 47 | 49 | 52 | 56 | 57 | 66 | 67 | 74 | 73 | 80 | 82 | 83 | 87 | 90 |

TABLE 6.17

CBO Values

CBO before Refactoring	CBO after Refactoring	Differences (d)
45	43	2
48	47	1
49	49	0
52	52	0
56	56	0
58	57	1
66	66	0
67	67	0
74	74	0
75	73	2
81	80	1
82	82	0
83	83	0
88	87	1
90	90	0
$\mu_{CBO1} = 67.6$	$\mu_{CBO2} = 67.07$	

Step 4: Define significance level.

As the computed p-value is 0.015, which is less than $\alpha = 0.05$. Thus, the result is significant at $\alpha = 0.05$.

Step 5: Derive conclusions.

Fifteen classes were selected and CBO metric is calculated for these classes. The mean CBO is found to be 67.6. With the aim to improve the quality of these classes, the software developer applied refactoring technique on these classes. The mean of CBO metrics after applying refactoring is found to be 67.06. The reduction in the mean is found to be statistically significant at 0.05 significance level (p-value < 0.05). Hence, the software developer can reject the null hypothesis and conclude that there is a statistically significant improvement in the mean value of CBO metric after refactoring technique is applied.

6.4.7 Chi-Squared Test

It is a nonparametric test, symbolically denoted as χ^2 (pronounced as Ki-square). This test is used when the attributes are categorical (nominal or ordinal). It measures the distance of the observed values from the null expectations. The purpose of this test is to

- Test the interdependence between attributes.
- Test the goodness-of-fit of models.
- Test the significance of attributes for attribute selection or attribute ranking.
- Test whether the data follows normal distribution or not.

The χ^2 calculates the difference between the observed and expected frequencies and is given as

$$\chi^2 = \sum \frac{\left(O_{ij} - E_{ij}\right)^2}{E_{ij}}$$

where:
   O_{ij} is the observed frequency of the cell in the ith row and jth column
   E_{ij} is the expected frequency of the cell in the ith row and jth column

The expected frequency is calculated as below:

$$E_{\text{row,column}} = \frac{N_{\text{row}} \times N_{\text{column}}}{N}$$

where:
   N is the total number of observations
   N_{row} is the total of all observations in a specific row
   N_{column} is the total of all observations in a specific column
   $E_{\text{row,column}}$ is the grand total of a row or column

The larger the difference of the observed and the expected values, the more is the deviation from the stated null hypothesis. The DOF is (row − 1) × (column − 1) for any given table. The expected values are calculated for each category of the categorical variable at each factor of the other categorical variable. Then, calculate the x^2 value for each cell. After calculating individual x^2 value, add the individual x^2 values of each cell to obtain an overall x^2 value. The overall x^2 value is compared with the tabulated value for (row − 1) × (column − 1) DOF. If the calculated x^2 value is greater than the tabulated x^2 value at critical value α, we reject the null hypothesis.

Example 6.5:

Consider Table 6.18 that consists of data for a particular software. It states the categorization of modules according to three maintenance levels (high, medium, and low) and according to the number of LOC (high and low). A researcher wants to investigate whether LOC and maintenance level are independent of each other or not.

   Step 1: Formation of hypothesis.
      In this step, null (H_0) and alternative (H_a) hypotheses are formed. The hypotheses for the example are given below:
         H_0: LOC and maintenance level are independent of each other.
         H_a: LOC and maintenance level are not independent of each other.
   Step 2: Select the appropriate statistical test.
      The attributes explored in the example "maintenance level" and "LOC" are ordinal. The data can be arranged in a bivariate table to investigate the

TABLE 6.18

Categorization of Modules

		Maintenance Level			
		High	Low	Medium	Total
LOC	High	23	40	22	85
	Low	17	30	20	67
Total		40	70	42	152

relationship between the two attributes. Thus, chi-square test is an appropriate test for checking the independence of the two attributes.

Step 3: Apply test and calculate p-value.

Calculate the expected frequency of each cell according to the following formula:

$$E_{\text{row,column}} = \frac{N_{\text{row}} \times N_{\text{column}}}{N}$$

Table 6.19 shows the calculated expected frequency of each cell.

Now, calculate the chi-square value for each cell according to the following formula as shown in Table 6.20:

$$\chi^2 = \sum \frac{\left(O_{ij} - E_{ij}\right)^2}{E_{ij}}$$

Finally, calculate the overall χ^2 value by adding all corresponding χ^2 values of each cell.

$$\chi^2 = 0.017 + 0.018 + 0.093 + 0.022 + 0.023 + 0.118 = 0.291$$

Step 4: Define significance level.

The DOF = (rows − 1) × (columns − 1) = (2 − 1) × (3 − 1) = 2. Given 2 DOF and referring the χ^2-distribution table, the obtained p-value is 0.862 and χ^2 value is 5.991. It can be seen that the results are not statistically significant at 0.05 significance value.

Step 5: Derive conclusions.

The results are not statistically significant at 0.05 significance level. Hence, we accept the null hypothesis, and the results show that the two attributes "maintenance level" and "LOC" are independent ($\chi^2 = 0.291$, $p = 0.862$).

TABLE 6.19

Calculation of Expected Frequency

		Maintenance Level		
		High	**Low**	**Medium**
LOC	High	$\frac{85 \times 40}{152} = 22.36$	$\frac{85 \times 70}{152} = 39.14$	$\frac{85 \times 42}{152} = 23.48$
	Low	$\frac{67 \times 40}{152} = 17.63$	$\frac{67 \times 70}{152} = 30.85$	$\frac{67 \times 42}{152} = 18.52$

TABLE 6.20

Calculation of χ^2 Values

		Maintenance Level		
		High	**Low**	**Medium**
LOC	High	$\frac{(23 - 22.36)^2}{23} = 0.017$	$\frac{(40 - 39.14)^2}{39.14} = 0.018$	$\frac{(22 - 23.48)^2}{23.48} = 0.093$
	Low	$\frac{(17 - 17.63)^2}{17.63} = 0.022$	$\frac{(30 - 30.85)^2}{30.85} = 0.023$	$\frac{(20 - 18.52)^2}{18.52} = 0.118$

Example 6.6

Analyze the performance of four algorithms when applied on a single data set as given in Table 6.21. Evaluate whether there is any significant difference in the performance of the four algorithms at 5% significance level.

Solution:

Step 1: Formation of hypothesis.

The hypotheses for the example are given below:

H_0: There is no significant difference in the performance of the algorithms.

H_a: There is significant difference in the performance of the algorithms.

Step 2: Select the appropriate statistical test.

To explore the "goodness-of-fit" of different algorithms when applied on a specific data set, we can effectively use chi-square test.

Step 3: Apply test and calculate *p*-value.

Calculate the expected frequency of each cell according to the following formula:

$$E = \frac{\sum_{i=1}^{n} O_i}{n}$$

where:

O_i is the observed value of *i*th observation

n is the total number of observations

$$E = \frac{81+61+92+43}{4} = 69.25$$

Next, we calculate individual χ^2 values as shown in Table 6.22.

TABLE 6.21

Performance Values of Algorithms

Algorithm	Performance
A1	81
A2	61
A3	92
A4	43

TABLE 6.22

Calculation of χ^2 Values

Algorithm	Observed Frequency O_{ij}	Expected Frequency E_{ij}	$\left(O_{ij}-E_{ij}\right)$	$\left(O_{ij}-E_{ij}\right)^2$	$\dfrac{\left(O_{ij}-E_{ij}\right)^2}{E_{ij}}$
A1	81	69.25	11.75	138.06	1.99
A2	61	69.25	−8.25	68.06	0.98
A3	92	69.25	22.75	517.56	7.47
A4	43	69.25	−26.25	689.06	9.95

Now

$$\chi^2 = \sum \frac{(O_{ij} - E_{ij})^2}{E_{ij}} = 20.393$$

The DOF would be $n - 1$, that is, $(4 - 1) = 3$. Given 3 DOF and referring the χ^2-distribution table, we get χ^2 value as 7.815 at $\alpha = 0.05$, and the obtained p-value is 0.0001.

Step 4: Define significance level.

It can be seen that the results are statistically significant at 0.05 significance value as the obtained p-value in Step 3 is less than 0.05.

Step 5: Derive conclusions.

The results are significant at 0.05 significance level. Hence, we reject the null hypothesis, and the results show that there is significant difference in the performance of four algorithms ($\chi^2 = 20.393$, $p = 0.0001$).

Example 6.7:

Consider a scenario where a researcher wants to find the importance of SLOC metric, in deciding whether a particular class having more than 50 source LOC (SLOC) will be defective or not. The details of defective and not defective classes are provided in Table 6.23. Test the result at 0.05 significance value.

Solution:

Step 1: Formation of hypothesis.

The null and alternate hypotheses are formed as follows:

H_0: Classes having more than 50 SLOC will not be defective.

H_a: Classes having more than 50 SLOC will be defective.

Step 2: Select the appropriate statistical test.

To investigate the importance of SLOC attribute in detection of defective and not defective classes, we can appropriately use chi-square test to find an attribute's importance.

Step 3: Apply test and calculate p-value.

Calculate the expected frequency of each cell according to the following formula:

$$E_{row,column} = \frac{N_{row} \times N_{column}}{N}$$

Table 6.24 shows the observed and the calculated expected frequency of each cell. We also then calculate the individual χ^2 value of each cell.

Now

$$\chi^2 = \sum \frac{(O_{ij} - E_{ij})^2}{E_{ij}} = 716.66$$

The DOF = (rows − 1) × (columns − 1) = (2 − 1) × (2 − 1) = 1. Given 1 DOF and referring the χ^2-distribution table, the obtained p-value is 0.00001.

TABLE 6.23

SLOC Values for Defective and Not Defective Classes

	Defective (D)	Not Defective (ND)	Total
Number of classes having SLOC ≥ 50	200	200	400
Number of classes having SLOC < 50	100	700	800
Total	300	900	1,200

TABLE 6.24

Calculation of Expected Frequency

Observed Frequency O_{ij}	Expected Frequency E_{ij}	$\left(O_{ij} - E_{ij}\right)$	$\left(O_{ij} - E_{ij}\right)^2$	$\dfrac{\left(O_{ij} - E_{ij}\right)^2}{E_{ij}}$
200	$\dfrac{400 \times 300}{1200} = 100$	100	10,000	100
200	$\dfrac{400 \times 900}{1200} = 300$	−100	10,000	33.33
100	$\dfrac{800 \times 300}{1200} = 200$	−100	10,000	50
700	$\dfrac{400 \times 900}{1200} = 300$	400	160,000	533.33

Step 4: Define significance level.

The tabulated χ^2 value is 3.841. It can be seen that the results are statistically significant at $\alpha = 0.05$ significance value as the computed p-value is 0.00001.

Step 5: Derive conclusions.

The results are significant at 0.05 significance level. Hence, we reject the null hypothesis, and the results show that classes having more than 50 SLOC value would be defective ($\chi^2 = 716.66$, $p = 0.00001$).

Example 6.8:

Consider a scenario where 40 students had developed the same program. The size of the program is measured in terms of LOC and is provided in Table 6.25. Evaluate whether the size values of the program developed by 40 students individually follows normal distribution.

Solution:

Step 1: Formation of hypothesis.

The null and alternative hypotheses are as follows:

H_0: The data follows a normal distribution.

H_a: The data does not follow a normal distribution.

Step 2: Select the appropriate statistical test.

In the case of the normal distribution, there are two parameters, the mean (μ) and the standard deviation (σ) that can be estimated from the data. Based on the data, $\mu = 793.125$ and $\sigma = 64.81$. To test the normality of data, we can use chi-square test.

Step 3: Apply test and calculate p-value.

We first need to divide data into segments in such a way that the segments have the same probability of including a value, if the data actually is normally

TABLE 6.25

LOC Values

641	672	811	770	741	854	891	792	753	876
801	851	744	948	777	808	758	773	734	810
833	704	846	800	799	724	821	757	865	813
721	710	749	932	815	784	812	837	843	755

distributed with mean μ and standard deviation σ. We divide the data into 10 segments. We find the upper and lower limits of all the segments. To find upper limit (x_i) of ith segment, the following equation is used:

$$P(X < x_i) = \frac{i}{10}$$

where:
  $i = 1$–9
  X is $N(\mu, \sigma^2)$

which in terms of the standard normal distribution corresponds to

$$P(X_s < z_i) = \frac{i}{10}$$

where:
  $i = 1$–9
  X_s is $N(0,1)$

$$z_i = \frac{x_i - \mu}{\sigma}$$

Using standard normal table, we can calculate the values of z_i. We can then calculate the value of x_i using the following equation:

$$x_i = \sigma z_i + \mu$$

The calculated values z_i and x_i are given in Table 6.26. Since, a normally distributed variable theoretically ranges from $-\infty$ to $+\infty$, the lower limit of segment 1 is taken as $-\infty$ and the upper limit of segment 10 is taken as $+\infty$. The number of values that fall in each segment are also shown in the table. They represent the observed frequency (O_i). The expected number of values (E_i) in each segment can be calculated as $40/10 = 4$.
Now,

$$\chi^2 = \sum \frac{(O_{ij} - E_{ij})^2}{E_{ij}} = 5$$

TABLE 6.26

Segments and χ^2 Calculation

Segment No.	z_i	Lower Limit	Upper Limit	O_i	E_i	$(O_i-E_i)^2$
1	−1.28	−∞	710.17	4	4	0
2	−0.84	710.17	738.68	3	4	1
3	−0.525	738.68	759.10	7	4	9
4	−0.255	759.10	776.60	2	4	4
5	0	776.60	793.13	3	4	1
6	0.255	793.13	809.65	4	4	0
7	0.525	809.65	827.15	6	4	4
8	0.84	827.15	847.56	4	4	0
9	1.28	847.56	876.08	4	4	0
10	–	876.08	+∞	3	4	1

DOF = $n - e - 1$, where e is the number of parameters that must be estimated (mean [μ] and standard deviation [σ]) and n is the number of segments. In our example, DOF = $10 - 2 - 1 = 7$. The computed p-value is 0.0499.

Step 4: Define significance level.

At significance value (0.05), χ^2 value from distribution table is 14.07. Since, the tabulated value of χ^2 is greater than the calculated value, the results are not significant. The obtained p-value is also almost equal to $\alpha = 0.05$.

Step 5: Derive conclusions.

The results are not significant at 0.05 significance level. Hence, we accept the null hypothesis, which means that the data follows a normal distribution ($\chi^2 = 5$).

6.4.8 *F*-Test

F-test is used to investigate the equality of variance for two populations. A number of assumptions need to be checked for application of *F*-test, which includes the following (Kothari 2004):

1. The samples should be drawn from normally distributed populations.
2. All the observations should be independent.
3. The samples are randomly drawn from respective populations and there is no measurement error.

We can formulate the following null and alternative hypotheses for the application of *F*-test on a given problem with two populations:

H_0: $\sigma_1^2 = \sigma_2^2$ (Variances of two populations are equal.)
H_a: $\sigma_1^2 \neq \sigma_2^2$ (Variances of two populations are not equal.)

To test the above stated hypothesis, we compute the *F*-statistic as follows:

$$F = \frac{\left(\sigma_{sample1}\right)^2}{\left(\sigma_{sample2}\right)^2}$$

The variance of a sample can be computed by the following formula:

$$\sigma_{sample} = \sqrt{\frac{\sum_{i=1}^{n}(x_i - \mu)^2}{n-1}}$$

where:
n represents the number of observations in a sample
x_i represents the ith observation of the sample
μ represents the mean of the sample observations

We also designate v_1 as the DOF in the sample having greater variance and v_2 as the DOF in the other sample. The DOF is designated as one less than the number of observations in the corresponding sample. For example, if there are 5 observations in a sample, then the DOF is designated as 4 (5 − 1). The calculated value of *F* is compared with tabulated $F_\alpha(v_1, v_2)$ value at the desired α value. If the calculated *F*-value is greater than F_α, we reject the null hypothesis (H_0).

TABLE 6.27

Runtime Performance of Learning Techniques

A1	11	16	10	4	8	13	17	18	5
A2	14	17	9	5	7	11	19	21	4

Example 6.9:

Consider Table 6.27 that shows the runtime performance (in seconds) of two learning techniques (A1 and A2) on several data sets. We want to test whether the populations have the same variances.

Solution:

Step 1: Formation of hypothesis.

In this step, null (H_0) and alternative (H_a) hypothesis are formed. The hypotheses for the example are given below:

$H_0: \sigma_1^2 = \sigma_2^2$ (Variances of two populations are equal.)

$H_a: \sigma_1^2 \neq \sigma_2^2$ (Variances of two populations are not equal.)

Step 2: Select the appropriate statistical test.

The samples belong to normal populations and are independent in nature. Thus, to investigate the equality of variances of two populations, we use F-test.

Step 3: Apply test and calculate p-value.

In this example, $n_1 = 9$ and $n_2 = 9$. The calculation of two sample variances is as follows:

We first compute the means of the two samples,

$$\mu_1 = 11.33 \text{ and } \mu_2 = 11.89$$

$$\sigma_1^2 = \frac{\sum_{i=1}^{9}(x_i - \mu)^2}{n_1 - 1} = \frac{(11 - 11.33)^2 + \cdots + (5 - 11.33)^2}{9 - 1} = 26$$

$$\sigma_2^2 = \frac{\sum_{i=1}^{9}(x_i - \mu)^2}{n_2 - 1} = \frac{(14 - 11.89)^2 + \cdots + (4 - 11.89)^2}{9 - 1} = 38.36$$

Now, compute the F-statistic,

$$F = \frac{\sigma_2^2}{\sigma_1^2} = \frac{38.36}{26} = 1.47 \text{ (because } \sigma_2^2 > \sigma_1^2)$$

DOF in Sample 1 (v_2) = 8.

DOF in Sample 2 (v_1) = 8.

The computed p-value is 0.299.

Step 4: Define significance level.

We look up the tabulated value of F-distribution with $v_1 = 8$ and $v_2 = 8$ at $\alpha = 0.05$, which is 3.44. The calculated value of F ($F = 1.47$) is lesser than the tabulated value and, as obtained in Step 3, the computed p-value is 0.299. The results are not significant at $\alpha = 0.05$.

Step 5: Derive conclusions.

Because the calculated value of F is less than the tabulated value, we accept the null hypothesis. Thus, we conclude that the variance in runtime performance of both the techniques do not differ significantly ($F = 1.47$, $p = 0.299$).

6.4.9 Analysis of Variance Test

Analysis of variance (ANOVA) test is a method used to determine the equality of sample means for three or more populations. The variation in data can be attributed to two reasons: chance or just specific causes (Harnett and Murphy 1980). ANOVA test helps in determining whether the cause of variance is "specific" or just by chance. It splits up the variance into "within samples" and "between samples." A "within sample" variance is attributed to just random effects and other influences that cannot be explained. However, a "between samples" variance is attributed to a "specific factor," which can also be termed as the "treatment effect" (Kothari 2004). This helps a researcher in drawing conclusions about different factors that can affect the dependent variable outcome. However, the ANOVA test only indicates that there is difference among different groups, but not which specific group is different. The various assumptions required for use of ANOVA test is as follows:

1. The populations from which samples (observations) are extracted should be normally distributed.
2. The variance of the outcome variable should be equal for all the populations.
3. The observations should be independent.

We also assume that all the other factors except the ones that are being investigated are adequately controlled, so that the conclusions can be appropriately drawn. One-way ANOVA, also called the single factor ANOVA, considers only one factor for analysis in the outcome of the dependent variable. It is used for a completely randomized design.

In general, we calculate two variance estimates, one "within samples" and the other "between samples." Finally, we compute the F-value with these two variance estimates as follows:

$$F = \frac{\text{Variance between samples}}{\text{Variance within samples}}$$

The computed F-value is then compared with the F-limit for specific DOF. If the computed F-value is greater than the F-limit value, then we can conclude that the sample means differ significantly.

6.4.9.1 One-Way ANOVA

This test is used to determine whether various sample means are equal for a quantitative outcome variable and a single categorical factor (Seltman 2012). The factor may have two or more number of levels. These levels are called "treatments." All the subjects are exposed to only one level of treatment at a time. For example, one-way ANOVA can be used to determine whether the performance of different techniques (factors) vary significantly from each other when applied on a number of data sets. It is analogous to two independent samples t-test and is applied when we want to investigate the equality of means of more than two samples; otherwise independent samples *t*-test is sufficient. We can formulate the following null and alternative hypotheses for application of one-way ANOVA on a given problem:

$H_0: \mu_1 = \mu_2 = \mu_3 = \ldots\ldots \mu_k$ (Means of all the samples are equal.)

$H_a: \mu_1 \neq \mu_2 \neq \mu_3 \neq \ldots.. \mu_k$ (Means of all the samples are not equal, i.e., at least mean value of one sample is different than the others.)

The steps for computing *F*-statistic is as follows. Here, we assume *k* is the number of samples and *n* is the number of levels:

Step a: Calculate the means of each of the samples: $\mu_1, \mu_2, \mu_3 \ldots \mu_k$.

Step b: Calculate the mean of sample means.

$$\mu = \frac{\mu_1 + \mu_2 + \mu_3 + \cdots + \mu_k}{\text{Number of samples } (k)}$$

Step c: Calculate the sum of squares of variance between the samples (SSBS).

$$SSBS = n_1(\mu_1 - \mu)^2 + n_2(\mu_2 - \mu)^2 + n_3(\mu_3 - \mu)^2 + \cdots + n_k(\mu_k - \mu)^2$$

Step d: Calculate the sum of squares of variance within samples (SSWS). To obtain SSWS, we find the deviation of each sample observation with their corresponding mean and square the obtained deviations. We then sum all the squared deviations values to obtain SSWS.

$$SSWS = \Sigma(x_{1i} - \mu_1)^2 + \Sigma(x_{2i} - \mu_2)^2 + \Sigma(x_{3i} - \mu_3)^2 + \cdots + \Sigma(x_{ki} - \mu_k)^2 \text{ for } i = 1,2,3\ldots$$

Step e: Calculate the sum of squares for total variance (SSTV).

$$SSTV = SSBS + SSWS$$

Step f: Calculate the mean square between samples (MSBS) and mean square within samples (MSWS), and setup an ANOVA summary as shown in Table 6.28.

The calculated value of *F* is compared with tabulated F_α ($k - 1, n - k$) value at the desired α value. If the calculated *F*-value is greater than F_α, we reject the null hypothesis (H_0).

TABLE 6.28

Computation of Mean Square and F-Statistic

Source of Variation	Sum of Squares (SS)	DOF	Mean Square (MS)	F-Ratio
Between sample	SSBS	$k - 1$	$MSBS = \dfrac{SSBS}{K-1}$	$F\text{-ratio} = \dfrac{MSBS}{MSWS}$
Within sample	SSWS	$n - k$	$MSWS = \dfrac{SSWS}{n-k}$	
Total	SSTV	$n - 1$		

TABLE 6.29

Accuracy Values of Techniques

	Techniques		
Data Sets	A1	A2	A3
D1	60 (x_{11})	50 (x_{12})	40 (x_{13})
D2	40 (x_{21})	50 (x_{22})	40 (x_{23})
D3	70 (x_{31})	40 (x_{32})	50 (x_{33})
D4	80 (x_{41})	70 (x_{42})	30 (x_{43})

Example 6.10:

Consider Table 6.29 that shows the performance values (accuracy) of three techniques (A1, A2, and A3), which are applied on four data sets (D1, D2, D3, and D4) each. We want to investigate whether the performance of all the techniques calculated in terms of accuracy (refer to Section 7.5.3 for definition of accuracy) are equivalent.

Solution:
The following steps are carried out to solve the example.

> **Step 1:** Formation of hypothesis.
> In this step, null (H_0) and alternative (H_a) hypotheses are formed. The hypotheses are given below:
>> H_0: $\mu_1 = \mu_2 = \mu_3$ (Means of all the samples are equal, i.e., all techniques work equally well.)
>> H_a: $\mu_1 \neq \mu_2 \neq \mu_3$ (Means of all the samples are not equal, i.e., at least mean value of one technique is different than the others.)
>
> **Step 2:** Select the appropriate statistical test.
> The given hypothesis checks the means of more than two sample populations. The data is normally distributed, and the homogeneity of variance of outcome variables is checked. The observations are independent, that is, at a time only one treatment is applied on a specific data set. Thus, we use one-way ANOVA to test the hypothesis as only one factor (technique) is used to determine the outcome (performance).
>
> **Step 3:** Apply test and calculate p-value.
> Step a: Calculate the means of each of the samples.

$$\mu_1 = \frac{60+40+70+80}{4} = 62.5 \; ; \; \mu_2 = \frac{50+50+40+70}{4} = 52.5 \; ; \; \mu_1 = \frac{40+40+50+30}{4} = 40$$

> Step b: Calculate the mean of sample means.

$$\mu = \frac{\mu_1 + \mu_2 + \mu_3 \ldots + \mu_k}{\text{Number of samples } (k)}$$

$$\mu = \frac{62.5 + 52.5 + 40}{3} = 51.67$$

> Step c: Calculate the SSBS.

$$SSBS = n_1(\mu_1 - \mu)^2 + n_2(\mu_2 - \mu)^2 + n_3(\mu_3 - \mu)^2 + \cdots + n_k(\mu_k - \mu)^2$$

$$SSBS = 4(62.5 - 51.67)^2 + 4(52.5 - 51.67)^2 + 4(40 - 51.67)^2 = 1016.68$$

Step d: Calculate the SSWS.

$$SSWS = \Sigma(x_{1i} - \mu_1)^2 + \Sigma(x_{2i} - \mu_2)^2 + \Sigma(x_{3i} - \mu_3)^2 + \cdots + \Sigma(x_{ki} - \mu_k)^2 \text{ for } i = 1,2,3\ldots$$

$$SSWS = \left[(60 - 62.5)^2 + \cdots + (80 - 62.5)^2\right] + \left[(50 - 52.5)^2 + \cdots + (70 - 52.5)^2\right]$$

$$+ \left[(40 - 40)^2 + \cdots + (30 - 40)^2\right] = 1550$$

Step e: Calculate the SSTV.

$$SSTV = SSBS + SSWS = 1016.68 + 1550 = 2566.68$$

Step f: Calculate MSBS and MSWS, and setup an ANOVA summary as shown in Table 6.30.

The DOF for between sample variance is 2 and that for within sample variance is 9. For the corresponding DOF, we compute the F-value using the F-distribution table and obtain the p-value as 0.103.

Step 4: Define significance level.

After obtaining the p-value in Step 3, we need to decide the threshold or α value. The calculated value of F at Step 3 is 2.95, which is less than the tabulated value of F (4.26) with DOF being $v_1 = 2$ and $v_2 = 9$ at 5% level. Thus, the results are not statistically significant at 0.05 significance value.

Step 5: Derive conclusions.

As the results are not statistically significant at 0.05 significance value, we accept the null hypothesis, which states that there is no difference in sample means and all the three techniques perform equally well. The difference in observed values of the techniques is only because of sampling fluctuations ($F = 2.95$, $p = 0.103$).

6.4.10 Wilcoxon Signed Test

Wilcoxon signed-ranks test is a nonparametric test that is used to perform pairwise comparisons among different treatments (Wilcoxon 1945). It is also called Wilcoxon matched pairs test and is used in the scenario of two related samples (Kothari 2004). The Wilcoxon signed-ranks test is based on the following hypotheses:

H_0: There is no statistical difference between the two treatments.

H_a: There exists a statistical difference between the two treatments.

TABLE 6.30

Computation of Mean Square and F-Statistic

Source of Variation	Sum of Squares (SS)	DOF	Mean Square (MS)	F-Ratio	F-Limit (0.05)
Between sample	1016.68	$3 - 1 = 2$	$MSBS = \dfrac{1016.68}{2} = 508.34$	$F = \dfrac{508.34}{172.22} = 2.95$	$F(2,9) = 4.26$
Within sample	1550	$12 - 3 = 9$	$MSWS = \dfrac{1550}{9} = 172.22$		
Total	2566.68	11			

To perform the test, we compute the differences among the related pair of values of both the treatments. The differences are then ranked based on their absolute values. We perform the following steps while assigning ranks to the differences:

1. Exclude the pairs where the absolute difference is 0. Let n_r be the reduced number of pairs.

2. Assign rank to the remaining n_r pairs based on the absolute difference. The smallest absolute difference is assigned a rank 1.

3. In case of ties among differences (more than one difference having the same value), each tied difference is assigned an average of tied ranks. For example, if there are two differences of data value 5 each occupying 7th and 8th ranks, we would assign the mean rank, that is, 7.5 ([7 + 8]/2 = 7.5) to each of the difference.

We now compute two variables R^+ and R^-. R^+ represents the sum of ranks assigned to differences, where the data instance in the first treatment outperforms the second treatment. However, R^- represents the sum of ranks assigned to differences, where the second treatment outperforms the first treatment. They can be calculated by the following formula (Demšar 2006):

$$R^+ = \sum_{d_i > 0} \text{rank}(d_i)$$

$$R^- = \sum_{d_i < 0} \text{rank}(d_i)$$

where:

d_i is the difference between performance measures of first treatment from the second treatment when applied on n different data instances

Finally, we calculate the Z-statistic as follows, where $Q = \text{minimum}(R^+, R^-)$.

$$Z = \frac{Q - (1/4) n_r (n_r + 1)}{\sqrt{(1/24) n_r (n_r + 1)(2n_r + 1)}}$$

If the Z-statistic is in the critical region with specific level of significance, then the null hypothesis is rejected and it is concluded that there is significant difference between two treatments, otherwise null hypothesis is accepted.

Example 6.11:

For example, consider an example where a researcher wants to compare the performance of two techniques (T1 and T2) on multiple data sets using a performance measure as given in Table 6.31. Investigate whether the performance of two techniques measured in terms of AUC (refer to Section 7.5.6 for details on AUC) differs significantly.

Solution:

Step 1: Formation of hypothesis.

The hypotheses for the example are given below:

H_0: The performance of the two techniques does not differ significantly.

H_a: The performance of the two techniques differs significantly.

TABLE 6.31

Performance Values of Techniques

Data Sets	Techniques	
	T1	T2
D1	0.75	0.65
D2	0.87	0.73
D3	0.58	0.64
D4	0.72	0.72
D5	0.60	0.70

Step 2: Select the appropriate statistical test.

The two techniques have matched pairs as they are evaluated on the same data sets. Moreover, the performance measurement scale is continuous. As we need to investigate the comparative performance of the two techniques, we use Wilcoxon signed test.

Step 3: Apply test and calculate p-value.

We assign ranks based on the basis of absolute difference between the performances of two techniques. Here, $n = 5$. For each pair, ranks are given in Table 6.32.

According to Table 6.32, we can see that $n_r = 4$. We now compute R^+ and R^- as follows:

$$R^+ = \sum_{d_i > 0} \text{rank}(d_i) = 1 + 2.5 = 3.5$$

$$R^- = \sum_{d_i < 0} \text{rank}(d_i) = 2.5 + 4 = 6.5$$

Thus, $Q = \text{minimum } (R^+, R^-) = 3.5$. The Z-statistic can be computed as follows:

$$Z = \frac{Q - (1/4)n_r(n_r + 1)}{\sqrt{(1/24)n_r(n_r + 1)(2n_r + 1)}} = \frac{3.5 - (1/4)4(4+1)}{\sqrt{(1/24)4(4+1)(2 \times 4 + 1)}} = -0.549$$

The obtained p-value is 0.581 with Z-distribution table, when DOF is $(n - 1)$, that is, 1.

Step 4: Define significance level.

The chi-square value is $\chi^2_{0.05} = 3.841$. As the test statistic value ($Z = -0.549$) is less than χ^2 value, we accept the null hypothesis. The obtained p-value in Step 3 is greater than $\alpha = 0.05$. Thus, the results are not significant at critical value $\alpha = 0.05$.

TABLE 6.32

Computing R^+ and R^-

Data Set	T1	T2	d_i	$|d_i|$	Rank(d_i)
D1	0.75	0.65	−0.10	0.10	2.5
D2	0.87	0.73	−0.14	0.14	4
D3	0.58	0.64	0.06	0.06	1
D4	0.72	0.72	0	0	–
D5	0.60	0.70	0.10	0.10	2.5

Step 5: Derive conclusions

As shown in Step 4, we accept the null hypothesis. Thus, we conclude that the performance of both the techniques do not differ significantly ($Z = -0.549$, $p = 0.581$).

6.4.11 Wilcoxon–Mann–Whitney Test (*U*-Test)

This test is used to ascertain the difference among two independent samples when the outcome variable is continuous or ordinal (Anderson et al. 2002). It is the nonparametric equivalent of the independent samples *t*-test. However, the underlying data does not need to be normal for the application of Wilcoxon–Mann–Whitney test. It is also commonly known as Wilcoxon rank-sum test or Mann–Whitney *U*-test. The test investigates whether the two samples drawn independently belong to the same population by checking the equality of the two sample means. It can be used when sample sizes are unequal. We can formulate the following null and alternative hypotheses for application of Wilcoxon–Mann–Whitney test on a given problem:

H_0: $\mu_1 - \mu_2 = 0$ (The two sample means belong to the same population and are identical.)

H_a: $\mu_1 - \mu_2 \neq 0$ (The two sample means are not equal and belong to different populations.)

To perform the test, we need to compute the rank-sum statistics for all the observations in the following manner. We assume that the number of observations in sample 1 is n_1 and the number of observations in sample 2 is n_2. The total number of observations is denoted by N ($N = n_1 + n_2$):

1. Arrange the data values of all the observations (both the samples) in ascending (low to high) order.

2. Assign ranks to all the observations. The lowest value observation is provided rank 1, the next to lowest observation is provided rank 2, and so on, with the highest observation given the rank N.

3. In case of ties (more than one observation having the same value), each tied observation is assigned an average of tied ranks. For example: if there are three observations of data value 20 each occupying 7th, 8th, and 9th ranks, we would assign the mean rank, that is, 8 ($[7 + 8 + 9]/3 = 8$) to each of the observation.

4. We then find the sum of all the ranks allotted to observations in sample 1 and denote it with T_1. Similarly, find the sum of all the ranks allotted to observations in sample 2 and denote it as T_2.

5. Finally, we compute the *U*-statistic by the following formula:

$$U = n_1.n_2 + \frac{n_1(n_1 + 1)}{2} - T_1$$

or

$$U = n_1.n_2 + \frac{n_2(n_2 + 1)}{2} - T_2$$

It can be observed that the sum of the *U*-values obtained by the above two formulas is always equal to the product of the two sample sizes ($n_1.n_2$; Hooda 2003). It should be noted that we should use the lower computed U-value as obtained by the two equations described above. Wilcoxon–Mann–Whitney test has two specific cases (Anderson et al. 2002; Hooda 2003): (1) when the sample sizes are small ($n_1 < 7$, $n_2 < 8$) or (2) when the sample sizes are large ($n_1 \geq 10$, $n_2 \geq 10$). The p-values for the corresponding computed U-values are interpreted as follows:

Case 1: When the sample sizes are small ($n_1 < 7$, $n_2 < 8$)

To decide whether we should accept or reject the null hypothesis, we should derive the p-value from the tables shown in Appendix I. For the given values of n_1 and n_2, we find a p-value that is less than or equal to the computed U-value. For example, if the value of n_1 and n_2 is 4 and 5, respectively, and the computed U-value is 3, then the p-value would be 0.056. For a two-tailed test, the U-value should be computed for the lesser of the two computed U-values.

Case 2: When the sample sizes are large ($n_1 \geq 10$, $n_2 \geq 10$)

For sample sizes, where each sample contains 10 or more data values, the sampling U-distribution can be approximated by the normal distribution. In this case, we can calculate the mean (μ_U) and standard deviation (σ_U) of the normal population as follows:

$$\mu_U = \frac{n_1 . n_2}{2} ; \sigma_U = \sqrt{\frac{n_1 . n_2 (n_1 + n_2 + 1)}{12}}$$

Thus, the Z-statistic can be defined as,

$$Z = \frac{U - \mu_u}{\sigma_u}$$

If the tabulated Z-value at a significance level α is greater than the computed Z-value, we reject the null hypothesis. Otherwise, we accept the alternate hypothesis.

Example 6.12:

Consider an example for comparing the coupling values of two different software (one open source and other academic software), to ascertain whether the two samples are identical with respect to coupling values (coupling of a module corresponds to the number of other modules to which a module is coupled).

Academic: 89, 93, 35, 43
Open source: 52, 38, 5, 23, 32

Solution:
Step 1: Formation of hypothesis.
In this step, null (H_0) and alternative (H_a) hypotheses are formed. The hypotheses for the example are given below:
$H_0 : \mu_1 - \mu_2 = 0$ (The two samples are identical in terms of coupling values.)
$H_a : \mu_1 - \mu_2 \neq 0$ (The two sample are not identical in terms of coupling values.)
Step 2: Select the appropriate statistical test.
The two samples of our study are independent in nature, as they are collected from two different software. Also, the outcome variable (amount of coupling) is continuous or ordinal in nature. The data may not be normal. Hence, we

TABLE 6.33

Computation of Rank Statistics for
Coupling Values of Two Software

Observations	Rank	Sample Name
5	1	Open source
23	2	Open source
32	3	Open source
35	4	Academic
38	5	Open source
43	6	Academic
52	7	Open source
89	8	Academic
93	9	Academic

use the Wilcoxon–Mann–Whitney test for comparing the differences among coupling values of an academic and open source software.

Step 3: Apply test and calculate p-value.

In this example, $n_1 = 4$, $n_2 = 5$, and $N = 9$. Table 6.33 shows the arrangement of all the observations in ascending order, and the ranks allocated to them.

Sum of ranks assigned to observations in Academic software $(T_1) = 4+6+8+9 = 27$. Sum of ranks assigned to observations in open source software $(T_2) = 1+2+3+5+7 = 18$.

The U-statistic is given below:

$$U = n_1.n_2 + \frac{n_1(n_1+1)}{2} - T_1$$

$$= 4.5 + \frac{4(4+1)}{2} - 27 = 3$$

$$U = n_1.n_2 + \frac{n_2(n_2+1)}{2} - T_2$$

$$= 4.5 + \frac{5(5+1)}{2} - 18 = 17$$

We compute the p-value to be 0.056 at $\alpha = 0.05$ for the values of n_1 and n_2 as 4 and 5, respectively, and the U-value as 3.

Step 4: Define significance level.

As the derived p-value of 0.056, in Step 3, is greater than $2\alpha = 0.10$, we accept the null hypothesis at $\alpha = 0.05$. Thus, the results are not significant at $\alpha = 0.05$.

Step 5: Derive conclusions.

As shown in Step 4, we accept the null hypothesis. Thus, we conclude that the coupling values of the academic and open source software do not differ significantly $(U = 3, p = 0.056)$.

Example 6.13:

Let us consider another example for large sample size, where we want to ascertain whether the two sets of observations (sample 1 and sample 2) are extracted from identical populations by observing the cohesion values of the two samples.

Sample 1: 55, 40, 71, 59, 48, 40, 75, 46, 71, 72, 58, 76
Sample 2: 46, 42, 63, 54, 34, 46, 72, 43, 65, 70, 51, 70

Solution:

Step 1: Formation of hypothesis.

In this step, null (H_0) and alternative (H_a) hypotheses are formed. The hypothesis for the example is given below:

H_0: $\mu_1 - \mu_2 = 0$ (The two samples are identical in terms of cohesion values.)

H_a: $\mu_1 - \mu_2 \neq 0$ (The two sample are not identical in terms of cohesion values.)

Step 2: Select the appropriate statistical test.

The two samples of our study are independent in nature as they are collected from two different software. Also, the outcome variable (amount of cohesion) is continuous or ordinal in nature. The data may not be normal. Hence, we use the Wilcoxon–Mann–Whitney test for comparing the differences among cohesion values of the two software.

Step 3: Apply test and calculate p-value.

In this example, $n_1 = 12$, $n_2 = 12$, and $N = 24$. Table 6.34 shows the arrangement of all the observations in ascending order, and the ranks allocated to them.

Sum of ranks assigned to observations in sample 1 (T_1) = 2.5 + 2.5 + 7 + 9 + 12 + 13 + 14 + 19.5 + 19.5 + 21.5 + 23 + 24 = 167.5.

Sum of ranks assigned to observations in sample 2 (T_2) = 1 + 4 + 5 + 7 + 7 + 10 + 11 + 15 + 16 + 17.5 + 17.5 + 21.5 = 132.5.

TABLE 6.34

Computation of Rank Statistics for Cohesion Values of Two Samples

Observations	Rank	Sample Name
34	1	Sample 2
40	2.5	Sample 1
40	2.5	Sample 1
42	4	Sample 2
43	5	Sample 2
46	7	Sample 1
46	7	Sample 2
46	7	Sample 2
48	9	Sample 1
51	10	Sample 2
54	11	Sample 2
55	12	Sample 1
58	13	Sample 1
59	14	Sample 1
63	15	Sample 2
65	16	Sample 2
70	17.5	Sample 2
70	17.5	Sample 2
71	19.5	Sample 1
71	19.5	Sample 1
72	21.5	Sample 1
72	21.5	Sample 2
75	23	Sample 1
76	24	Sample 1

The U-statistic is given below:

$$U = n_1.n_2 + \frac{n_1(n_1+1)}{2} - T_1$$

$$= 12 \cdot 12 + \frac{12(12+1)}{2} - 167.5 = 54.5$$

$$U = n_1.n_2 + \frac{n_2(n_2+1)}{2} - T_2$$

$$= 12 \cdot 12 + \frac{12(12+1)}{2} - 132.5 = 89.5$$

As the sample size is large, we can calculate the mean (μ_U) and standard deviation (σ_U) of the normal population as follows:

$$\mu_U = \frac{n_1.n_2}{2} = \frac{12 \cdot 12}{2} = 72; \sigma_U = \sqrt{\frac{n_1.n_2(n_1+n_2+1)}{12}} = \sqrt{\frac{12 \cdot 12(12+12+1)}{12}} = 17.32$$

Thus, the Z-statistic can be computed as,

$$Z = \frac{U - \mu_u}{\sigma_u} = \frac{54.5 - 72}{17.32} = -1.012$$

The obtained p-value from the normal table is 0.311.

Step 4: Define significance level.

As computed in Step 3, the obtained p-value is 0.311. This means that the results are not significant at $\alpha = 0.05$. Thus, we accept the null hypothesis.

Step 5: Derive conclusions.

As shown in Step 4, we accept the null hypothesis. Thus, we conclude that the cohesion values of two software samples do not differ significantly ($U = 54.5$, $p = 0.311$).

6.4.12 Kruskal–Wallis Test

This test is used to investigate whether there is any significant difference among three or more independent sample distributions (Anderson et al. 2002). It is a nonparametric test that extends the Wilcoxon–Mann–Whitney test on k sample distributions. We can formulate the following null and alternative hypothesis for application of Kruskal–Wallis test on a given problem:

H_0: $\mu_1 = \mu_2 = \ldots \mu_k$ (All samples have identical distributions and belong to the same population.)

H_a: $\mu_1 \neq \mu_2 \neq \ldots \mu_k$ (All samples do not have identical populations and may belong to different populations.)

The steps to compute the Kruskal–Wallis test statistic H are very similar to that of Wilcoxon–Mann–Whitney test statistic U. Assuming there are k samples of size $n_1, n_2, \ldots n_k$, respectively, and the total number of observations N ($N = n_1 + n_2 + \ldots n_k$), we perform the following steps:

1. Organize and sort the data values of all the observations (belonging to all the samples) in an ascending (low to high) order.

2. Next, allocate ranks to all the observations from 1 to N. The observation with the lowest data value is assigned a rank of 1, and the observation with the highest data value is assigned rank N.

3. In case of two or more observations of equal values, assign the average of the ranks that would have been assigned to the observations. For example, if there are two observations of data value 40 each occupying 3rd and 4th ranks, we would assign the mean rank, that is, 3.5 ($[3+4]/2=3.5$) to each of the 3rd and 4th observations.

4. We then compute the sum of ranks allocated to observations in each sample and denote it as $T_1, T_2 \ldots T_k$.

5. Finally, the H-statistic is computed by the following formula:

$$H = \frac{12}{N(N+1)} \sum_{i=1}^{k} \frac{T_i^2}{n_i} - 3(N+1)$$

The calculated H-value is compared with the tabulated χ_α^2 value at $(k-1)$ DOF at the desired α value. If the calculated H-value is greater than χ_α^2 value, we reject the null hypothesis (H_0).

Example 6.14:

Consider an example (Table 6.35) where three research tools were evaluated by 17 different researchers and were given a performance score out of 100. Investigate whether there is a significant difference in the performance rating of the tools.

Solution:
Step 1: Formation of hypothesis.

In this step, null (H_0) and alternative (H_a) hypotheses are formed. The hypotheses for the example are given below:

$H_0: \mu_1 = \mu_2 = \mu_3$ (The performance rating of all tools does not differ significantly.)

$H_a: \mu_1 \neq \mu_2 \neq \mu_3$ (The performance rating of all tools differ significantly.)

Step 2: Select the appropriate statistical test.

The three samples are independent in nature as they are rated by 17 different researchers. The outcome variable is continuous. As we need to compare more than two samples, we use Kruskal–Wallis test to investigate whether there is a significant difference in the performance rating of the tools.

TABLE 6.35

Performance Score of Tools

Tools		
Tool 1	Tool 2	Tool 3
30	65	55
75	25	75
65	35	65
90	20	85
100	45	95
95		75

TABLE 6.36

Computation of Rank Kruskal–Wallis Test for Performance Score of Research Tools

Observations	Rank	Sample Name
20	1	Tool 2
25	2	Tool 2
30	3	Tool 1
35	4	Tool 2
45	5	Tool 2
55	6	Tool 3
65	8	Tool 1
65	8	Tool 2
65	8	Tool 3
75	11	Tool 1
75	11	Tool 3
75	11	Tool 3
85	13	Tool 3
90	14	Tool 1
95	15.5	Tool 1
95	15.5	Tool 3
100	17	Tool 1

Step 3: Apply test and calculate *p*-value.

In this example, $n_1 = 6$, $n_2 = 5$, $n_3 = 6$, and $N = 17$. The arrangement of all the performance rating observations in ascending order and their corresponding ranks are shown in Table 6.36.

Sum of ranks assigned to performance rating observations of Tool 1 $(T_1) = 3 + 8 + 11 + 14 + 15.5 + 17 = 68.5$.

Sum of ranks assigned to performance rating observations of Tool 2 $(T_2) = 1 + 2 + 4 + 5 + 8 = 20$.

Sum of ranks assigned to performance rating observations of Tool 3 $(T_3) = 6 + 8 + 11 + 11 + 13 + 15.5 = 64.5$.

The *H*-statistic can be computed as follows:

$$H = \frac{12}{N(N+1)} \sum_{i=1}^{k} \frac{T_i^2}{n_i} - 3(N+1)$$

$$= \frac{12}{17(17+1)} \left[\frac{(68.5)^2}{6} + \frac{(20)^2}{5} + \frac{(64.5)^2}{6} \right] - 3(17+1) = 7$$

The *p*-value obtained at 2 DOF is 0.029.

Step 4: Define significance level.

We compute chi-square distribution with 2 $(k - 1)$ DOF at $\alpha = 0.05$. The chi-square value is $\chi^2_{0.05} = 5.99$. As the test statistic value $(H = 7)$ is greater than χ^2 value, we reject the null hypothesis. Thus, the results are significant with a *p*-value of 0.029.

Step 5: Derive conclusions.

As shown in Step 4, we reject the null hypothesis. Thus, we conclude that the performance rating of all tools differ significantly $(H = 7, p = 0.029)$.

6.4.13 Friedman Test

Friedman test is a nonparametric test, which can be used to rank a set of k treatments over multiple data instances or subjects (Friedman 1940). The test can be used to investigate the existence of any statistical difference between various treatments. It is generally used in a scenario where same set of treatments (techniques/methods) are repeatedly applied over n independent data instances or subjects. A uniform measure is required to compute the performance of different treatments on n data instances. However, Friedman test does not require that the samples should be drawn from normal populations. To proceed with the test, we must compute the ranks based on the performance of different treatments on n data instances. The Friedman test is based on the assumption that the measures over data instances are independent of each other. The hypotheses can be formulated as follows:

H_0: There is no statistical difference between the performances of various treatments.

H_a: There is statistical significant difference between the performances of various treatments.

The steps to compute the Friedman test statistic χ^2 are as follows. Assuming there are k treatments that are applied on n independent data instances each.

1. Organize and sort the data values of all the treatments for a specific data instance or data set in descending (high to low) order. Allocate ranks to all the observations from 1 to k, where rank 1 is assigned to the best performing treatment value and rank k to the worst performing treatment. In case of two or more observations of equal values, assign the average of the ranks that would have been assigned to the observations.

2. We then compute the total of ranks allocated to a specific treatment on all the data instances. This is done for all the treatments and the rank total for k treatments is denoted by $R_1, R_2, \ldots R_k$.

3. Finally, the χ^2-statistic is computed by the following formula:

$$\chi^2 = \frac{12}{nk(k+1)} \sum_{i=1}^{k} R_i^2 - 3n(k+1)$$

where:

R_i is the individual rank total of the ith treatment

n is the number of data instances

The value of Friedman measure χ^2 is distributed over $k - 1$ DOF. If the value of Friedman measure is in the critical region (obtained from chi-squared table with specific level of significance, i.e., 0.01 or 0.05 and $k - 1$ DOF), then the null hypothesis is rejected and it is concluded that there is difference among performance of different treatments, otherwise the null hypothesis is accepted.

Example 6.15:

Consider Table 6.37, where the performance values of six different classification methods are stated when they are evaluated on six data sets. Investigate whether the performance of different methods differ significantly.

TABLE 6.37

Performance Values of Different Methods

Data Sets	M1	M2	M3	M4	M5	M6
D1	83.07	75.38	73.84	72.30	56.92	52.30
D2	66.66	75.72	73.73	71.71	70.20	45.45
D3	83.00	54.00	54.00	77.00	46.00	59.00
D4	61.93	62.53	62.53	64.04	56.79	53.47
D5	74.56	74.56	73.98	73.41	68.78	43.35
D6	72.16	68.86	63.20	58.49	60.37	48.11

Solution:

Step 1: Formation of hypothesis.

In this step, null (H_0) and alternative (H_a) hypotheses are formed. The hypotheses for the example are given below:

H_0: There is no statistical difference between the performances of various methods.

H_a: There is statistical significant difference between the performances of various methods.

Step 2: Select the appropriate statistical test.

As we need to evaluate the difference between the performances of different methods when they are evaluated using six data sets, we are evaluating different treatments on different data instances. Moreover, there is no specific assumption for data normality. Thus, we can use Friedman test.

Step 3: Apply test and calculate p-value.

We compute the rank total allocated to each method on the basis of performance ranking of each method on different data sets as shown in Table 6.38. Now, compute the Friedman statistic,

$$\chi^2 = \frac{12}{nk(k+1)} \sum R^2 - 3n(k+1)$$

$$= \frac{12}{6 \times 6 \times (6+1)} \left(13.5^2 + 13.5^2 + 18^2 + 19^2 + 29^2 + 33^2\right) - 3.6(6+1) = 16.11$$

$$DOF = k - 1 = 5$$

TABLE 6.38

Computation of Rank Totals for Friedman Test

Data Sets	M1	M2	M3	M4	M5	M6
D1	1	2	3	4	5	6
D2	5	1	2	3	4	6
D3	1	4.5	4.5	2	6	3
D4	4	2.5	2.5	1	5	6
D5	1.5	1.5	3	4	5	6
D6	1	2	3	5	4	6
Rank total	13.5	13.5	18	19	29	33

We look up the tabulated value of χ^2-distribution with 5 DOF, and find the tabulated value as 15.086 at $\alpha = 0.01$. The p-value is computed as 0.007.
Step 4: Define significance level.
   The calculated value of χ^2 ($\chi^2 = 16.11$) is greater than the tabulated value. As the computed p-value in Step 3 is <0.01, the results are significant at $\alpha = 0.01$.
Step 5: Derive conclusions.
   Since the calculated value of χ^2 is greater than the tabulated value, we reject the null hypothesis. Thus, we conclude that the performance of six methods differ significantly ($\chi^2 = 16.11$, $p = 0.007$).

6.4.14 Nemenyi Test

Nemenyi test is a post hoc test that is used to compare multiple subjects (techniques/tools/other experimental design settings) when the sample sizes are equal. It can be used after the application of a Kruskal–Wallis test or a Friedman test, if the null hypothesis of the corresponding test is rejected. Nemenyi test is applicable when we compare all the subjects with each other and want to investigate whether the performance of two subjects differ significantly (Demšar 2006). We compute the critical distance (CD) value as follows:

$$CD = q_\alpha \sqrt{\frac{k(k+1)}{6n}}$$

Here k corresponds to the number of subjects and n corresponds to the number of observations for a subject. The critical values (q_α) are studentized range statistic divided by $\sqrt{2}$. The computed CD value is compared with the difference between average ranks allocated to two subjects. If the difference is at least equal to or greater than the CD value, the two subjects differ significantly at the chosen significance level α.

Example 6.16:

Consider an example where we compare four techniques by analyzing the performance of the models predicted using these four techniques on six data sets each. We first apply Friedman test to obtain the average ranks of all the methods. The computed average ranks are shown in Table 6.39. The result of the Friedman test indicated the rejection of null hypothesis. Evaluate whether there are significant differences among different methods using pairwise comparisons.

Solution:
Step 1: Formation of hypothesis.
   In this step, null (H_0) and alternative (H_a) hypotheses are formed. The hypotheses for the example are given below:
   H_{01}: The performance of T1 and T2 techniques do not differ significantly.
   H_{a1}: The performance of T1 and T2 techniques differ significantly.
   H_{02}: The performance of T1 and T3 techniques do not differ significantly.
   H_{a2}: The performance of T1 and T3 techniques differ significantly.
   H_{03}: The performance of T1 and T4 techniques do not differ significantly.

TABLE 6.39

Average Ranks of Techniques after Applying Friedman Test

	T1	T2	T3	T4
Average rank	3.67	2.67	1.92	1.75

H_{a3}: The performance of $T1$ and $T4$ techniques differ significantly.
H_{04}: The performance of $T2$ and $T3$ techniques do not differ significantly.
H_{a4}: The performance of $T2$ and $T3$ techniques differ significantly.
H_{05}: The performance of $T2$ and $T4$ techniques do not differ significantly.
H_{a5}: The performance of $T2$ and $T4$ techniques differ significantly.
H_{06}: The performance of $T3$ and $T4$ techniques do not differ significantly.
H_{a6}: The performance of $T3$ and $T4$ techniques differ significantly.

Step 2: Select the appropriate statistical test.

The evaluation of different techniques is performed using Friedman test, and the result led to rejection of the null hypothesis. To analyze whether there are any significant differences among pairwise comparisons of all the techniques, we need to apply a post hoc test. The number of data sets for evaluating each technique is same (six each, i.e., equal sample sizes), thus we use Nemenyi test.

Step 3: Apply test and calculate CD.

In this example, $k = 4$ and $n = 6$. The value of q_α for four subjects at $\alpha = 0.05$ is 2.569. The CD can be calculated by the following formula:

$$CD = q_\alpha \sqrt{\frac{k(k+1)}{6n}} = 2.569 \sqrt{\frac{4.(4+1)}{6.6}} = 1.91$$

We now find the differences among ranks of each pair of techniques as shown in Table 6.40.

Step 4: Define significance level.

Table 6.41 shows the comparison results of critical difference and actual rank differences among different techniques. The rank difference of only $T1$–$T4$ pair is higher than the computed critical difference. The rank differences of all other

TABLE 6.40

Computation of Pairwise Rank
Differences among Techniques
for Nemenyi Test

Pair	Difference
$T1$–$T2$	$3.67 - 2.67 = 1.00$
$T1$–$T3$	$3.67 - 1.92 = 1.75$
$T1$–$T4$	$3.67 - 1.75 = 1.92$
$T2$–$T3$	$2.67 - 1.92 = 0.75$
$T2$–$T4$	$2.67 - 1.75 = 0.92$
$T3$–$T4$	$1.92 - 1.75 = 0.17$

TABLE 6.41

Comparison of Differences
for Nemenyi Test

Pair	Difference
$T1$–$T2$	$1.00 < 1.91$
$T1$–$T3$	$1.75 < 1.91$
$T1$–$T4$	**$1.92 > 1.91$**
$T2$–$T3$	$0.75 < 1.91$
$T2$–$T4$	$0.92 < 1.91$
$T3$–$T4$	$0.17 < 1.91$

technique pairs is not significant at $\alpha = 0.05$. The rank difference of only $T1$–$T4$ pair (shown in bold) is higher than the computed critical difference.

Step 5: Derive conclusions.

As the rank difference of only $T1$–$T4$ pair is higher than the computed critical difference, we conclude that the $T4$ technique significantly outperforms $T1$ technique at significance level $\alpha = 0.05$. The difference in performance of all other techniques is not significant. We accept all the null hypotheses H_{01}–H_{06}, except H_{03}.

6.4.15 Bonferroni–Dunn Test

Bonferroni–Dunn test is a post hoc test that is similar to Nemenyi test. It can be used to compare multiple subjects, even if the sample sizes are unequal. It is generally used when all subjects are compared with a control subject (Demšar 2006). For example, all techniques are compared with a specific control technique A for evaluating the comparative pairwise performance of all techniques with technique A. Bonferroni–Dunn test is also called Bonferroni correction and is used to control family-wise error rate. A family-wise error may occur when we are testing a number of hypotheses referred to as family of hypotheses, which are performed on a single set of data or samples. The probability that at least one hypothesis may be significant just because of chance (Type I error) needs to be controlled in such a case (Garcia et al. 2007). Bonferroni–Dunn test is mostly used after a Friedman test, if the null hypothesis is rejected. To control family-wise error, the critical value α is divided by the number of comparisons. For example, if we are comparing $k - 1$ subjects with a control subject then the number of comparisons is $k - 1$. The formula for new critical value is as follows:

$$\alpha_{New} = \frac{\alpha}{\text{Number of comparisons}}$$

There is another method for performing the Bonferroni–Dunn's test by computing the CD (same as Nemenyi test). However, the α values used are adjusted to control family-wise error. We compute the CD value as follows:

$$CD = q_\alpha \sqrt{\frac{k(k+1)}{6n}}$$

Here k corresponds to the number of subjects and n corresponds to the number of observations for a subject. The critical values (q_α) are studentized range statistic divided by $\sqrt{2}$. Note that the number of comparisons in the Appendix table includes the control subject. We compare the computed CD with difference between average ranks. If the difference is less than CD, we conclude that the two subjects do not differ significantly at the chosen significance level α.

Example 6.17:

Consider an example where we compare four techniques by analyzing the performance of the models predicted using these four techniques on six data sets each. We first apply Friedman test to obtain the average ranks of all the methods. The computed average ranks are shown in Table 6.42. The result of the Friedman test indicated the rejection of the null hypothesis. Evaluate whether there are significant difference among $M1$ and all the other methods.

TABLE 6.42

Average Ranks of Techniques

	T1	T2	T3	T4
Average rank	3.67	2.67	1.92	1.75

Solution:

Step 1: Formation of hypothesis.

In this step, null (H_0) and alternative (H_a) hypotheses are formed. The hypotheses for the example are given below:

H_{01}: The performance of T1 and T2 techniques do not differ significantly.

H_{a1}: The performance of T1 and T2 techniques differ significantly.

H_{02}: The performance of T1 and T3 techniques do not differ significantly.

H_{a2}: The performance of T1 and T3 techniques differ significantly.

H_{03}: The performance of T1 and T4 techniques do not differ significantly.

H_{a3}: The performance of T1 and T4 techniques differ significantly.

Step 2: Select the appropriate statistical test.

The example needs to evaluate the comparison of T1 technique with all other techniques. Thus, T1 is the control technique. The evaluation of different techniques is performed using Friedman test, and the result led to rejection of the null hypothesis. To analyze whether there are any significant differences among the performance of the control technique and other techniques, we need to apply a post hoc test. Thus, we use Bonferroni–Dunn's test.

Step 3: Apply test and calculate CD.

In this example, $k = 4$ and $n = 6$. The value of q_α for four subjects at $\alpha = 0.05$ is 2.394. The CD can be calculated by the following formula:

$$CD = q_\alpha \sqrt{\frac{k(k+1)}{6n}} = 2.394 \sqrt{\frac{4.(4+1)}{6.6}} = 1.79$$

We now find the differences among ranks of each pair of techniques, as shown in Table 6.43.

Step 4: Define significance level.

Table 6.44 shows the comparison results of critical difference and actual rank differences among different techniques. The rank difference of only T1–T4 pair is higher than the computed critical difference. However, the rank difference of T1–T3 is quite close to the critical difference. The difference in performance of T1–T2 is not significant.

Step 5: Derive conclusions.

As the rank difference of only T1–T4 pair is higher than the computed critical difference. We conclude that the T4 technique significantly outperforms T1 technique at significance level $\alpha = 0.05$. We accept the null hypothesis

TABLE 6.43

Computation of Pairwise Rank Differences among Techniques for Bonferroni–Dunn Test

Pair	Difference
T1–T2	3.67 − 2.67 = 1.00
T1–T3	3.67 − 1.92 = 1.75
T1–T4	3.67 − 1.75 = 1.92

TABLE 6.44

Comparison of Differences
for Bonferroni–Dunn Test

Pair	Difference
T1–T2	1.00 < 1.79
T1–T3	1.75 < 1.79
T1–T4	**1.92 > 1.79**

H_{03} and reject hypotheses H_{01} and H_{02}. As the rank difference of only $T1$–$T4$ pair (shown in bold) is higher than the computed critical difference.

6.4.16 Univariate Analysis

Univariate LR may be defined as a statistical method that works by formulating a mathematical model to depict the relationship between dependent variable and each of the independent variables, taken one at a time. As discussed in 6.2, one of the purposes of the univariate analysis is to screen out the independent variables that are not significantly related to the dependent variables. The other goal is to test the hypothesis about the relationship of independent variables with the dependent variable. The choice of methods in the univariate analysis depends on the type of dependent variables being used. The formula for univariate LR is given below:

$$\text{prob}(X_1) = \frac{e^{(A_0 + A_1 X_1)}}{1 + e^{(A_0 + A_1 X_1)}}$$

where:
X_1 is an independent variable
A_1 is the weight
A_0 is a constant

The sign of the weight indicates the direction of effect of the independent variable on the dependent variable. The positive sign indicates that independent variable has positive effect on the dependent variable, and negative sign indicates that the independent variable has negative effect on the dependent variable. The significance statistic is employed to test the hypothesis.

In linear regression, t-test is used to find the significant independent variables and, in LR, Wald test is used for the same purpose.

6.5 Example—Univariate Analysis Results for Fault Prediction System

We treat a class as faulty, if it contained at least one fault. Tables 6.45 through 6.48 provide the coefficient (B), standard error (SE), statistical significance (sig), odds ratio [exp (B)], and R^2 statistic for each measure. The statistical significance estimates the importance or the significance level for each independent variable. Odd ratio represents the probability of occurrence of an event divided by the probability of nonoccurrence of an event. The R^2 statistic depicts the variance in the independent variable caused by the variance in the independent variable. A higher value of R^2 means high accuracy. The metrics with a significant relationship to fault proneness, that is, below or at the significance (named as Sig. in Tables 6.45 through 6.48) threshold of 0.01 are shown in bold (see Tables 6.45 through 6.48). Table 6.45 presents the

TABLE 6.45

Univariate Analysis Using LR Method for HSF

Metric	B	SE	Sig.	Exp(B)	R²
CBO	0.145	0.028	**0.0001**	1.156	0.263
WMC	0.037	0.011	**0.0001**	1.038	0.180
RFC	0.016	0.004	**0.0001**	1.016	0.160
SLOC	0.003	0.001	**0.0001**	1.003	0.268
LCOM	0.015	0.006	**0.0170**	1.015	0.100
NOC	−18.256	5903.250	0.9980	0.000	0.060
DIT	0.036	0.134	0.7840	1.037	0.001

TABLE 6.46

Univariate Analysis Using LR Method for MSF

Metric	B	SE	Sig.	Exp(B)	R²
CBO	0.276	0.030	**0.0001**	1.318	0.375
WMC	0.065	0.011	**0.0001**	1.067	0.215
RFC	0.025	0.004	**0.0001**	1.026	0.196
SLOC	0.010	0.001	**0.0001**	1.110	0.392
LCOM	0.009	0.003	**0.0050**	1.009	0.116
NOC	−1.589	0.393	**0.0001**	0.204	0.090
DIT	0.058	0.092	0.5280	1.060	0.001

TABLE 6.47

Univariate Analysis Using LR Method for LSF

Metric	B	SE	Sig.	Exp(B)	R²
CBO	0.175	0.025	**0.0001**	1.191	0.290
WMC	0.050	0.011	**0.0001**	1.052	0.205
RFC	0.015	0.004	**0.0001**	1.015	0.140
SLOC	0.004	0.001	**0.0001**	1.004	0.338
LCOM	0.004	0.003	0.2720	1.004	0.001
NOC	−0.235	0.192	0.2200	0.790	0.002
DIT	0.148	0.099	0.1340	1.160	0.005

TABLE 6.48

Univariate Analysis Using LR Method for USF

Metric	B	SE	Sig.	Exp(B)	R²
CBO	0.274	0.029	**0.0001**	1.315	0.336
WMC	0.068	0.019	**0.0001**	1.065	0.186
RFC	0.023	0.004	**0.0001**	1.024	0.127
SLOC	0.011	0.002	**0.0001**	1.011	0.389
LCOM	0.008	0.003	**0.0100**	1.008	0.013
NOC	−0.674	0.185	**0.0001**	0.510	0.104
DIT	0.086	0.091	0.3450	1.089	0.001

results of univariate analysis for predicting fault proneness with respect to high-severity faults (HSF). From Table 6.45, we can see that five out of seven metrics were found to be very significant (Sig. < 0.01). However, NOC and DIT metrics are not found to be significant. The LCOM metric is significant at 0.05 significance level. The value of R^2 statistic is highest for SLOC and CBO metrics.

Table 6.46 summarizes the results of univariate analysis for predicting fault proneness with respect to medium-severity faults (MSF). Table 6.46 shows that the values of R^2 statistic is the highest for SLOC metric. All the metrics except DIT are found to be significant. NOC has a negative coefficient, which implies that classes with higher NOC value are less fault prone.

Table 6.47 summarizes the results of univariate analysis for predicting fault proneness with respect to low-severity faults (LSF). Again, it can be seen from Table 6.47 that the value of R^2 statistic is highest for SLOC metric. The results show that four out of seven metrics are found to be very significant. LCOM, NOC, and DIT metrics are not found to be significant.

Table 6.48 summarizes the results of univariate analysis for predicting fault proneness. The results show that six out of seven metrics were found to be very significant when the faults were not categorized according to their severity, that is, ungraded severity faults (USF). The DIT metric is not found to be significant and the NOC metric has a negative coefficient. This shows that the NOC metric is related to fault proneness but in an inverse manner.

Thus, the SLOC metric has the highest R^2 value at all the severity of faults, which shows that it is the best predictor. The CBO metric has the second highest R^2 value. The values of R^2 statistic are more important as compared to the value of sig. as they show the strength of the correlation.

Exercises

6.1 Describe the measures of central tendency? Discuss the concepts with examples.

6.2 Consider the following data set on faults found by inspection technique for a given project. Calculate mean, median, and mode.

100, 160, 166, 197, 216, 219, 225, 260, 275, 290, 315, 319, 361, 354, 365, 410, 416, 440, 450, 478, 523

6.3 Describe the measures of dispersion. Explain the concepts with examples.

6.4 What is the purpose of collecting descriptive statistics? Explain the importance of outlier analysis.

6.5 What is the difference between attribute selection and attribute extraction techniques?

6.6 What are the advantages of attribute reduction in research?

6.7 What is CFS technique? State its application with advantages.

6.8 Consider the data set consisting of lines of source code given in exercise 6.2. Calculate the standard deviation, variance, and quartile.

6.9 Consider the following table presenting three variables. Determine the normality of these variables.

Fault Count	Cyclomatic Complexity	Branch Count
332	25	612
274	24	567
212	23	342
106	12	245
102	10	105
93	09	94
63	05	89
23	04	56
09	03	45
04	01	32

6.10 What is outlier analysis? Discuss its importance in data analysis. Explain univariate, bivariate, and multivariate.

6.11 Consider the table given in exercise 6.7. Construct box plots and identify univariate outliers for all the variables given in the data set.

6.12 Consider the data set given in exercise 6.7. Identify bivariate outliers between dependent variable fault count and other variables.

6.13 Consider the following data with the performance accuracy values for different techniques on a number of data sets. Check whether the conditions of ANOVA are met. Also apply ANOVA test to check whether there is significant difference in the performance of techniques.

Data Sets	Techniques		
	Technique 1	Technique 2	Technique 3
D1	84	71	59
D2	76	73	66
D3	82	75	63
D4	75	76	70
D5	72	68	74
D6	85	82	67

6.14 Evaluate whether there is significant difference between different algorithms evaluated on three data sets on the runtime performance (in seconds) of the model using appropriate statistical test.

Algorithms	Data Sets #		
	1	2	3
Algorithm 1	9	7	9
Algorithm 2	19	20	20
Algorithm 3	18	15	14
Algorithm 4	13	7	6
Algorithm 5	10	9	8

6.15 A software company plans to adopt a new programming paradigm, that will ease the task of software developers. To assess its effectiveness, 50 software developers used the traditional programming paradigm and 50 others used the new one. The productivity values per hour are stated as follows. Perform a *t*-test to assess the effectiveness of the new programming paradigm.

Statistic	Old Programming Paradigm	New Programming Paradigm
Mean	1.5	2.21
Standard Deviation	0.4	0.36

6.16 A company deals with development of certain customized software products. The following data lists the proposed cost and the actual cost of 10 different software products. Evaluate whether the company makes a good estimate of the proposed cost using a paired sample *t*-test.

Software Product	Proposed Cost	Actual Cost
P1	1,739	1,690
P2	2,090	2,090
P3	979	992
P4	997	960
P5	2,750	2,650
P6	799	799
P7	980	1,000
P8	1,099	1,050
P9	1,225	1,198
P10	900	943

6.17 The software team needs to determine average number of methods in a class for a particular software product. Twenty-two classes were chosen at random and the number of methods in these classes were analyzed. Evaluate whether the hypothesized mean of the chosen sample is different from 11 methods per class for the whole population.

Class No.	No. of Methods	Class No.	No. of Methods	Class No.	No. of Methods
C1	11.5	C9	9	C17	11.5
C2	12	C10	14	C18	12.5
C3	10	C11	11.5	C19	14
C4	13	C12	7.5	C20	8.5
C5	9.5	C13	11	C21	12
C6	14	C14	6	C22	9.5
C7	11.5	C15	12		
C8	12	C16	12.5		

6.18 A software organization develops software tools using five categories of programming languages. Evaluate a goodness-of-fit test on the data given below to test whether the organization develops equal proportion of software tools using the five different categories of programming languages.

Programming Language Category	Number of Software Tools
Category 1	35
Category 2	30
Category 3	45
Category 4	44
Category 5	28

6.19 Twenty-five students developed the same program and the cyclomatic complexity values of these 25 programs are stated. Evaluate whether the cyclomatic complexity values of the program developed by the 25 students follows normal distribution.

6, 11, 9, 14, 16, 10, 13, 9, 15, 12, 10, 14, 15, 10, 8, 11, 7, 12, 13, 17, 17, 19, 9, 20, 26, 6, 11, 9, 14, 16,

6.20 A software organization uses either OO methodology or procedural methodology for developing software. It also uses effective verification techniques at different stages to obtain errors. Given the following data, evaluate whether the two attributes, software development stage for verification and methodology, are independent.

		Methodology		
		OO	Procedural	Total
Software Development Stage	Requirements	80	100	180
	Initial design	50	110	160
	Detailed design	75	65	140
Total		205	275	480

6.21 The coupling values of a number of classes are provided below for two different samples. Test the hypothesis using F-test whether the two samples belong to the same population.

Sample 1	32	42	33	40	42	44	42	38	32
Sample 2	31	31	31	35	35	32	30	36	

6.22 Two training programmes were conducted for software professionals by an organization. Nine participants were asked to rate the training programmes on a scale of 1 to 100. Using Wilcoxon signed-rank test, evaluate whether one program is favorable over the other.

Participant No.	Program A	Program B
1	25	45
2	15	55
3	25	65
4	15	65
5	5	35
6	35	15
7	45	45
8	5	75
9	55	85

6.23 A researcher wants to compare the performance of two learning algorithms across multiple data sets using receiver operating characteristic (ROC) values as shown below. Investigate whether there is a statistical difference among the performance of two learning algorithms.

	Algorithms	
Data Sets	A1	A2
D1	0.65	0.55
D2	0.78	0.85
D3	0.55	0.70
D4	0.60	0.60
D5	0.89	0.70

6.24 Two attribute selection techniques were analyzed to check whether they have any effect on model's performance. Seven models were developed using attribute selection technique X and nine models were developed using attribute selection technique Y. Use Wilcoxon–Mann–Whitney test to evaluate whether there is any significant difference in the model's performance using the two different attribute selection techniques.

Attribute Selection Technique X	Attribute Selection Technique Y
57.5	58.9
58.6	58.0
59.3	61.5
56.9	61.2
58.4	62.3
58.8	58.9
57.7	60.0
	60.9
	60.4

6.25 A researcher wants to find the effect of the same learning algorithm on three data sets. For every data set, a model is predicted using the same learning algorithm with a specific performance measure area under the ROC curve.

Evaluate whether there is statistical difference in the performance of learning algorithm on different data sets.

Data Set	ROC Values
1	0.76
2	0.85
3	0.66

6.26 A market survey is conducted to evaluate the effectiveness of three text editors by 20 probable customers. The customers assessed the text editors on various criteria and provided a score out of 300. Test the hypothesis whether there is any significant differences among the three text editors using Kruskal–Wallis test.

Text Editor A	Text Editor B	Text Editor C
200	110	260
60	200	290
150	60	240
190	70	150
150	140	250
270	30	280
	210	230

6.27 A researcher wants to compare the performance of four learning techniques on multiple data sets (five) using the performance measure, area under the ROC curve. The data for the scenario is given below. Determine whether there is any statistical difference in the performance of different learning techniques.

Data Sets	Methods			
	A1	A2	A3	A4
D1	0.65	0.56	0.72	0.55
D2	0.79	0.69	0.69	0.59
D3	0.65	0.65	0.62	0.60
D4	0.85	0.79	0.66	0.76
D5	0.71	0.61	0.61	0.78

6.28 What is the purpose of Bonferroni–Dunn correction? Consider data given in Exercise 6.27. Evaluate the pairwise differences using Wilcoxon test with Bonferroni–Dunn correction.

6.29 A researcher wants to evaluate the effectiveness of four tools by analyzing the performances of different models as given below. Evaluate using Friedman test whether the performance of tools is significantly different. If the difference is significant, evaluate the pairwise differences using Nemenyi test.

Data Sets	Tools			
	T1	T2	T3	T4
Model 1	69	60	83	73
Model 2	70	68	81	69
Model 3	73	54	75	67
Model 4	71	61	91	79
Model 5	77	59	85	69
Model 6	73	56	89	77

6.30 Explain a scenario where application of Nemenyi test is advisable?

6.31 Which test is used to control family-wise error?

6.32 What is type-I and type-II errors? Why are they important to be identified?

6.33 Compare and contrast various statistical tests with respect to their assumptions and normality conditions of the underlying data.

6.34 Differentiate between:

 (a) Wrapper and filter methods

 (b) Nemenyi and Bonferroni–Dunn

 (c) One-tailed and two-tailed tests

 (d) Independent sample and Wilcoxon–Mann–Whitney tests

6.35 Discuss two applications of univariate analysis.

Further Readings

The following books provide details on summarizing data:

D. D. Boos, and C. Brownie, "Comparing variances and other measures of dispersion," *Statistical Science*, vol. 19, pp. 571–578, 2004.

J. I. Marden, *Analysing and Modeling Rank Data*, Chapman and Hall, London, 1995.

H. Mulholland, and C. R. Jones, "Measures of dispersion," In: *Fundamentals of Statistics*, Springer, New York, chapter 6, pp. 93–110, 1968.

R. R. Wilcox, and H. J. Keselman, "Modern robust data analysis methods: Measures of central tendency," *Psychological Methods*, vol. 8, no. 3, pp. 254–274, 2003.

There are several books on research methodology and statistics in which various concepts and statistical tests are explained:

W. G. Hopkins, *A New View of Statistics*, Sportscience, 2003. http://sportsci.org/resource/stats

C. R. Kothari, *Research Methodology: Methods and Techniques*, New Age International Limited, New Delhi, India, 2004.

The details on outlier analysis can be obtained from:

V. Barnett, and T. Price, *Outliers in Statistical Data*, John Wiley & Sons, New York, 1995.

The concept of principal component analysis is explained in the following:

H. Abdi, and L. J. Williams, "Principal component analysis," *Wiley Interdisciplinary Reviews: Computational Statistics*, vol. 2, no. 4, pp. 433–459, 2010.

The basic concept of univariate analysis are presented in:

F. Hartwig, and B. E. Dearing, *Exploratory Data Analysis*, Sage Publications, Beverly Hills, CA, 1979.

H. M. Park, "Univariate analysis and normality test using SAS, Stata, and SPSS," The University Information Technology Services, Indiana University, Bloomington, IA, 2008.

The details about the CFS technique are provided in:

M. A. Hall, "Correlation-based feature selection for machine learning," PhD dissertation, The University of Waikato, Hamilton, New Zealand, 1999.

M. A. Hall, and L. A. Smith, "Feature subset selection: A correlation based filter approach," In proceedings of International Conference of Neural Information Processing and Intelligent Information Systems, pp. 855–858, 1997.

A detailed description of various wrapper and filter methods can be found in:

A. L. Blum, and P. Langley, "Selection of relevant features and examples in machine learning," Artificial Intelligence, vol. 97, pp. 245–271, 1997.

N. Sánchez-Maroño, A. Alonso-Betanzos, and M. Tombilla-Sanromán, "Filter methods for feature selection, a comparative study," In: *Proceedings of the 8th International Conference on Intelligent Data Engineering and Automated Learning*, H. Yin, P. Tino, W. Byrne, X. Yao, E. Corchado (eds.), Springer-Verlag, Berlin, Germany, pp. 178–187.

Some of the useful facts and concepts of significance tests are presented in:

P. M. Bentler, and D. G. Bonett, "Significance tests and goodness of fit in the analysis of covariance structures," *Psychological Bulletin*, vol. 88, no. 3, pp. 588–606, 1980.

J. M. Bland, and D. G. Altman, "Multiple significance tests: The Bonferroni method," *BMJ*, vol. 310, no. 6973, pp. 170, 1995.

L. L. Harlow, S. A. Mulaik, and J. H. Steiger, *What If There Were No Significance Tests*, Psychology Press, New York, 2013.

The one-tailed and two-tailed tests are described in:

J. Hine, and G. B. Wetherill (eds.), "One-and Two-Tailed Tests," In: *A Programmed Text in Statistics Book 4: Tests on Variance and Regression,* Springer, Amsterdam, the Netherlands, pp. 6–11, 1975.

D. B. Pillemer, "One-versus two-tailed hypothesis tests in contemporary educational research," *Educational Researcher,* vol. 20, no. 9, pp. 13–17, 1991.

Frick provides an excellent use of hypothesis testing based on null hypothesis:

R. W. Frick, "The appropriate use of null hypothesis testing," *Psychological Methods,* vol. 1, no. 4, 379–390, 1996.

The following books provide details on parametric and nonparametric tests:

D. J. Sheskin, *Handbook of Parametric and Nonparametric Statistical Procedures,* CRC Press, Boca Raton, FL, 2003.

D. J. Sheskin (ed.), "Parametric versus nonparametric tests," In: *International Encyclopedia of Statistical Science,* Springer, Berlin, Germany, pp. 1051–1052, 2011.

The following is an excellent and widely used book on hypothesis testing:

E. L. Lehmann, and J. P. Romano, *Testing Statistical Hypotheses: Springer Texts in Statistics,* Springer, New York, 2008.

7

Model Development and Interpretation

In Chapter 6, we presented data preprocessing techniques, feature reduction methods, and statistical tests for hypothesis testing. In software engineering research, the researcher may relate the software metrics (independent variables) with quality attributes (dependent variable) such as fault proneness, maintainability, reliability, or testability. The relationship between software metrics and quality attributes can be analyzed using statistical or machine learning (ML) techniques. The models are created to predict the quality attributes using performance measures or analyzers. After obtaining the values of performance measures, the hypothesis may be applied to analyze the difference between the techniques over multiple data sets. The results are then interpreted and assessed, and final conclusions are derived.

In this chapter, we present various statistical and ML techniques for model development. The performance measures for measuring the accuracy of the developed models are described, and the guidelines for interpreting the obtained results are presented.

7.1 Model Development

Models are constructed using historical data for development of prediction systems. These prediction systems can be used in the early phases of software development by developers and managers to obtain insight about the quality of the systems.

Software quality prediction helps in identification of weak portions of a software. Thus, it aids in efficient utilization of limited resources like cost, time, and effort by focusing these resources on the identified weak parts to improve the quality of the system. For example, if we can determine the classes that are more prone to faults, we can focus our resources on these classes so that minimum faults propagate to later phases of software development life cycle.

Model prediction involves the following three main elements:

1. Independent variables
2. Dependent variables
3. Learning technique

Before development of models, the experiment design is constructed, data is processed, and the attributes are reduced.

The data is divided into two parts: training and validation. The training data is used for model development where the model is trained by learning from the relationship between the independent variables and dependent variable. The model development process consists of the following steps (Figure 7.1):

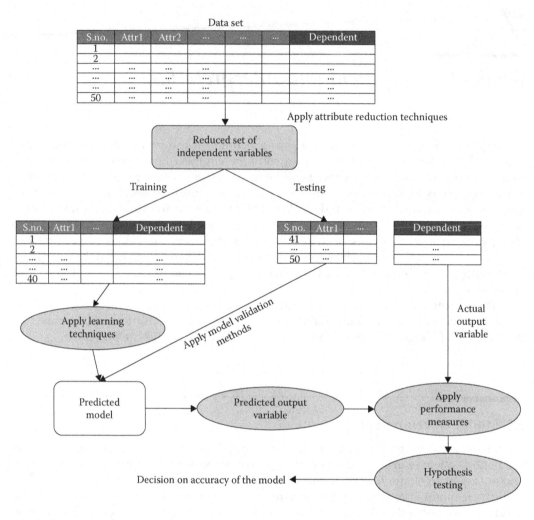

FIGURE 7.1
Steps in model prediction.

1. Dividing data set into two parts (training and testing)
2. Selecting relevant attributes (independent variables)
3. Developing model using learning technique
4. Validating the predicted model
5. Applying hypothesis testing using statistical tests, if required
6. Interpreting the results

7.1.1 Data Partition

There can be serious problems, if the model is tested using the same data set from which it is trained. In other words, a learning technique might perform well on the training data but poorly on future unseen test data. To overcome this problem, cross-validation techniques explained in Section 7.7 should be used for evaluating and comparing the

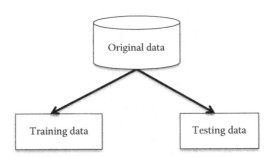

FIGURE 7.2
Data partition.

techniques. Figure 7.2 shows that the original data can be divided into training and testing data samples. The model is developed using training data and validated on the unseen test data. In cross-validation, the data is split into two independent parts, one for training and the other for testing. The study may also divide the data into training, validation, and testing samples in empirical studies where the data set available is very large. The validation data can be used to choose the correct architecture as an optional step. However, in this book, we describe the concepts in terms of training and testing samples.

During model development, the data must be randomly divided into training and testing data samples.

7.1.2 Attribute Reduction

The independent variables, also known as attributes, are reduced using the attribute subselection techniques explained in Section 6.2. The purpose is to extract the best and relevant attributes, and these attributes are provided as input to the algorithm or technique in the next step.

7.1.3 Model Construction using Learning Algorithms/Techniques

After the data has been partitioned, the next step is to train the model using the training data. In this step, the learning technique(s) selected in experimental design must be used for creation of the model. The independent variables such as software metrics are used to predict the dependent variable. In software engineering research, the dependent variables are generally software quality attributes. The techniques may be statistical or ML and are briefly explained in Sections 7.2 and 7.3.

7.1.4 Validating the Model Predicted

The model created in the previous step is validated using testing data by computing the values of various performance analyzers. The performance analyzers are selected on the basis of the following:

1. Type of dependent variable
2. Nature of dependent variable

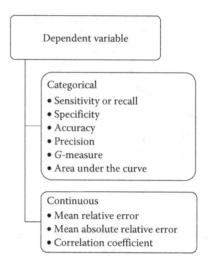

FIGURE 7.3
Performance measures for dependent variable.

Figure 7.3 shows the performance measures that may be used depending on the type of dependent variable. The type of dependent variable can be either categorical or continuous and the nature of dependent variable is determined by the distribution of the outcome variable (ratio of positive and negative samples). The guidelines on the selection of performance measures on the basis of nature of dependent variables are given in Section 7.5.7. The cross-validation method is applied for model validation. Depending on the size of data, the appropriate cross-validation method is selected.

7.1.5 Hypothesis Testing

On the basis of performance measures, the prediction capability of the models predicted using various learning techniques could be compared using the statistical tests given in Section 6.4. The performance of more than two models can be compared using Friedman or Kruskal and Wallis tests, and further post hoc analysis is also recommended to compare the pairwise performance of learning techniques. The model comparison tests are summarized in Section 7.8.

7.1.6 Interpretation of Results

Finally, the results of model prediction are interpreted and discussed. The interpretation of results of hypothesis testing is also done. The answers to research questions, practical application of the work, and limitations of the work are summarized. The commonalities and differences of the results produced by the current study in view of the current literature work are also presented.

7.1.7 Example—Software Quality Prediction System

Figure 7.4 presents the general framework for software quality prediction. The inputs to the classification algorithm can be either process or product metrics, and the outputs are

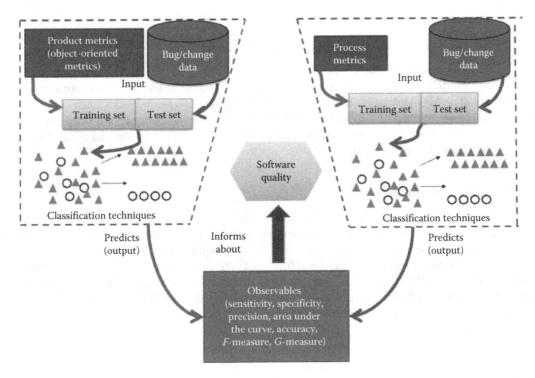

FIGURE 7.4
Software quality assessment framework.

the various observables such as maintainability, fault proneness, and reliability that can be used to provide information about the software quality.

A software quality prediction system takes as an input the various product-oriented metrics that define various characteristics of a software. The focus of this study is object-oriented (OO) software. Thus, OO metrics are used, throughout the software process, to determine and quantify various aspects of an OO application. A single metric alone is not sufficient to uncover all characteristics of an application under development. Several metrics must be used together to gauge a software product. The metrics used in this study are the ones that are most commonly used by various researchers to account for software characteristics like size, coupling, cohesion, inheritance, and so on. Along with metrics, the collection of fault/change-prone data of a class is also an essential input to create an efficient and intelligent classifier prediction system. The prediction system learns to distinguish and identify fault/change-prone classes of the software data set under study.

Development of a software quality prediction system helps in ascertaining software quality attributes and focused use of constraint resources. It also guides researchers and practitioners to perform preventive actions in the early phases of software development and commit themselves for creation of better quality software. Once a software quality prediction system is trained, it can be used for quality assessment and for predictions on future unseen data. These predictions are then utilized for assessing software quality processes and procedures as we evaluate the software products, which are a result of these processes.

7.2 Statistical Multiple Regression Techniques

In the present scenario, we have a variety of learning techniques that can be used for development of models for predicting software quality attributes. A learning technique could be statistical or ML. Unlike univariate analysis, multiple regression works with combination of variables to predict a model. The multiple regression techniques are very popular techniques for model prediction where the combined effect of the independent variables is found on the dependent variable.

7.2.1 Multivariate Analysis

There are various techniques used in multivariate analysis depending on the type of dependent variable. If the dependent variable is continuous, linear regression is used; whereas, if the dependent variable is categorical then logistic regression (LR) is used.

In multiple linear regression, the weighted linear combination of independent variables is identified to optimally predict the dependent variable. Each predictor is assigned a weight, and the result of the product is summed up together with the constant to predict the outcome. The equation is given below:

$$y = a + b_1 x_1 + b_2 x_2 + \cdots + b_n x_n$$

where:
   a is constant
   $b_1 \ldots b_n$ are weights
   $x_1 \ldots x_n$ are independent variables

The weights are generated in such a way that the predicted values are closest to the actual value. Closeness of predicted values to the actual value can be measured using ordinary linear squares where the sum squared difference between predicted and actual value is kept to a minimum. The difference between the actual and observed predicted values is known as prediction errors. Thus, the linear regression model that best fits the data for predicting dependent variable is such that the sum of squared errors are minimum.

LR is used to predict the dependent variable from a set of independent variables to determine the percentage of variance in the dependent variable explained by the independent variable (a detailed description is given by Basili et al. [1996], Hosmer and Lemeshow [1989], and Aggarwal et al. [2009]). The multivariate LR formula can be defined as (Aggarwal et al. 2009):

$$prob\left(X_1, X_2, \ldots, X_n\right) = \frac{e^{\left(A_0 + A_1 X_1 + \cdots + A_n X_n\right)}}{1 + e^{\left(A_0 + A_1 X_1 + \cdots + A_n X_n\right)}}$$

where:
   $X_i, i = 1, 2, \ldots, n$ are the independent variables
   "Prob" is the probability of occurrence of an event

7.2.2 Coefficients and Selection of Variables

The coefficients are assigned weights to the independent variables in the multivariate analysis. The importance of the predictive capability of the independent variables is specified using the coefficient. The higher the value of the coefficients, more is the impact of the independent variables.

In multivariate analysis two stepwise selection methods—forward selection and backward elimination—are used (Hosmer and Lemeshow 1989). The forward stepwise procedure examines the variables that are selected one at a time for entry at each step. The backward elimination method includes all the independent variables in the model. Variables are deleted one at a time from the model, until a stopping criteria is fulfilled.

The statistical significance defines the significance level of a coefficient. Larger the value of statistical significance, lesser is the estimated impact of an independent variable on the dependent variable. Usually, the value of 0.01 or 0.05 is used as threshold cutoff value.

7.3 Machine Learning Techniques

The goal of ML is to develop programs that learn from experience, automatically improve the performance, and adapt to new environment over time. ML techniques are well suited for real-life problems that use methods to extract useful information from complex and intractable problems in less time. They can be tolerant to data that is inaccurate, partially incorrect, or uncertain. These methods can be used to construct models and make predictions. Thus, ML techniques have the following benefits:

- Complex relationships can be modeled
- Easily adaptable to changing environment as new knowledge is discovered

ML techniques can be divided into two categories—supervised and unsupervised learning. Supervised learning makes predictions when the outcome variable is available while training models, whereas unsupervised learning attempts to search for relevant patterns when the outcome variable is unknown. Examples of use of supervised learning in software engineering are fault prediction, prediction of change-prone modules, and reliability prediction. In supervised learning, classification techniques such as decision tree (DT), neural networks (NN), and support vector machines (SVM) are used. In unsupervised learning, clustering methods are used to identify patterns from unlabeled samples.

7.3.1 Categories of ML Techniques

The ML is categorized by Malhotra (2015) into eight broad categories as given below. The classification taxonomy of ML techniques is presented in Figure 7.5. A brief summary of ML broad categories is provided in sections below.

- DT
- Bayesian learners (BL)
- Ensemble learners (EL)
- NN
- SVM
- Rule-based learning (RBL)
- Search-based techniques (SBT)
- Miscellaneous

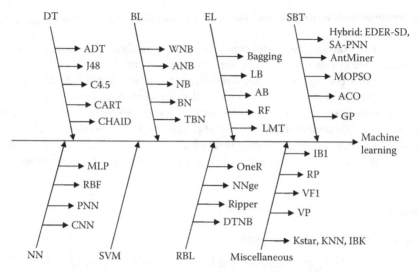

FIGURE 7.5
Classification of ML techniques. CHAID: chi-squared automatic interaction detection, CART: classification and regression trees, ADT: alternating decision tree, RF: random forest, NB: naïve Bayes, BN: Bayesian networks, ABN: augmented Bayesian networks, WNB: weighted Bayesian networks, TNB: transfer Bayesian networks, MLP: multilayer perceptron, PNN: probabilistic neural network, RBF: radial basis function, LB: Logit boost, AB: AdaBoost, NNge: neighbor with generalization, GP: genetic programming, ACO: ant colony optimization, SVM: support vector machines, RP: recursive partitioning, AIRS: artificial immune system, KNN: K-nearest neighbor, VFI: Voting Feature Intervals, EDER-SD: evolutionary decision rules for subgroup discovery, SA-PNN: simulated annealing probabilistic neural network, VP: voted perceptron, DTNB: decision tree naive bayes, LMT: logistic model trees. (Data from Malhotra, 2015.)

7.3.2 Decision Trees

The DT technique begins at the root node, and at each step the best variable is found to split the given node into two child nodes. The best variable is found by checking the possibilities of all the variables while making the decision of split at each node. To select the best variable at each node, the algorithm used (such as classification and regression tree) works with the aim to decrease the average impurity at a given split (Quinlan 1993; Breiman et al. 1994). Figure 7.6 shows the basic steps in DT algorithm. The attribute selection can be made using information gain, gini index, or gain ratio measures.

7.3.3 Bayesian Learners

BL are used to predict probabilities of data sample belonging to a given class or category. Bayesian learning is based on Bayes theorem. A Bayesian network (BN) is an interconnected network of nodes, where each node represents a random variable and all directed edges connecting these nodes represent probabilistic dependencies among nodes (Witten and Frank 2005). BN helps in computing joint probability distribution among a set of random variables. BN can easily handle incomplete data sets and also allows to investigate casual relationships. Bayesian belief networks and naïve Bayes are two popularly used BL techniques.

7.3.4 Ensemble Learners

In ensemble learning, multiple training sets are created and multiple ML techniques are applied to these sets. The individual predictions of the ML techniques are combined to

DT (records, outcome_variable, list_of_attributes)

Create Head node for the tree

if *the records in the data set belong to the positive class then*

        return *Node Head as leaf node and label it with the positive class*

if *the records in the data set belong to the negative class then*

        return *Node Head as leaf node and label it with the negative class*

if *list_of_attributes is empty then*

        return *Node Head as leaf node and label it with most common value of outcome_*
        variable in records

otherwise call attribute_selection_method(list_of_attributes)

        select best attribute using the splitting criteria and initialise A ← splitting_attribute

        list_of_attributes ← list_of_attributes A

        for *each value i for attribute A*

            Add new tree branch below Head node

            if *records_i is empty* **then**

                Add leaf node and label it with most common value of outcome_variable
                in records

            else *add new subtree*

                call *DT (records_i, outcome_variable, list_of_attributes)*

    return Head

FIGURE 7.6
DT algorithm.

obtain the final outcome by taking a vote. Rather than using a single ML technique, this approach aims to improve the accuracy of model by combining the results obtained by multiple ML techniques. It is proved that the multiple ML techniques give more accurate results, rather than using individual ML technique. Figure 7.7 depicts the concept of EL. There are various techniques based on EL such as boosting, bagging, and RF. Table 7.1 summarizes the widely used EL techniques.

7.3.5 Neural Networks

The NN repetitively adjusts different weights so that the difference between desired output from the network and actual output from the NN is minimized. The network learns by finding a vector of connection weights that minimizes the sum of squared errors on the training data set. The NN is trained by standard error back propagation algorithm at a given learning rate (e.g., 0.005), having the minimum square error as the training stopping criterion.

The input layer has one unit for each input variable. Each input value in the data set is normalized within the interval [0, 1] using min–max normalization. Min–max normalization performs a linear transformation on the original data (Han and Kamber 2001). Suppose that $\min_A$ and $\max_A$ are the minimum and maximum values of an attribute A. It maps value v of A to v' in the range 0–1 using the formula:

$$v' = \frac{v - \min_A}{\max_A - \min_A}$$

In NN, first random weights are assigned to the nodes and then back propagation algorithm is applied to update the weights using multiple epochs (iterations) through the

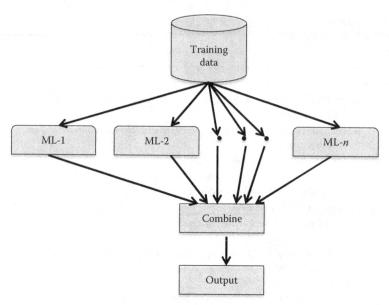

FIGURE 7.7
Ensemble learning.

TABLE 7.1

Ensemble Learning Techniques

Technique	Description
RF	RF was proposed by Breiman (2001) and constructs a forest of multiple trees and each tree depends on the value of a random vector. For each of the tree in the forest, this random vector is sampled with the same distribution and independently. Hence, RF is a classifier that consists of a number of decision trees.
Boosting	Boosting uses DT algorithm for creating new models. Boosting assigns weights to models based on their performance. There are many variants of boosting algorithms available in the literature. There are two variants of boosting technique—AdaBoost (Freund and Schapire 1996) and LogitBoost (Friedman et al. 2000).
Bagging	Bagging or bootstrap aggregating improves the performance of classification models by creating various sets of the training sets.

training sets. During this process, the architecture is determined, such as the number of hidden layers and number of nodes in the hidden layer (Figure 7.8). Usually, one hidden layer is used in research as what can be achieved in function approximation, with more than one hidden layer also achieved by one hidden layer (Khoshgaftaar 1997).

The weights between the jth hidden node and input nodes are represented by W_{ji}, while the weights between the jth hidden node and output node are represented by α_j. The threshold of the jth hidden node is represented by β_j, while the threshold of the output layer is represented by β. If x represents the input vector to the network, the net input to the hidden node j is given by (Haykin 1994):

$$\text{net}_j = \sum_{i=1}^{M} W_{ji}x_i + B_j; \quad j = 1, 2, \ldots N$$

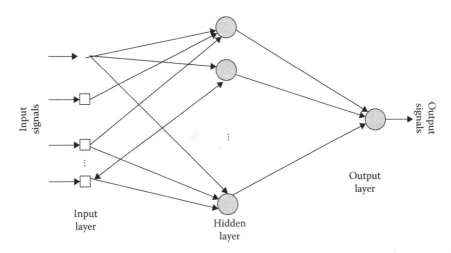

FIGURE 7.8
Architecture of NN.

The output from the jth hidden node is:

$$\sigma_j = \sigma\left(net_j\right)$$

The output from the network is given by:

$$y = \sigma\left[\sum_{j=1}^{N} \alpha_j \sigma_j + \beta\right]$$

7.3.6 Support Vector Machines

SVM are useful tools for performing data classification, and have been successfully used in various applications such as face identification, medical diagnosis, text classification, and pattern recognition. SVM constructs an N-dimensional hyperplane that optimally separates the data set into two categories. The purpose of SVM modeling is to find the optimal hyperplane that separates clusters of vector in such a way that cases with one category of the dependent variable on one side of the plane and the cases with the other category on the other side of the plane (Sherrod 2003). The support vectors are the vectors near the hyperplane. The SVM modeling finds the hyperplane that is oriented so that the margin between the support vectors is maximized. When a nonlinear region separates the points, SVM handles this by using a kernel function to map the data into a different space when a hyperplane can be used to do the separation. Details on SVM can be found in Cortes and Vapnik (1995) and Cristianini and Shawe-Taylor (2000).

The most commonly used functions in the literature are: linear, polynomial, radial basis function (RBF), and sigmoid. The recommended kernel function is the RBF (Sherrod 2003). The RBF kernel nonlinearly maps data into a higher dimensional space, so it can handle nonlinear relationships between dependent and independent variables. Figure 7.9 shows the RBF kernel. One category of dependent variable is shown as rectangles and the other as circles. The shaded circles and rectangles are support vectors.

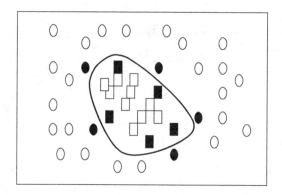

FIGURE 7.9
Radial basis function.

Given a set of $(x_i, y_i),..., (x_m, y_m)$ and $y_i \varepsilon \{-1, +1\}$ training samples. $\alpha_i = (i = 1,..., m)$ is a lagrangian multiplier. $K(x_i, y_i)$ is called a kernel function and b is a bias. The discriminant function D of two class SVM is given below (Zhao and Takagi 2007):

$$D(x) = \sum_{i=1}^{m} y_i \alpha_i K(x_i, x) + b$$

Then, an input pattern x is classified as (Zhao and Takagi 2007):

$$x = \begin{cases} +1, \text{ if } D(x) > 0 \\ -1, \text{ if } D(x) < 0 \end{cases}$$

7.3.7 Rule-Based Learning

RBL is a ML technique that creates if–then rules. To illustrate RBL, a simple covering algorithm is shown in Figure 7.10. Let d be the total data samples for a value v of a given attribute A of which p are positive samples that classifies a given category C. For example, consider the case where we want to predict classes as faulty and not faulty.
    If ? then faulty
The above rule consists an empty (?) left-hand side, and the rule for predicting faulty category of dependent variable is extracted. The new attribute–value pair is extracted based on the maximum correct predictions for category C of dependent variable.
    There are various rule-based techniques such as RIPPER, OneR, and NNge.

```
For each value of category C
    Initialize I to the data samples
    While I contains data samples in category C
        Create a rule R with an empty left-hand side that predicts category C
        Until R is complete do
            For each attribute A not mentioned in R, and each value v,
                Consider adding the condition A = v to the left-hand side of R
                Select A and v to maximize the accuracy p/d
            Add A = v to rule R
        Remove the data samples covered by R from I
```

FIGURE 7.10
Basic algorithm for rule-based learning

7.3.8 Search-Based Techniques

Search-based techniques (SBT) are inspired from the process of biological evolution (Eiben and Smith 2003). SBT are metaheuristic procedures that are capable of searching for an optimized solution in a large search space of potential solutions. These techniques are modeled on the basis of a fitness function, which is evaluated to ascertain the goodness of a specific solution, thus guiding the search process (Harman et al. 2012d). The usefulness of these techniques in the field of software test data generation, requirement analysis, and so on has been well established (Harman et al. 2012c).

SBT are population-based algorithms that undergo a series of iterations to find candidates with the desired characteristics. The working of SBT starts from a set of candidates that is called the initial population. These candidates are actually competitors, which compete against one another to achieve the tag of "best" candidates among the solution. The ranking among candidates is based on fitness function. The "best" candidates are used in future generations to produce new candidate solutions. These new candidate solutions may replace the candidates displaying the worst performance in the initial population, and the whole iteration process starts again (Grosan and Abraham 2007). The whole process of iteration stops either if maximum number of iterations has been performed or the candidates produced are the ones with the desired quality and fitness parameters. Figure 7.11 shows a diagrammatic representation of the process followed by SBT.

The process of production of new candidates may involve a series of operators like selection, mutation, crossover, and replacement. A brief explanation of these operators is stated below:

- Selection: It refers to the process of determination of those candidates that can be included in the next generation based on their fitness parameters. Before incorporation in the next generation, the selected candidates may undergo mutation and crossover.

- Mutation: It is an operator to incorporate a degree of alteration in the population from the previous generation to the next. It diversifies or changes one or more values of the parent candidate. In SBT, the probability of mutation basically refers to the degree of randomness, while traversing the solution space.

- Crossover: This operator couples two parent candidates to formulate a child with the assumption that the child candidate would be superior to both the parent candidates, as it would involve the good attributes of both the parents.

- Replacement: This operator involves identification of those candidates that may be replaced from the current population by some better candidates. The most popularly used strategy is to replace "worst" candidates and to replace the "most similar" candidates.

Harman and Jones (2001) advocated the application of the SBT for predictive modeling work, as SBT will allow software engineers to balance constraints and conflicts in search space because of noisy, partially inaccurate, and incomplete data sets. A systematic review of studies was performed on software quality prediction which reported that there are few studies that assess the predictive performance of SBT for defect prediction and change prediction (Malhotra 2014a; Malhotra and Khanna 2015). Thus, future studies should employ SBT to evaluate their capability in the area of defect and change model prediction. An important factor while developing a prediction model is its runtime

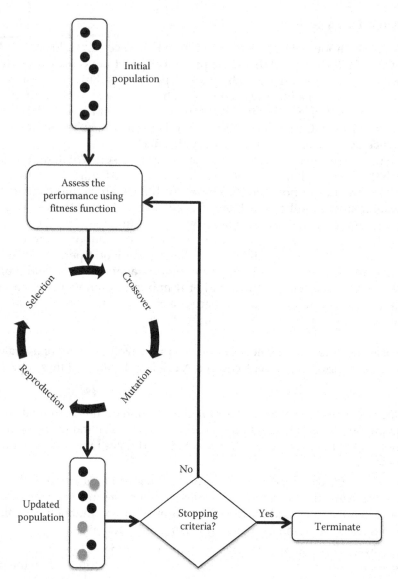

FIGURE 7.11
Process of search-based techniques.

(speed). The systematic review also revealed that the SBT require higher running time for model development as compared to the ML techniques, but parallel or cloud search-based software engineering (SBSE) can lead to promising results and significant time reduction (Di Geronimo 2012; White 2013).

A summary of characteristics of ML techniques is given in Table 7.2.

TABLE 7.2

Characteristics of ML Techniques

Technique Name	Characteristics
DT	The model developed is cost-effective and simple.
	Easy to build and apply.
	Comprehensive capability.
	May overfit data.
	Fast and are not based on any assumption.
	Can handle missing values.
	No effect of outliers.
RF	It can handle large data and are consistent performers.
	The technique is robust to noisy and missing data.
	Fast to train, robust toward parameter setting.
	Provide understandable model.
	Comprehensive capability.
	Runs efficiently on large data sets.
	Helps in identifying most important independent variables.
Bagging	Uses an ensemble of independently trained classifiers.
	Reduce the variance associated with prediction.
	Helps to avoid overfitting.
	Improved stability and accuracy.
	Improve the performance of weak learners.
Boosting	Uses a weighted average of results obtained from applying a prediction technique to various samples.
	Used to improve the accuracy of classification or regression techniques.
	Reduces bias primarily and also variance.
NB	It is robust in nature.
	It is easy to interpret and construct.
	Computationally efficient.
	Does not consider attribute correlation.
	Performance of model dependent on attribute selection technique used.
	Assumes normal distribution of numeric attributes.
	Unable to discard irrelevant attributes.
SVM	Good tolerance for high-dimension space and redundant features.
	Robust in nature specifically to outliers.
	It can handle complex functions.
	It can handle nonlinear problems.
	Less overfitting.
NN	It can infer complex nonlinear I/O transformation.
	Simple to implement.
RBL	Computation based on specified minimum coverage.
SBT	Provides optimal solution.
	Consumes more memory.
	High running time.
	Handles noisy data.

7.4 Concerns in Model Prediction

The prediction models must be carefully developed. Before and during model prediction, there are many issues and concerns that must be addressed.

- Data must be preprocessed using outlier analysis, normality tests and so on. It may help in increasing the accuracy of the models. Section 6.1 presents the preprocessing techniques.
- The model must be checked for multicollinearity effects (see Section 7.4.2).
- Dealing with imbalanced data (see Section 7.4.3).
- A suitable learning technique must be selected for model development (see Section 7.4.4).
- The training and test data must be as independent as possible, as new data is expected to be applied for model validation.
- The parameter setting of ML techniques may be adjusted (not over adjusted) and should be carefully documented so that repeatable studies can be conducted (see Section 7.4.5).

7.4.1 Problems with Model Prediction

Overfitting: In learning, overfitting occurs when the model learns noise rather than depicting the relationship. When the training error is much lower than the generalization or testing error, the model predicted is said to be overfitted. Generally, overfitting occurs when the model has too many parameters as compared to number of data samples. Increasing the number of data samples may reduce overfitting.

Model error rates: Empirical error occurs when the actual values do not match the predicted values. In some cases, although the empirical error is low but the testing or generalization error are high. Generalization error represents the degree to which the model estimates new or unseen data.

Bias versus variance: In supervised learning, creating a model that learns the relevant patterns or relationships and do not overfit the training data is very difficult. High bias means underfitting, that is, model predicted is too simple to capture the relationships, and high variance signifies that the model is too complex and captures noise along with relevant patterns or relationships. Thus, obtaining balance between bias and variance (or reducing them both simultaneously) is difficult.

Table 7.3 summarizes the possible remedies for issues or problems encountered during model prediction.

7.4.2 Multicollinearity Analysis

Multicollinearity refers to the degree to which any variable effect can be predicted by the other variables in the analysis. As multicollinearity rises, the ability to define any variable's effect is diminished. Thus, the interpretation of the model becomes difficult, as the impact of individual variables on the dependent variable can no longer be judged independently from the other variables (Belsley et al. 1980; Aggarwal et al. 2009). Thus, a test of multicollinearity can be performed on the model predicted. Let $X_1, X_2, \ldots, X_n$ be the covariates of the

TABLE 7.3

Recommended Solution to Learning Problems

Issue	Remedy
High bias	Increase attributes or independent variables
High variance	Increase training data samples
High variance	Reduce attributes
High bias	Decrease regularization parameter (lamda)
High variance	Increase regularization parameter

model predicted. Principal component method (or P.C. method) is a standard technique used to find the interdependence among a set of variables. The factors summarize the commonality of these variables, and factor loadings represent the correlation between the variables and the factor. P.C. method maximizes the sum of squared loadings of each factor extracted in turn (Aggarwal et al. 2009). The P.C. method is applied to these variables to find the maximum eigenvalue, e_{max}, and minimum eigenvalue, e_{min}. The conditional number is defined as $\lambda = \sqrt{e_{min}/e_{max}}$. If the value of the conditional number exceeds 30 then multicollinearity is not tolerable (Belsley et al. 1980).

Variance inflation factor (VIF) is used to estimate the degree of multicollinearity in predicted models. R^2s are calculated using ordinary least square regression method and VIF is defined below:

$$VIF = \frac{1}{1 - R^2}$$

According to literature, VIF value less than 10 is tolerable.

7.4.3 Guidelines for Selecting Learning Techniques

In binary model prediction, the data set that has skewed values of either positive or negative instances is known as imbalanced data. For example, given a data set with 100 instances, if 85 instances (majority class) are of faulty classes and only 15 instances (minority class) are of non-faulty classes, then the data is highly imbalanced or skewed. When a model is developed with imbalanced data it tends to be strongly biased toward the majority class as the learning technique tries to maximize the prediction accuracy of the model. There are several methods available to deal with the imbalancing learning issue. One method is to use an appropriate performance measure to estimate the model's prediction accuracy (refer Section 7.5.7). Another method uses a sampling technique such as undersampling or oversampling. In undersampling method, the samples of the majority class are removed to balance the distribution of classes. In oversampling method, the samples of the minority class are duplicated to increase the proportion of the minority class.

7.4.4 Dealing with Imbalanced Data

As explained in Section 4.9, the data analysis techniques are selected in experiment design phase. Figure 7.12 provides the factors that must be considered for selection of learning techniques. Accuracy is a desirable feature of any technique, but apart from that the technique should be easily understandable and simple. The technique should be fast in training and testing. The reason behind the good performance of a given technique must also be interpretable. The technique should also be scalable to large data sets.

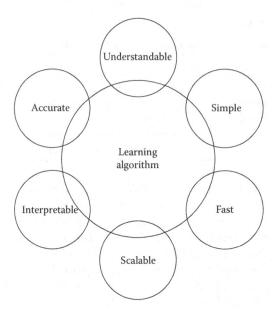

FIGURE 7.12
Properties of learning algorithms.

7.4.5 Parameter Tuning

While applying a technique, researchers require to select an optimum set of parameters among a large number of possible parameters for achieving effective and high quality results. A researcher may use his knowledge, experience, certain rules of thumb, or brute force search for optimum parameter selection (Snoek et al. 2012). Manual search of effective parameters is a very time-consuming process. Although, use of "default" parameter settings gives quite effective results, they are not close to optimum parameter settings on specific problem instances (Arcuri and Fraser 2013). However, care should be taken as overtuning the parameters of a technique leads to biased results as the technique overfits a specific training and testing data. Such models are discouraged in practice.

7.5 Performance Measures for Categorical Dependent Variable

The measures for evaluating the performance of the models are different for categorical and continuous dependent variable. This section presents the various performance measures used to evaluate the performance of the prediction models when the dependent variable is of categorical type.

7.5.1 Confusion Matrix

Confusion matrix is used to depict the accuracy of the predictions made by the model. In other words, it is used to evaluate the performance of the predicted model. Confusion matrix is a table consisting of two rows and two columns. The rows correspond to the actual (known) outputs and columns correspond to the output predicted by the model.

The predictions made by the model are with respect to the classes of the outcome variable (also referred as dependent variable) of a problem, which is under consideration. For example, if the outcome variable of the problem has two classes, then that problem is referred to as a binary problem. Similarly, if the outcome variable has three classes, then that problem is known as a three-class problem, and so on.

Consider the confusion matrix given in Table 7.4 for a two-class problem, where the outcome variable consists of positive and negative values.

The following measures are used in the confusion matrix:

- True positive (TP): Refers to the number of correctly predicted positive instances
- False negative (FN): Refers to the number of incorrectly predicted positive instances
- False positive (FP): Refers to number of incorrectly predicted negative instances
- True negative (TN): Refers to number of correctly predicted negative instances

Now, consider a three-class problem where an outcome variable consists of three classes, C_1, C_2, and C_3, as shown in Table 7.5.

From the above confusion matrix, we will get the values of TP, FN, FP, and TN corresponding to each of the three classes, C_1, C_2, and C_3, as shown in Figures 7.6 through 7.8.

Table 7.6 depicts the confusion matrix corresponding to class C_1. This table is derived from Table 7.5, which shows the confusion matrix for all the three classes C_1, C_2, and C_3. In Table 7.6, the number of TP instances are "a," where "a" are the class C_1 instances that are correctly classified as belonging to class C_1. The "b" and "c" are the class C_1 instances that are incorrectly labeled as belonging to class C_2 and class C_3, respectively. Therefore, these instances come under the category of FN. On the other hand, d and g are the instances belonging to class C_2 and class C_3, respectively, and they have been incorrectly marked as belonging to class C_1 by the prediction model. Hence, they are FP instances. The e, f, h, and i are all the remaining samples that are correctly classified as nonclass C_1 instances.

TABLE 7.4

Confusion Matrix for Two-Class Outcome Variables

		Predicted	
		Positive	Negative
Actual	Positive	TP	FN
	Negative	FP	TN

TABLE 7.5

Confusion Matrix for Three-Class Outcome Variables

		Predicted		
		C_1	C_2	C_3
Actual	C_1	a	b	c
	C_2	d	e	f
	C_3	g	h	i

TABLE 7.6

Confusion Matrix for Class "C_1"

		Predicted	
		C_1	Not C_1
Actual	**C_1**	TP = a	FN = b + c
	Not C_1	FP = d + g	TN = e + f + h + i

TABLE 7.7

Confusion Matrix for Class "C_2"

		Predicted	
		C_2	Not C_2
Actual	**C_2**	TP = e	FN = d + f
	Not C_2	FP = b + h	TN = a + c + g + i

TABLE 7.8

Confusion Matrix for Class "C_3"

		Predicted	
		C_3	Not C_3
Actual	**C_3**	TP = i	FN = g + h
	Not C_3	FP = c + f	TN = a + b + d + e

Therefore, they are referred to as TN instances. Similarly, Tables 7.7 and 7.8 depict the confusion matrix for classes C_2 and C_3.

7.5.2 Sensitivity and Specificity

Sensitivity is defined as the ratio of correctly classified positive instances to the total number of actual positive instances. It is also referred to as recall or true positive rate (TPR). If we get a sensitivity value of 1.0 for a particular class C, then this means that all the instances that belong to class C are correctly classified as belonging to class C. Sensitivity is given by the following formula:

$$\text{Sensitivity or recall(Rec)} = \frac{TP}{TP+FN} \times 100$$

But, the important point to note here is that this value comments nothing about the other instances, which do not belong to class C, but are still incorrectly classified as belonging to class C.

Specificity is defined as the ratio of correctly classified negative instances to the total number of actual negative instances. It is given by the following formula:

$$\text{Specificity} = \frac{TN}{FP+TN} \times 100$$

Ideally, the value of both sensitivity and specificity should be as high as possible. Low value of sensitivity specifies that there are many high-risk classes (positive classes) that are incorrectly classified as low-risk classes. Low value of specificity specifies that there are many low-risk classes (negative classes) that are incorrectly classified as high-risk classes (Aggarwal et al. 2009). For example, consider a two-class problem in a software organization where a module may be faulty or not faulty. In this case, low sensitivity would result in delivery of software with faulty modules to the customer, and low specificity would result in the wastage of the organization's resources in testing the software.

7.5.3 Accuracy and Precision

Accuracy is used to measure the correctness of the predicted model and is defined as the ratio of the number of correctly classified classes to the total number of classes. It is given by the following formula:

$$\text{Accuracy} = \frac{\text{TP+TN}}{\text{TP+FN + FP+TN}} \times 100$$

Precision measures how many positive predictions are correct. It is defined as the ratio of actual correctly predicted positives instances to the total predicted positive instances. In a classification task, a precision of 100% for a class C means that all the instances that belong to class C are correctly classified as belonging to class C. But, the value comments nothing about the other instances that belong to class C and are not correctly predicted.

$$\text{Precision(Pre)} = \frac{\text{TP}}{\text{TP+FP}} \times 100$$

7.5.4 Kappa Coefficient

Kappa coefficient is used to measure the degree of agreement between two given variables. Its values lie in the range of −1 to 1 (Briand et al. 2000). The higher the value of kappa coefficient, the better is the agreement between two variables. A kappa of zero indicates that the agreement is no better than what can be expected from chance.

7.5.5 *F*-measure, *G*-measure, and *G*-mean

F-measure is defined as the weighted harmonic mean of precision and recall. Therefore, the value of *F*-measure is dependent on the value of precision and recall. The *F*-measure value is less if the value of either precision or recall is less. This is the most important property of *F*-measure. It is defined as follows:

$$F\text{-measure} = \frac{2 \times \text{Pre} \times \text{Rec}}{\text{Pre} \times \text{Rec}}$$

G-measure represents the harmonic mean of recall and (100-false positive rate [FPR]) and is defined as given below:

$$G\text{-measure} = \frac{2 \times \text{Recall} \times (100 - \text{FPR})}{\text{Recall} + (100 - \text{FPR})}$$

where:

FPR is defined as the ratio of incorrectly predicted positive instances that are actually negative instances to the total actually negative instances and is given below

$$FPR = \frac{FP}{FP + TN} \times 100$$

G-mean is popularly used in an imbalanced data set, where the effect of negative cases prevails. It is the combination of two evaluations, namely, the accuracy of positives (a+) and the accuracy of negatives (a−) (Shatnawi 2010). Therefore, it keeps a balance between both these accuracies and is high if both the accuracies are high. It is defined as follows:

$$a+ = \frac{TP}{TP + FP} ; a- = \frac{TN}{TN + FN} ; g = \sqrt{(a+) \times (a-)}$$

Example 7.1:

Let us consider an example system consisting of 1,276 instances. The independent variables of this data set are the OO metrics belonging to the popularly used Chidamber and Kemerer (C&K) metric suite. The dependent variable has two values, namely, faulty or not faulty. In other words, this data set depicts whether a particular module of software contains a fault or not. If a module is containing a fault, then the value of the outcome variable corresponding to that module is 1. On the other hand, if a module is not faulty, then the value of the outcome variable for that module is 0. Now, this data set is used to predict the model. The observed and the predicted values of the outcome variable thus obtained are then used to construct the confusion matrix to evaluate the performance of the model. Confusion matrix obtained from the results is thus given in Table 7.9. Compute values of performance measures sensitivity, specificity, accuracy, precision, *F*-measure, *G*-measure, and *G*-mean based on Table 7.9.

Solution:
The values of different measures to evaluate the performance of the prediction model when the dependent variable is of categorical type are shown below in Table 7.10.

Example 7.2: An Example for a Three-Class Problem

Consider an example where dependent variable has three classes, namely, high, medium, and low. These three categories high, medium, and low are the type of severity levels of a fault associated with a module. If a module is containing a fault of high severity, then the value of the outcome variable corresponding to that class is 1. On the other hand, if a particular module is containing a fault having medium severity, then the value of the outcome variable corresponding to that class is 2. Finally, if a module is containing a low-severity fault (LSF), then the value of the outcome variable corresponding to that class is 3. So, we have rated high, medium, and low-severity faults in terms of 1, 2, and 3, respectively. There are in total sixty instances depicted in this example. The confusion matrix is shown in Table 7.11. Compute all the values of performance measures based on Table 7.11.

TABLE 7.9

Confusion Matrix for Binary Categorical Variable

		Predicted	
		Faulty (1)	Not Faulty (0)
Actual	Faulty (1)	TP = 516	FN = 25
	Not faulty (0)	FP = 10	TN = 725

TABLE 7.10

Performance Measures for Confusion Matrix given in Table 7.7

Performance Measures	Formula	Values Obtained	Results
Sensititvity or recall (Rec)	$\dfrac{TP}{TP+FN}\times100$	$\dfrac{516}{516+25}\times100$	95.37
Specificity	$\dfrac{TN}{FP+TN}\times100$	$\dfrac{725}{10+725}\times100$	98.63
Accuracy	$\dfrac{TP+TN}{TP+FN+FP+TN}\times100$	$\dfrac{516+725}{516+25+10+725}\times100$	97.25
Precision (Pre)	$\dfrac{TP}{TP+FP}$	$\dfrac{516}{516+10}$	0.981
F-measure	$\dfrac{2\times Pre\times Rec}{Pre+Rec}$	$\dfrac{2\times0.981\times0.954}{0.981+0.954}$	0.967
a+	$\dfrac{TP}{TP+FP}$	$\dfrac{516}{516+10}$	0.981
a−	$\dfrac{TN}{TN+FN}$	$\dfrac{725}{725+25}$	0.967
FPR	$\dfrac{FP}{FP+TN}\times100$	$\dfrac{10}{10+5+6}\times100$	1.90
G-measure	$\dfrac{2\times Recall\times(100-FPR)}{Recall+(100-FPR)}$	$\dfrac{2\times95.4\times(100-1.90)}{95.4+(100-1.90)}$	96.73
G-mean	$\sqrt{(a+)\times(a-)}$	$\sqrt{0.981\times0.967}$	0.973

TABLE 7.11

Confusion Matrix for Three-Class Outcome Variable

		Predicted		
		High (1)	Medium (2)	Low (3)
Actual	High (1)	3	9	0
	Medium (2)	3	34	1
	Low (3)	1	4	5

Solution:
From the confusion matrix given in Table 7.11, the values of TP, FN, FP, and TN are derived and corresponding to each of the three classes high (1), medium (2), and low (3), and are shown in Tables 7.12 through 7.14.

The value of different performance measures at each severity level, namely, high, medium, and low on the basis of Tables 7.12 through 7.14 are given in Table 7.15.

TABLE 7.12

Confusion Matrix for Class "High"

		Predicted	
		High	Not High
Actual	High	TP = 3	FN = 9
	Not high	FP = 4	TN = 44

TABLE 7.13

Confusion Matrix for Class "Medium"

		Predicted	
		Medium	Not Medium
Actual	Medium	TP = 34	FN = 4
	Not medium	FP = 13	TN = 9

TABLE 7.14

Confusion Matrix for Class "Low"

		Predicted	
		Low	Not Low
Actual	Low	TP = 5	FN = 5
	Not low	FP = 1	TN = 49

7.5.6 Receiver Operating Characteristics Analysis

Receiver operating characteristic (ROC) analysis is one of the most popular techniques used to measure the accuracy of the prediction model. It determines how well the model has worked on the test data. It is used in situations where a decision between two possible outcomes is to be made. For example, whether a sample belongs to change-prone or not change-prone class.

ROC analysis is well suited for problems with binary outcome variable, that is, when the outcome variable has two possible outcome values. In case of multinomial outcome variable, that is, when the outcome variable has three or more possible groups of outcome values, then separate prediction functions are generated for each of the groups, and then ROC analysis is done for each of the group individually.

Often the values of outcome variables are referred as target and reference group (Meyers et al. 2013). Now, the question arises as to what should be considered as the target group. For example, if the outcome variable has two classes A and B. Then, which outcome class should be considered as target group and reference group? Usually, the target group is the one that would satisfy the condition that we need to identify or predict. Therefore, it is referred to as the group of positive outcomes. The remaining instances correspond to the alterative group referred to as the "reference group." These instances are referred to as the negative outcomes, as shown in Table 7.4.

7.5.6.1 ROC Curve

One of the most important characteristics of the ROC analysis is the curve. ROC curve is the visual representation that is used to picture the overall accuracy of the prediction model. The ROC curve is defined as a plot between sensitivity on the y-coordinate and 1-specificity on the x-coordinate (Hanley and McNeil 1982; El Emam et al. 1999). So, we can say that ROC curve is a plot of the TP rate against the FP rate at different possible threshold values (cutoff points). It is represented by a graph together with a diagonal line, as shown in Figure 7.13. This diagonal line represents a random model that has no predictive power. We can also interpret the curve by saying that there is a tradeoff between sensitivity and specificity in the sense that any increase in the value of sensitivity will

TABLE 7.15

Performance Measures at Each Severity Level

Category (Severity of Fault)	Sensitivity/ Recall (Rec) $\dfrac{TP}{TP+FN}\times100$	Specificity $\dfrac{TN}{FP+TN}\times100$	Accuracy $\dfrac{TP+TN}{TP+FN+FP+TN}\times100$	Precision (Pre) $\dfrac{TP}{TP+FP}$	F-Measure $\dfrac{2\times Pre\times Rec}{Pre+Rec}$	a+ $\dfrac{TP}{TP+FP}$	a− $\dfrac{TN}{TN+FN}$	G-Mean $\sqrt{(a+)\times(a-)}$	FPR $\dfrac{FP}{FP+TN}\times100$	G-Measure $\dfrac{2\times Recall\times(100-FPR)}{Recall+(100-FPR)}$
High	25.0	91.66	78.33	0.428	0.316	0.428	0.830	0.595	8.3	49.86
Medium	89.47	40.90	71.66	0.723	0.799	0.723	0.692	0.707	59.09	56.14
Low	50.0	98.0	90.0	0.833	0.625	0.833	0.907	0.868	2.00	66.21

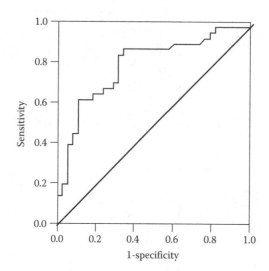

FIGURE 7.13
Example of an ROC curve.

lead to a decrease in the value of specificity. ROC curve starts from the origin and moves toward the upper-right portion of the graph, as can be seen from Figure 7.13. The ends of the curve meet the end points of the diagonal line. The closer the curve is toward the left-hand border and the top border of the ROC graph, the more accurate is the prediction capability of the model. In contrast, the closer the curve comes to the 45-degree diagonal of the ROC graph, the less accurate is the model prediction.

7.5.6.2 Area Under the ROC Curve

The prediction capability of the model depends on the degree to which the ROC curve bends away from the random model projection. In other words, we can say that the accuracy of the model depends on how well it is able to separate the instances being tested as positives and negatives. The area under the ROC curve (AUC) measures this accuracy of the predicted model. An AUC value of 1 represents that the model prediction is 100% accurate and an area of 0.5 represents that the performance of the model is worthless in predicting the unknown instances. An AUC value in the range of 0.5 will arise in the case where the ROC curve lies very close to the diagonal line. In other words, the AUC provides a measure of how much better, than the random model, a given prediction model is able to differentiate between positive and negative outcomes (target and reference group).

7.5.6.3 Cutoff Point and Co-Ordinates of the ROC Curve

We should not select arbitrary cutoff points in the analysis to calculate sensitivity and specificity. Another important use of ROC analysis is to provide optimal threshold value. The optimal threshold value provides balance between sensitivity and specificity values and can be obtained by ROC analysis. The ROC curve depicts the overall accuracy of the model, but the success of correctly predicting group membership depends on the location of a particular decision threshold value on the ROC curve. This threshold value is also known as the cutoff point. Determining the threshold value is very important in model prediction because the basis of our classification is a quantitative measure, that is,

TABLE 7.16

Co-Ordinates of the ROC Curve

Cutoff Point	Sensitivity	1-Specificity
0.450	0.890	0.600
0.500	0.800	0.400
0.550	0.740	0.220
0.900	0.100	0.020

the predicted probability of an instance as being a positive or negative outcome. In other words, we can say that based on the decision criterion selected, we can identify whether a particular instance of the test data will belong to the positive or negative outcome on the basis of its predicted probability. An instance whose predicted probability falls below the selected cutoff point would be classified as negative outcome, and an instance whose predicted probability falls at or above the selected threshold value would be classified as positive outcome.

This crucial decision regarding the selection of the threshold point can be made based on the results of the ROC analysis. In addition to obtaining the AUC value and the ROC curve, a range of co-ordinates that define the ROC curve are also obtained and shown in Table 7.16. This range of co-ordinates is a combination of TP (sensitivity) and FP (1-specificity).

Along with each set of sensitivity and 1-specificity, we have the predicted probability of an instance as being classified as positive outcome. Now, based on this predicted probability, the value of an instance can be decided. The remaining two columns of Table 7.16 represent the TP rate (sensitivity) and FP rate (1-specificity) that are plotted as the data points for the ROC curve on *y*- and *x*-axes, respectively. For example, consider the second row in Table 7.16, where the predicted probability of an instance being in the target group is 0.500 as the threshold value (cutoff point). The TP rate is 0.800 and FP rate is 0.400 corresponding to 0.500 cutoff value. This means that 80% of the positive values are correctly classified, and 60% of the negative values are correctly classified. When we plot these data points together, then the ROC curve is depicted as shown in Figure 7.13.

Example 7.2:

Consider an example to compute AUC using ROC analysis. In this example, OO metrics are taken as independent variables and fault proneness is taken as the dependent variable. The model is predicted by applying an ML technique. Table 7.17 depicts actual and predicted dependent variable. Use ROC analysis for the following:

1. Identify AUC.
2. Based on the AUC, determine the predicted capability of the model.

Solution:

The value of the dependent variable is 0 if the module does not contain any fault, and its value is 1 if the module contains a fault. On the basis of this input, the ROC curve obtained using Statistical Package for the Social Sciences (SPSS) tool is shown in Figure 7.14, the value of AUC is 0.642 and the coordinates of the curve are depicted in Table 7.18. Table 7.18 shows the values of sensitivity and 1-specificity along with their corresponding cutoff points. The results show that AUC is 0.642. Hence, the model performance is not good (for interpretation of performance measures refer Section 7.9.1).

TABLE 7.17

Example for ROC Analysis

Actual	Predicted Probability	Actual	Predicted Probability
1	0.055	0	0.061
1	0.124	0	0.254
1	0.124	0	0.191
1	0.964	0	0.024
1	0.124	0	0.003
1	0.016	0	0.123
0	0.052	1	0.123
0	0.015	1	0.024
0	0.125	1	0.169
0	0.123	1	0.169
1	0.052	1	0.169

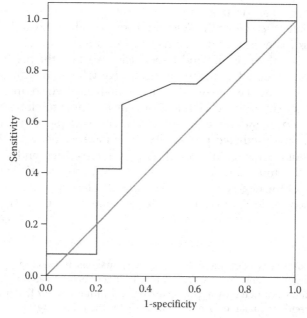

FIGURE 7.14
The obtained ROC curve.

7.5.7 Guidelines for Using Performance Measures

There are a number of different measures that are used to evaluate the performance of the prediction model. These measures have already been explained in the previous sections. Each performance measure has its own advantages and, therefore, is applicable only in selected situations. In this section, we will discuss the suitability of performance measures and provide the guidelines that can be followed in selecting a suitable performance measure.

Accuracy and error rate (1-accuarcy) are the simplest of the measures that are used to evaluate the performance of the prediction model. However, these measures are highly

TABLE 7.18

Co-Ordinates of the ROC Curve

Cutoff Point	Sensitivity	1-Specificity
0	1	1
0.009	1	0.9
0.015	1	0.8
0.020	0.917	0.8
0.038	0.833	0.7
0.056	0.750	0.6
0.092	0.750	0.5
0.123	0.667	0.3
0.124	0.417	0.3
0.147	0.417	0.2
0.180	0.083	0.2
0.222	0.083	0.1
0.609	0.083	0
1	0	0

sensitive to the distributions in the data. In other words, accuracy is very sensitive to the imbalances in a given data set. Any data set that exhibits unequal distribution of positive and negative instances is considered as an imbalanced data (Malhotra 2015). Therefore, as the class distribution will vary, the performance of the measure will also change even though the performance of the learning technique remains the same. As a result, the accuracy measure will not be a true representative of the model performance. For example, if data set contains maximum negative samples and all the samples are predicted as negative, the accuracy will be very high but the predicted model is useless. Hence, this measure is not recommended to be used when there is a need to compare the performance of two learning techniques over different data sets.

Therefore, other measures popularly used in learning are precision, recall, *F*-measure, and *G*-measure. We will first discuss precision and recall and see their behavior with respect to imbalanced data. As we know, precision is a measure of exactness that determines the number of instances which are labeled correctly out of the total number of instances labeled as positive. In contrast, recall is a measure of completeness that determines the number of positive class instances, which are labeled correctly. By these definitions, it is clear that both precision and recall have an inverse relationship with each other and precision is sensitive to data distributions, whereas recall is not. But recall is not able to give any information regarding the number of instances that are incorrectly labeled as positive. Similarly, precision does not tell anything about the number of positive instances that are labeled incorrectly. Therefore, precision and recall are often combined together to form a measure referred to as *F*-measure. *F*-measure is considered as an effective measure of classification that provides an insight into the functionality of a classifier, unlike the accuracy metric. However, *F*-measure is also sensitive to data distributions. Another metric, the *G*-measure is also one of the popularly used evaluation measure that is used to evaluate the degree of inductive bias, in terms of a ratio of positive accuracy and negative accuracy. Although *F*-measure and *G*-measure are much better than the accuracy measure, they are still not suitable to compare the performance of different classifiers over a range of sample distributions.

To overcome the above issues, AUC curve generated by the ROC analysis is widely used as the performance measure specifically for imbalanced data. AUC computed from ROC analysis is widely used in medical diagnosis for the past many years, and its use is increasing in the field of data mining research. Carvalho et al. (2010) advocated AUC to be the relevant criterion for dealing with unbalanced and noisy data, as AUC is insensitive to the changes in distribution of class. He and Garcia (2009) have recommended the use of AUC for dealing the issues of imbalanced data with regard to class distributions, it provides a visual representation of the relative tradeoffs between the advantages (represented by TP) and costs (represented by FP) of classification. In addition to, ROC curves for data sets that are highly skewed, a researcher may use precision–recall (PR) curves. The PR curve is expressed as a plot of precision rate and the recall rate (He and Garcia 2009). The ROC curves achieve maximum model accuracy in the upper left-hand of the ROC space. However, a PR curve achieves maximum model accuracy in the upper right-hand of the PR space. Hence, PR space can be used as an effective mechanism for predicted model's accuracy assessment when the data is highly skewed.

Another shortcoming of ROC curves is that they are not able to deduce the statistical significance of different model performance over varying class probabilities or misclassification costs. To address these problems, another solution suggested by He and Garcia (2009) is to use cost curves. A cost curve is an evaluation method that, like ROC curve, visually depicts the model's performance over varying misclassification of costs and class distributions (He and Garcia 2009).

In general, given the limitations of each performance measures, the researcher may use multiple measures to increase the conclusion validity of the empirical study.

7.6 Performance Measures for Continuous Dependent Variable

This section highlights on the various performance measures used to evaluate the performance of the prediction models when the dependent variable is of continuous type.

7.6.1 Mean Relative Error

The mean relative error (MRE) is a measure that is used to find out how far are the estimated values from actual values of the instances in a given data set. It is applicable to any two sets of values. Here, by two sets we mean that one set consists of the actual values and the other set consists of the estimated (predicted) values. It is defined by the following formula:

$$\text{MRE} = \frac{1}{N} \sum_{i=1}^{N} \frac{P_i - A_i}{A_i}$$

where:
N is the total number of instances in a given data set
P_i is the predicted value of an instance i
A_i is the actual value of an instance i

7.6.2 Mean Absolute Relative Error

The mean absolute relative error (MARE) is the most frequently used measure to evaluate the performance of the model when the dependent variable is continuous. It is given by the following formula:

$$MARE = \frac{1}{N} \sum_{i=1}^{N} \frac{|P_i - A_i|}{A_i}$$

where:

N refers to the total number of instances in a given data set

P_i refers to the predicted value of an instance i

A_i refers to the actual value of an instance i

7.6.3 PRED (A)

It is calculated from the relative error. It is defined as the ratio of the instances that have an error value (absolute relative error [ARE]) less than or equal to "A" error divided by the total number of instances in the data set. Given the lowest value of A, the higher the value of Pred (A), the better it is as the prediction model. Generally, the value of "A" that is used by most of the studies is 25%, 50%, or 75%. PRED (A) is given by the following formula:

$$PRED(A) = \frac{d}{N}$$

where:

N refers to the total number of instances in a given data set

d is the number of instances having value of error less than or equal to "A" error

Example 7.3:

Consider an example to assess the performance of model predicted with lines of code (LOC) as outcome. Table 7.19 presents a data set consisting of ten instances that depict the LOC of a given software. The table shows the actual values of LOC and values of LOC that are predicted once the model has been trained. Calculate all the performance measures for the data given in Table 7.19.

Solution:

The difference between the predicted and the actual values has been shown in Table 7.20. Table 7.21 shows the values of the performance measures MRE, MARE, and Pred (A). MRE is the average of the values obtained after dividing the difference of the predicted and the actual values with the actual values. Similarly, MARE is the average of the values obtained after dividing the absolute difference of the predicted and the actual values with

TABLE 7.19

Actual and Predicted Values of Model Predicted

Module #	Actual (A_i)	Predicted (P_i)
1	100	90
2	76	35
3	45	60
4	278	300
5	360	90
6	240	250
7	520	500
8	390	800
9	50	45
10	110	52

TABLE 7.20

Actual and Predicted Values of Model Predicted

Module #	Actual (A_i)	Predicted (P_i)	$P_i - A_i$	$(P_i - A_i)/A_i$	$|(P_i - A_i)|/A_i$
1	100	90	−10	−0.1	0.100
2	76	35	−41	−0.539	0.539
3	45	60	15	0.333	0.333
4	278	300	22	0.079	0.079
5	360	90	−270	−0.75	0.750
6	240	250	10	0.047	0.042
7	520	500	−20	−0.038	0.038
8	390	800	410	1.051	1.051
9	50	45	−5	−0.1	0.100
10	110	52	−58	−0.527	0.527

TABLE 7.21

Performance Measures

Performance Measure	Values Obtained	Result
$\text{MRE} = \dfrac{1}{N}\sum_{i=1}^{N}\dfrac{P_i - A_i}{A_i}$	$\text{MRE} = \dfrac{-0.55}{10}$	−0.055
$\text{MARE} = \dfrac{1}{N}\sum_{i=1}^{N}\dfrac{|P_i - A_i|}{A_i}$	$\text{MARE} = \dfrac{3.56}{10}$	0.356
$\text{PRED}(25) = \dfrac{d}{N}$	$\text{PRED}(25) = \dfrac{5}{10}$	50%
$\text{PRED}(50) = \dfrac{d}{N}$	$\text{PRED}(50) = \dfrac{6}{10}$	60%
$\text{PRED}(75) = \dfrac{d}{N}$	$\text{PRED}(75) = \dfrac{9}{10}$	90%

the actual values. Pred (A) is obtained by dividing the instances that have an error value (MRE) less than or equal to "A" error by the total number of instances in the data set. The Pred value is calculated at 25%, 50%, and 75% levels, and the results are shown in Table 7.21. The results show that 50% of instances have error less than 25%, 60% of instances have error less than 50% and 90% of instances have error less than 75%.

7.7 Cross-Validation

The accuracy obtained by using the data set from which the model is build is quite optimistic. Cross-validation is a model evaluation technique that divides the given data set into training and testing data in a given ratio and proportion. The training data is used to train the model using any of the learning techniques available in the literature. This trained model is then used to make new predictions for data it has not already seen, that is, the testing data. The division of the data set into two parts is essential, as it will provide information about how well the learner performs on the new data. The ratio by which the data set is divided is decided on the basis of the cross-validation method used.

7.7.1 Hold-Out Validation

Hold-out validation is the simplest kind of cross-validation in which the given data set is randomly partitioned into two independent data sets. Generally, two-third of the data is used as the training set and remaining one-third of the data is used as the testing set. Figure 7.15 shows the concept of hold-out validation. Training set is used for model construction, and testing set is used for the estimation of model accuracy. In other words, testing set is used to determine how well the model is trained to make new predictions. The advantage of this method is that it takes less time to compute. However, its evaluation can have a high variance. This is so because here the evaluation depends greatly on which data points end up in the training set and which end up in the testing set. Thus, the evaluation may be quite different depending on how the division is made, and we may get a misleading estimate of error rate if we happen to get an "unfortunate" split. This problem can be overcome by using multiple runs of hold-out validation and then averaging the result or using majority voting. Hold-out method is applicable for problems with large data set and is not suitable for small data sets as, in this case, a portion of the data set for testing cannot be set aside.

7.7.2 *K*-Fold Cross-Validation

In k cross-validation, the data set is divided into k parts where $k - 1$ parts are used for training the model and one part is used for validation purpose. Thus, this procedure is repeated k-times and the results of each run are combined together. This validation method is widely used in empirical studies and mostly the value of k is ten. Figure 7.16 depicts the ten cross-validation procedure.

7.7.3 Leave-One-Out Validation

Leave-one-out (LOO) cross-validation is a K-fold cross-validation with K equal to 1. This means that for a data set with N instances, N separate experiments are conducted. For each experiment, $N - 1$ instances are used for training and the remaining 1 instance is used for testing. Thus, the prediction is made for that one particular data point. In this validation method, the average error is also computed, which is used to evaluate the model. The evaluation given by LOO cross-validation error is good, but at first glance it seems very expensive to compute. We also have the concept of locally weighted learners that can make LOO predictions just as easily as they make regular predictions. This means that computing the LOO cross-validation error is a much better way to evaluate models. However, when the available number of data samples is severely limited, one may use this form of cross-validation. The procedure is presented in Figure 7.17. Given the stochastic nature of learning techniques, specifically SBT, multiple runs maybe combined with cross-validation method being used by a researcher to increase the generalizability of results.

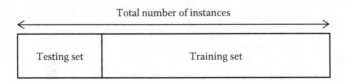

FIGURE 7.15
Hold-out validation.

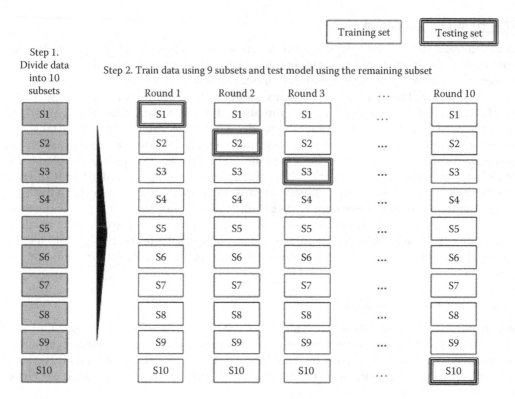

FIGURE 7.16
Tenfold cross-validation.

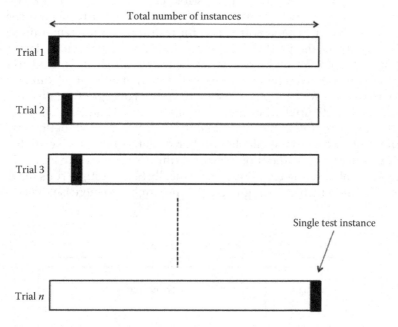

FIGURE 7.17
Leave-one-out validation.

7.8 Model Comparison Tests

The performance of the models predicted using various techniques and on multiple data sets can be compared using statistical tests. These tests have been explained in Chapter 6. The models predicted using various techniques over multiple data sets can be compared using one of the performance measures. The paired *t*-test, Friedman test, or Kruskal–Wallis test can be used for comparing the performance of models predicted. The selection of the appropriate test depends on the assumptions of the test. Figure 7.18 depicts the procedure of comparing predicted models. The figure shows that each ML technique is applied to multiple data sets resulting in predicted models. The statistical tests such as paired *t*-test, Friedman test, or Kruskal–Wallis test are applied on the outcome produced (such as recall, AUC, accuracy) by these predicted models.

Consider an example given in Table 7.22, where the researcher intends to compare the predictive performance of two models predicted using Bagging and LR technique. We can apply paired *t*-test to compare the performance of these two techniques with the following hypothesis.

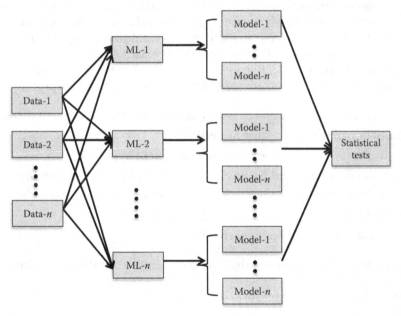

FIGURE 7.18
Model comparison using statistical tests.

TABLE 7.22

AUC of Models Predicted

Data Set	Bagging	LR
Data-1	0.74	0.69
Data-2	0.72	0.67
Data-3	0.75	0.65
Data-4	0.77	0.61
Data-5	0.71	0.69

Null hypothesis: There is no significant difference between the performance of model predicted using Bagging technique and the model predicted using LR technique.

Alternative hypothesis: There is a significant difference between the performance of model predicted using Bagging technique and the model predicted using LR technique.

After applying paired t-test using the procedure given in Section 6.4.6.3, the t-statistic is 3.087 (p-value = 0.037) and the test is significant at 0.05 significance level. Hence, null hypothesis is rejected and the alternative hypothesis is accepted. The example above demonstrates how statistical tests can be used for model comparison. The empirical study in Section 7.11 describes the practical example of comparison of models using statistical tests.

7.9 Interpreting the Results

The results are merely the facts and findings of the statistical analysis and hypothesis testing. The results are generally presented using tables and figures. For example, specifying that the two OO metrics are related to each other is a finding. However, discussing why the variables are related to each other is part of discussion of results or results interpretation portion. Thus, the meaning of the results is presented in the results interpretation section of the study. The following issues should be addressed while interpreting the results:

1. Answers to research questions or issues identified in the experimental design.
2. Determination of reasons related to findings.
3. Discussing the reasons of acceptance or rejection of hypothesis.
4. Identification of generalized findings in view of findings in the literature.
5. Acknowledging the weaknesses or limitations of the empirical study.
6. Determination of new lessons learned from the findings.
7. Identification of target audience of the study.

Figure 7.19 depicts the list of questions that must be addressed while interpreting the results.

7.9.1 Analyzing Performance Measures

The high values of recall represents whether the samples are correctly predicted as positive or not. Ideally, both the sensitivity and specificity should be high. For example, for predicting fault-prone classes, a low specificity means that there are many low-risk classes that are classified as faulty. Therefore, the organization would waste resources in focusing additional testing efforts on these classes. A low sensitivity means that there are many high-risk classes that are classified as not faulty. Therefore, the organization would be passing high-risk classes to customers.

The performance measures can be evaluated and analyzed as follows in different cases:

Case 1: Consider a scenario where the number of samples is 1,000 with 10 positive samples and rest negative samples, if all the samples are predicted as negative then the accuracy is 99% and specificity is 100%. However, the model is useless as

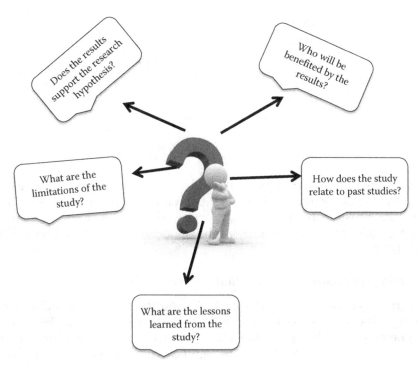

FIGURE 7.19
Issues to be addressed while result interpretation.

sensitivity or recall is computed its value is 0%. Hence, in this case, recall is the most appropriate performance measure that represents the model accuracy.

Case 2: Consider a scenario where the number of instances is 1,000 with 10 positive samples and rest negative instances, if all the samples are predicted as positive then precision is 1%, accuracy is 1%, specificity is 0%, and sensitivity is 100%. Hence, in this case, sensitivity is the most inappropriate performance measure that represents the model accuracy.

Case 3: Another situation is if most of the instances are positive. Consider a scenario where the number of instances is 1,000 with 990 positive instances and rest negative instances, if all the instances are predicted as positive then precision is 99%, accuracy is 99%, specificity is 0%, and sensitivity is 100%. Hence, in this case, specificity is the most appropriate performance measure that represents the model accuracy.

Case 4: Consider a scenario where the number of instances is 1,000 with 990 positive instances and rest negative instances, if all the instances are predicted as negative then precision is 0%, accuracy is 1%, specificity is 100%, and sensitivity is 0%. Hence, in this case, sensitivity, precision, and accuracy are the most appropriate performance measures that represent the model accuracy.

Case 5: Consider 1,000 samples with 60 positive instances and 940 negative instances, where 50 instances are predicted correctly as positive, 40 are incorrectly predicted as positive, and 900 are correctly predicted as negative. For this example, sensitivity is 83.33%, specificity is 95.74%, precision is 55.56%, and accuracy is 95%. Hence, in this case, precision is the most appropriate performance measure that represents the model accuracy.

TABLE 7.23

AUC Values

AUC Range	Guideline
0.50–0.60	No discrimination
0.60–0.70	Poor
0.70–0.80	Acceptable/good
0.80–0.90	Very good
0.90 and higher	Excellent

Hence, the problem domain has major influence on the values of performance measures, and the models can be interpreted in the light of more than one performance measures. AUC is another measure that provides a complete view of the accuracy of the model. Guidelines for interpreting the accuracy of the prediction model based on the AUC are given in Table 7.23.

7.9.2 Presenting Qualitative and Quantitative Results

The quantitative results are presented using tables and charts. The readability of the tables should be high. The readers would want to know the precise values of the results. There are many graphs such as box plots, line charts, scatter plots, and pie charts that can be used to present the quantitative results. The significance of the numerical results must be interpreted.

Qualitative research involves presenting the data that is non-numerical in nature. It presents people's reactions. The researcher must present the quotes, reactions, or texts that represent most significant results of the study.

7.9.3 Drawing Conclusions from Hypothesis Testing

The statistical test begins with the assumption that the null hypothesis is true, however, the researcher wants to reject the null hypothesis. When the null hypothesis is not rejected, it does not necessarily means that there is no difference. It means there might be a difference, but it is not detected by the sample data used in the hypothesis testing. Thus, a difference might exist but the result does not detect it.

When the null hypothesis is rejected, it means that a statistical significance has been obtained. However, the researcher has to decide whether this result is of any practical significance.

7.9.4 Example—Discussion of Results in Hypothesis Testing Using Univariate Analysis for Fault Prediction System

In this section, we validate the hypothesis set A stated in Section 4.7.6 and the results of univariate analysis presented in Section 6.6. In addition to the results of univariate analysis using LR technique provided in Section 6.6, the FPS study also conducts univariate analysis using two ML techniques: ANN and DT. In the FPS study, the values of performance measures (sensitivity, specificity, accuracy) are calculated for each individual metric using ANN and DT techniques. Thus, while reporting the results of hypothesis testing, the univariate results obtained using LR, ANN, and DT techniques are shown. While providing the results of the hypothesis formed in this work, we will also compare the results with those of previous studies till date shown in Table 7.24.

TABLE 7.24

Results of Different Validations

Metric	Our Results				Basili et al. (1996)	Tang et al. (1999)	Briand et al. (2000)	Briand et al. (2001)	El Emam et al. (2001a)		Yu et al. (2002)	Gyimothy et al. (2005)	Zhou and Leung (2006)		Olague et al. (2007)	
	HSF	MSF	LSF	USF					#1	#2			LSF/USF	HSF		
Language used	C++				C++	C++	C++	C++	C++	C++	Java	C++	C++	C++	Java	
Technique used	LR, ML (DT, ANN)				LR	LR	LR	LR	LR	LR	OLS	LR, ML (DT, ANN)	LR, ML (NNage, RF, NB)		LR	
Type of data	NASA data set				Univ.	Comm.	Univ.	Comm.	Comm.		Comm.	Open source	NASA data set		Open source	
Fault severity taken?	Yes				No	No	No	No	No		No	No	Yes		No	
WMC	++	++	++	++	+	+	+	++	+	0	++	++	++	++	++	++
DIT	0	0	0	0	++	0	++	-	0	0	0	+	0	0	0	-
RFC	++	++	++	++	++	+	++	++	++	0	+	++	++	++	++	++
NOC	-	0	0	-	--	0	-	0	0		++	0	-		0	0
CBO	++	++	++	++	+	0	++	++	+	0	+	++	++	++	++	++
LCOM	++	0	++	++	0							+	+	++	++	++
LOC	++	++	++	++			++		++	++		++	++	++	++	++

++ denotes metric is significant at 0.01, + denotes metric is significant at 0.05, -- denotes metric is significant at 0.01 but in an inverse manner, - denotes metric is significant at 0.05 but in an inverse manner, 0 denotes that metric is not significant. A blank entry means that our hypothesis is not examined or the metric is calculated in a different way. LR: logistic regression, OLS: ordinary least square, ML: machine learning, DT: decision tree, ANN: artificial neural network, RF: random forest, NB: naïve Bayes, LSF: low-severity fault, USF: ungraded severity fault, HSF: high-severity fault, MSF: medium-severity fault, #1: without size control, #2: with size control, comm.: commercial, univ.: university.

Coupling between objects (CBO) hypothesis is found to be significant in the LR analysis for all severities of faults. Sensitivity of both the LR and artificial NN (ANN) models are same. The ML techniques confirmed the findings of the regression analysis, as the values of sensitivity, accuracy, and correctness for the CBO metric is high. It is also found to be the significant predictor in all studies except Tang et al. (1999) and El Emam et al. (2001a).

All of the models found the CBO metric to be the significant predictor of fault proneness. Hence, the null hypothesis is rejected and the alternative hypothesis is accepted.

Response for a class (RFC) hypothesis is found to be significant in the LR analysis for all severities of faults. Sensitivity of the ANN and DT models is also high. Similar results are shown by Basili et al. (1996), Briand et al. (2000), El Emam et al. (2001a), Zhou and Leung (2006), Olague et al. (2007), and Gyimothy et al. (2005). Tang et al. (1999) found it significant at 0.05. Yu et al. (2002) also found the RFC metric as the significant predictor, but their method of calculating the RFC metric was different. Hence, the null hypothesis is rejected and the alternative hypothesis for the RFC metric is accepted.

Lack of cohesion in methods (LCOM) hypothesis is found to be significant in the LR analysis of this study (except for faults predicted with respect to low severity), contradicting the results of Basili et al. (1996), where the LCOM was shown to be insignificant. Zhou and Leung (2006), Olague et al. (2007), and Gyimothy et al. (2005) also found the LCOM metric to be a very significant predictor of fault proneness. Yu et al. (2002) calculated the LCOM in a totally different way; therefore, the study could not compare the results with theirs. Hence, the null hypothesis is rejected and the alternative hypothesis is accepted.

In the number of children (NOC) hypothesis, it was found that the NOC metric was not significant with respect to the LSF and high-severity fault (HSF) in the LR analysis. However, the NOC metric is found inversely related to fault proneness with respect to medium-severity fault (MSF) and ungraded severity fault (USF), that is, larger the value of the NOC, the lesser is the probability of fault detection. The results of the DT and ANN also predict all classes to be nonfaulty for all severity levels of faults. Braind et al. (2001), Gyimothy et al. (2005), and Tang et al. (1999) found the NOC not to be a significant predictor of fault proneness. Basili et al. (1996), Briand et al. (2000), and Zhou and Leung (2006) found the NOC metric to be significant, but they found that the larger the value of NOC, the lower the probability of fault proneness. According to Yu et al. (2002), NOC metric is a significant predictor of fault proneness, and they found that more the NOC in a class, the more fault prone it is.

The null hypothes is accepted for the NOC metric and alternative hypothesis. Most of the studies that examined this metric found either the NOC metric to be not related to fault proneness or negatively related to fault proneness. The conclusion is that the NOC metric is a bad predictor of fault proneness. Hence, perhaps more attention (e.g., through walkthroughs and inspections) is given during development to the classes on which other classes depend (Briand et al. 2000).

Depth of inheritance tree (DIT) hypothesis is not found to be significant in the univariate LR analysis. On the other hand, Briand et al. (2001) found it to be significant but in inverse manner. This finding is similar to those given by Yu et al. (2002), Tang et al. (1999), El Emam et al. (2001a), and Zhou and Leung (2006). Basili et al. (1996) and Briand et al. (2000) found DIT metric to be a significant predictor of fault proneness. Gyimothy et al. (2005) found the DIT metric to be a less significant predictor of fault proneness. The DT results show very less values of sensitivity and accuracy for medium, low, and ungraded severities of faults. For high severity of faults, the DT and ANN techniques predicted all classes as nonfaulty. The ANN technique shows low values of sensitivity for medium, low, and

TABLE 7.25

Summary of Hypothesis

Metric	Hypothesis Accepted/Rejected			
Severity of Faults	HSF	MSF	LSF	USF
RFC	√	√	√	√
CBO	√	√	√	√
LCOM	√	√	×	√
DIT	×	×	×	×
NOC	×	×	×	×
WMC	√	√	√	√
LOC	√	√	√	√

√ means the hypothesis is accepted, × means that the hypothesis is rejected.

ungraded severities of faults. The completeness value of the DIT is worse with respect to all the severities of faults. Table 7.24 shows that most of the studies found that the DIT metric is not related to fault proneness. The class may have less number of ancestors in most of the studies and is one of the reasons for nonrelation of the DIT metric with fault proneness, and further investigation is needed. The null hypothesis for the DIT metric is accepted and the alternative hypothesis is rejected.

Weighted methods per class (WMC) hypothesis is found to be significant in the LR and ANN analysis. On the other hand, Basili et al. (1996) found it to be less significant. In the study conducted by Yu et al. (2002), WMC metric was found to be a significant predictor of fault proneness. Similar to the regression and DT and ANN results in this study, Briand et al. (2000), Gyimothy et al. (2005), Olague et al. (2007), and Zhou and Leung (2006) also found the WMC metric as one of the best predictors of fault proneness. Rest of the studies found it to be a significant predictor, but at 0.05 significance level.

All of the three models found the WMC metric to be significant predictor of fault proneness. Hence, the null hypothesis is rejected and the alternative hypothesis is accepted.

LOC hypothesis is found to be significant in the LR analysis. It was also found significant in all the studies that examined them. Hence, the null hypothesis is rejected and the alternative hypothesis is accepted.

In Table 7.25, summary of the results of the hypothesis stated in Section 4.7.6 with respect to each severity of faults. The LCOM metric was found significant at 0.05 level in the study conducted by Zhou and Leung (2006) with regard to low and ungraded severity levels of faults. However, in this study, the LCOM metric is found significant at 0.01 level with respect to the HSF, MSF, and USF and is not found to be significant with respect to LSF.

7.10 Example—Comparing ML Techniques for Fault Prediction

In this study, the performance of 18 ML techniques on six releases of "MMS" application package of Android operating system (OS) is compared. This enables the investigation whether one technique outperforms others, and also provides insights on the selection of a particular ML technique. The 18 ML techniques are a subset of the 22 ML techniques used by Lessmann et al. (2008) to assess relationship between static code metrics (traditional, procedural, and module-based metrics [Malhotra 2014b]) and fault proneness. However, in this work, we predict the fault-prone classes using the OO metrics design suite given by Bansiya and Davis (2002) and Chidamber and Kemerer (1994), instead of static code metrics. The results are evaluated using AUC obtained from ROC analysis. Figure 7.20 presents the basic elements of the study (for details refer [Malhotra and Raje 2014]).

This section presents the evaluation results of the various ML techniques for fault prediction using selected OO metrics given in Table 7.26.

The results are validated using six releases of the "MMS" application package of the Android OS. The six releases of Android OS have been selected with three code names,

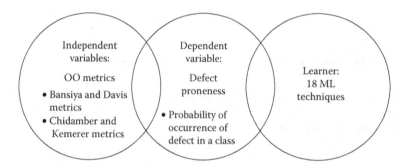

FIGURE 7.20
Elements of empirical study.

TABLE 7.26

Description of OO Metrics Used in the Study

Abb.	Metric	Definition
WMC	Weighted methods per class	Count of sum of complexities of the number of methods in a class.
NOC	Number of children	Number of subclasses of a given class.
DIT	Depth of inheritance tree	Provides the maximum steps from the root to the leaf node.
LCOM	Lack of cohesion in methods	Null pairs not having common attributes.
CBO	Coupling between objects	Number of classes to which a class is coupled.
RFC	Response for a class	Number of external and internal methods in a class.
DAM	Data access metric	Ratio of the number of private (and/or protected) attributes to the total number of attributes in a class.
MOA	Measure of aggression	Percentage of data declarations (user defined) in a class.
MFA	Method of functional abstraction	Ratio of total number of inherited methods to the number of methods in a class.
CAM	Cohesion among the methods of a class	Computes method similarity based on their signatures.
AMC	Average method complexity	Computed using McCabe's cyclomatic complexity method.
LCOM3	Lack of cohesion in methods	Revision of LCOM metric given by Henderson-Sellers
LOC	Line of code	Number of lines of source code of a given class.

(Continued)

TABLE 7.26 (*Continued*)

Description of OO Metrics Used in the Study

Abb.	Metric	Definition
NPM	Number of public methods	Number of public methods in a given class.
Ca	Afferent couplings	Number of classes calling a given class.
Ce	Efferent couplings	Number of other classes called by a class.
IC	Inheritance coupling	Number of parent classes to which a class is coupled.
Faults	Fault count	Binary variable indicating the presence or absence of the faults.

Source: S. R. Chidamber and C. F. Kemerer, *IEEE Trans. Softw. Eng.*, 20, 476–493, 1994.

namely—Ginger Bread, Ice Cream Sandwich, and Jelly Bean. The source code of these releases has been obtained from Google's Git repository (https://android.googlesource. com). The source code of Android OS is available in various application packages.

The results of the ML techniques are compared by first applying the Friedman test, followed by post hoc Wilcoxon signed-rank test if the results in the Friedman test are significant. The predicted models are validated using tenfold cross-validation. Further, the predictive capabilities of the ML techniques are evaluated using the across-release validation. To answer the research questions given below, an empirical validation is done using various techniques on the six releases of the Android OS using the following steps:

1. Preprocessing of collected data sets
2. Selection of various ML techniques for fault prediction
3. Selection of performance measures and model validation techniques (for analyzing the performance of the models developed using Android data sets
4. Selection of relevant OO metrics using correlation-based feature subselection (CFS) method
5. Model development for fault prediction using ML techniques in step 2.
6. Model validation using two validations methods: tenfold cross-validation and across-release validation
7. Testing whether the difference between the performances of ML techniques is statistically significant using Friedman test and post hoc analysis

The models are generated using all the independent variables selected using the CFS technique. The results obtained using the reduced set of variables are slightly better as compared to the results obtained using all the independent variables. Table 7.27 presents the relevant metrics found in each release of Android data set after applying the CFS technique. The results show that Ce, LOC, LCOM3, cohesion among methods (CAM), and data access metric (DAM) are the most commonly selected OO metrics over the six releases of the Android data sets.

After this, the ML techniques are empirically compared, and the results are evaluated in terms of the AUC. The AUC has been advocated as a primary indicator of comparative performance of the predicted models (Lessmann et al. 2008). The AUC measure can deal with noisy and unbalanced data and is insensitive to the changes in the class distributions (De Carvalho et al. 2008). Table 7.28 reports the tenfold cross-validation results of 18 ML techniques on six releases of Android OS. The ML technique yielding best AUC for a given

TABLE 7.27

Relevant OO Metrics

Release	Relevant Features
Android 2.3.7	Ce, LCOM3, LOC, DAM, MOA, CAM, AMC
Android 4.0.2	WMC, RFC, LCOM, LCOM3, DAM
Android 4.0.4	Ce, NPM, LOC, LCOM3, DAM, CAM
Android 4.1.2	Ce, CAM
Android 4.2.2	DAM
Android 4.3.1	Ce, LOC, DAM, MOA

TABLE 7.28

Tenfold Cross-Validation Results of 18 ML Techniques with Respect to AUC

	Android Data Set Release						
ML Tech.	**2.3.7**	**4.0.2**	**4.0.4**	**4.1.2**	**4.2.2**	**4.3.1**	**Avg.**
LR	0.81	0.66	**0.85**	0.73	0.56	0.68	0.72
NB	0.81	0.73	0.84	**0.76**	0.62	**0.80**	**0.76**
BN	0.79	0.46	0.84	0.73	0.43	0.52	0.63
MLP	0.79	0.71	**0.85**	0.71	0.61	0.76	0.74
RBF	0.77	0.76	0.80	0.74	**0.76**	0.74	**0.77**
SVM	0.64	0.50	0.76	0.50	0.50	0.50	0.57
VP	0.66	0.59	0.67	0.56	0.50	0.50	0.58
CART	0.77	0.45	0.75	0.74	0.43	0.45	0.60
J48	0.71	0.48	0.78	0.67	0.43	0.52	0.60
ADT	0.81	0.72	0.83	0.72	0.62	0.74	0.74
Bag	0.81	0.68	0.84	0.74	0.68	0.72	0.75
RF	0.79	0.65	0.82	0.67	0.70	0.73	0.73
LMT	0.77	0.66	0.83	0.75	0.56	0.73	0.72
LB	**0.83**	**0.75**	0.82	0.71	0.70	0.65	0.75
AB	0.81	0.70	0.81	0.69	0.70	0.65	0.73
NNge	0.69	0.53	0.75	0.66	0.65	0.51	0.64
DTNB	0.76	0.46	0.81	0.71	0.43	0.68	0.65
VFI	0.77	0.72	0.70	0.62	0.75	0.74	0.72

release is depicted in bold. The results show that the model predicted using the NB, AB, RBF, Bag, ADT, MLP, LB, and RF techniques have AUC greater than 0.7 corresponding to most of the releases of the Android data set.

RQ1: What is the overall predictive capability of various ML techniques on Android "MMS" data set?

A1: The AUC of most of the models predicted using the ML techniques is 0.7, which highlights the predictive capability of the ML techniques.

To confirm that the performance difference among the ML models is not random, Friedman test is used to evaluate the superiority of one ML technique over the other

TABLE 7.29

Friedman Test Results

ML Tech.	Mean Rank	ML Tech.	Mean Rank
NB	3.58	RF	8.58
Bag	5.67	VFI	9.08
RBF	5.67	BN	10.83
ADT	5.92	DTNB	12.83
LB	7.17	NNge	13.58
MLP	7.25	CART	13.92
LR	7.08	J48	14.42
LMT	7.92	SVM	15.67
AB	8.17	VP	15.67

ML techniques. The Friedman test resulted in significant value of zero. The results are significant at the 0.05 level of significance over 17 degrees of freedom. Thus, the null hypothesis that all the ML techniques have similar performance in terms of AUC is rejected. The results given in Table 7.29 show that NB technique is the best for predicting fault proneness of a class using OO metrics. The result supports the finding of Menzies et al. (2007) that NB is the best technique for building fault prediction models. It can be also seen that the models predicted using SVM-based techniques, SVM and VP, performed worst.

RQ2: Which is the best ML technique for fault prediction using OO metrics?

A2: The outcome of the Friedman test indicates that the performance of the NB technique for fault prediction is the best. The performance of the Bagging and RBF techniques for fault prediction are the second best among the 18 ML techniques that were compared.

After obtaining significant results using the Friedman test, post hoc analysis was performed using the Wilcoxon test. The Wilcoxon test is used to examine the statistical difference between the pairs of different ML techniques (see Section 6.4.10). The results of the pairwise comparisons of the ML techniques are shown in Table 7.30.

The results of Wilcoxon test show that out of the 18 ML techniques, the NB model is significantly better than the models predicted using 17 ML techniques such as LMT, BN, DTNB, NNge, CART, J48, SVM, and VP. Similarly, the VP model is significantly worse than models developed using NB, Bag, RBF, ADT, LB, MLP, LR, LMT, AB, RF, and VFI techniques, worse than the BN, DTNB, NNge, CART, and J48 techniques, and better than the SVM model.

Figure 7.21 shows the number of ML techniques from which the performance of a given ML technique is either superior, significantly superior, inferior, or significantly inferior. For example, from the bar chart shown in Figure 7.21, it can be seen that the performance of the NB technique is significantly superior to eight other ML techniques and nonsignificantly superior to nine other techniques. Similarly, the performance of the Bagging technique is significantly superior to seven other ML techniques, nonsignificantly superior to eight other techniques, and nonsignificantly inferior to two other ML techniques.

TABLE 7.30

Wilcoxon Test Results

	NB	Bag	RBF	ADT	LB	MLP	LR	LMT	AB	RF	VFI	BN	DTNB	NNge	CRT	J48	SVM	VP
NB		↑	↑	↑	↑	↑	↑	↑	↑	↑	↑	↑	↑	↑	↑	↑	↑	↑
Bag	↓		↑	↑	↑	↑	↑	↑	↑	↑	↑	↑	↑	↑	↑	↑	↑	↑
RBF	↓	↓		↑	↑	↑	↑	↑	↑	↑	↑	↑	↑	↑	↑	↑	↑	↑
ADT	↓	↓	↓		↑	↑	↑	↑	↑	↑	↑	↑	↑	↑	↑	↑	↑	↑
LB	↓	↓	↓	↓		↑	↑	↑	↑	↑	↑	↑	↑	↑	↑	↑	↑	↑
MLP	↓	↓	↓	↓	↓		↑	↑	↑	↑	↑	↑	↑	↑	↑	↑	↑	↑
LR	↓	↓	↓	↓	↓	↓		↑	↑	↑	↑	↑	↑	↑	↑	↑	↑	↑
LMT	↓	↓	↓	↓	↓	↓	↓		↑	↑	↑	↑	↑	↑	↑	↑	↑	↑
AB	↓	↓	↓	↓	↓	↓	↓	↓		↑	↑	↑	↑	↑	↑	↑	↑	↑
RF	↓	↓	↓	↓	↓	↓	↓	↓	↓		↑	↑	↑	↑	↑	↑	↑	↑
VFI	↓	↓	↓	↓	↓	↓	↓	↓	↓	↓		↑	↑	↑	↑	↑	↑	↑
BN	↓	↓	↓	↓	↓	↓	↓	↓	↓	↓	↓		↑	↑	↑	↑	↑	↑
DTNB	↓	↓	↓	↓	↓	↓	↓	↓	↓	↓	↓	↓		↑	↑	↑	↑	↑
NNge	↓	↓	↓	↓	↓	↓	↓	↓	↓	↓	↓	↓	↓		↑	↑	↑	↑
CART	↓	↓	↓	↓	↓	↓	↓	↓	↓	↓	↓	↓	↓	↓	=	=	↑	↑
J48	↓	↓	↓	↓	↓	↓	↓	↓	↓	↓	↓	↓	↓	↓	=		↑	↑
SVM	↓	↓	↓	↓	↓	↓	↓	↓	↓	↓	↓	↓	↓	↓	↓	↓		↑
VP	↓	↓	↓	↓	↓	↓	↓	↓	↓	↓	↓	↓	↓	↓	↓	↓	↓	

↑ implies performance of ML technique is significantly better than the compared ML technique, ↑ implies performance of ML technique is better than the compared ML technique, ↓ implies performance of ML technique is significantly worse than the compared ML technique, ↓ implies performance of ML technique is worse than the compared ML technique, = implies performance is equal.

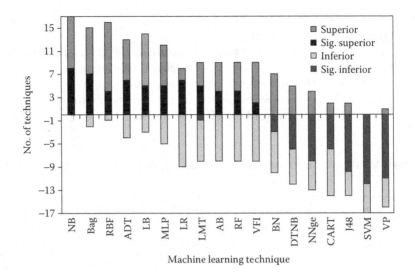

FIGURE 7.21
Results of Wilcoxon test.

The AUC values of the NB model are between 0.73 and 0.85 in five releases of the Android data sets. The results in this study confirm the previous findings that the NB technique is effective in fault prediction and may be used by researchers and practitioners in future applications. The NB technique is based on the assumption that the attributes are independent and unrelated. One of the reasons that the NB technique showed the best performance is that the features are reduced using the CFS method before applying the model prediction techniques in this work. The CFS method removes the features that are correlated with each other and retains the features that are correlated with the dependent variable. Hence, OO metrics selected by the CFS method for each data set are less correlated with each other and more correlated with the fault variable. The NB technique is easy to understand and interpret (linear model can be obtained as a sum of logs) and is also computationally efficient (Friedman 1940; Zhou and Leung 2006). The NB technique is not able to retain the results in one release of the Android data set (Android 4.2.2). This may be because of the reason that the NB technique is not able to make accurate predictions of faults on the basis of only one OO metric (DAM).

RQ3: Which pairs of ML techniques are significantly different from each other for fault prediction?

A3: There are 112 pairs of ML techniques that yield significantly different performance results in terms of AUC. The results show that the performance of the NB model is significantly better than BN, LMT, BN, DTNB, NNge, CRT, J48, SVM, and VP. Similarly, significant pairs of performance of the other ML techniques are given in Table 7.30.

To evaluate the accuracy of the predicted models, across-release cross-validation is also performed. The performance of the model derived from each release on the immediate subsequent release is validated. For example, the model trained using Android 2.3.7 data set is validated on the Android 4.0.2 data set, and so on. The results of the across-release cross-validation in terms of AUC are shown in Table 7.31. The results of across-release validation show that the AUC of NB, RBF, ADT, Bagging, LMT, and AB are greater than 0.7 in most of the releases of Android.

Figure 7.22 depicts the comparison of overall results of 18 ML techniques in terms of the average AUC using both tenfold and across-release validation over all the Android releases. The chart shows that the overall performance results obtained from the across-release validation are better or comparable than the results obtained from the tenfold cross-validation, except when Android 4.0.4 is validated using Android 4.0.2. One possible

TABLE 7.31

Across-Release Validation Results of 18 ML Techniques with Respect to AUC

ML Tech.	Android					
	2.3.7 on 4.0.2	4.0.2 on 4.0.4	4.0.4 on 4.1.2	4.1.2 on 4.2.2	4.2.2 on 4.3.1	Avg.
LR	0.81	0.80	0.80	0.66	0.58	0.73
NB	0.82	0.80	0.79	0.68	0.70	0.76
BN	0.85	0.50	0.79	0.63	0.50	0.66
MLP	0.84	0.82	0.81	0.66	0.60	0.75
RBF	0.82	0.76	0.78	0.72	0.80	0.78
SVM	0.68	0.50	0.71	0.50	0.50	0.58
VP	0.72	0.50	0.58	0.50	0.50	0.56
CART	0.80	0.50	0.70	0.63	0.50	0.63
J48	0.84	0.50	0.77	0.63	0.50	0.65
ADT	0.83	0.69	0.81	0.74	0.74	0.77
Bag	0.85	0.73	0.79	0.72	0.81	0.78
RF	0.84	0.57	0.80	0.71	0.65	0.72
LMT	0.81	0.77	0.80	0.74	0.58	0.74
LB	0.85	0.69	0.81	0.69	0.80	0.77
AB	0.82	0.78	0.78	0.66	0.80	0.77
NNge	0.78	0.54	0.71	0.65	0.56	0.65
DTNB	0.80	0.51	0.77	0.63	0.50	0.65
VFI	0.85	0.79	0.73	0.59	0.78	0.75

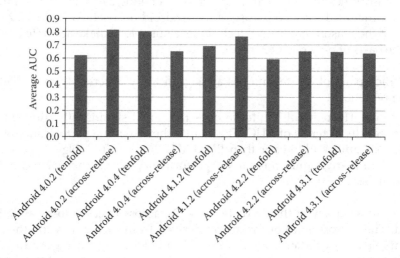

FIGURE 7.22

Comparison between AUC results of tenfold and across-release validation for five releases of android data set.

explanation to this is that the values of OO metrics in the Android releases are informative enough to predict faults in the subsequent releases. The reason for the low AUC values for across-release validation as compared to the AUC values for tenfold cross-validation in case of Android 4.0.4 could be that the faulty class percentage in Android 4.0.2 is very less (5.47%) as compared to the faulty class percentage in Android 4.0.4 (33.01%).

RQ4: What is the performance of ML techniques when across-release validation is used for predicting postrelease faults?

A4: The performance of the ML techniques when across-release validation is used is comparable (and even better) to the performance of the ML techniques when tenfold cross-validation is used for fault prediction.

Exercises

7.1 Briefly outline the steps of model prediction.

7.2 What is multicollinearity? How can it be removed?

7.3 What is ML? Define various categories of ML technique?

7.4 Discuss the guidelines for selecting ML techniques.

7.5 It is difficult to assess the accuracy of a model where most of the outcomes are negatives. In such cases, what criteria will you use to determine the accuracy of the model?

7.6 Consider two models predicted using tenfold cross-validation. The error rate produced by model1 is 32, 15, 14, 20, 35, 45, 48, 52, 27, and 29, and model2 is 20, 14, 10, 8, 15, 20, 25, 17, 19, and 7. We want to determine which model performance is significantly better than the other at 0.01 significance level. Apply appropriate statistical test and provide interpretation of the results.

7.7 How can bias and variance be reduced for a given model?

7.8 What is the difference between underfitting and overfitting?

7.9 Which measures are useful in predicting model performance when data is imbalanced?

7.10 How will a researcher decide on the selection of learning technique?

7.11 Consider the model with following predicted values. Given the actual values, comment on the performance of the model.

Actual	Predicted	Actual	Predicted
0	0.34	1	0.34
1	0.78	1	0.82
0	0.23	0	0.21
0	0.46	1	0.56
0	0.52	0	0.61
1	0.86	0	0.21
1	0.92	1	0.76
1	0.68	1	0.56
0	0.87	0	0.10

7.12 How can the results of hypothesis testing be interpreted?

7.13 What is the purpose of confusion matrix? How can confusion matrix for a three-class problem derived?

7.14 Explain steps in search-based techniques. What are the advantages and disadvantages of these techniques?

7.15 Define multivariate analysis. Multivariate analysis plays a vital role in software engineering research. Justify.

7.16 List the steps involved in conducting research by applying search-based techniques in software quality prediction. Take your research problem for illustration.

7.17 What is the significance of constructing ROC curves? What is the use of area under the curve metric?

7.18 Explain the *K*-fold , hold-out and leave-one-out cross-validation methods.

Further Readings

The statistical methods and concepts are effectively addressed in:

G. K. Bhattacharyya, and R. A. Johnson, *Statistical Methods and Concepts*, John Wiley & Sons, New York, 1977.

This book emphasizes problem-solving strategies that address the many issues arising when developing multivariable models using real data with examples:

F. H. Harrell, *Regression Modeling Strategies: With Applications to Linear Models, Logistic Regression, and Survival Analysis*, Springer Series in Statistics, Springer, Berlin, Germany, 2010.

Mertler and Vannatta provide description on specific methods on advanced multivariate statistics:

C. A. Mertler, and R. A. Vannatta, *Advanced and Multivariate Statistical Methods*, Pyrczak, Los Angeles, CA, 2002.

Marsland presents the algorithms of ML techniques.

S. Marsland, *Machine Learning: An Algorithmic Perspective*, 2nd Edition, Chapman & Hall, Boca Raton, FL, 2014.

The following book helps to explore the benefits in data mining that DTs offer:

L. Rokach, and O. Maimon, *Data Mining with Decision Trees: Theory and Applications*, Series in Machine Perception and Artificial Intelligence, World Scientific, Singapore, 2007.

The variations of DT techniques are proposed in:

Y. Freund, and L. Mason, "The alternating decision tree learning algorithm," *ICML*, vol. 99, pp. 124–133, 1999.

D. A. Hill, L. M. Delaney, and S. Roncal, "A chi-square automatic interaction detection (CHAID) analysis of factors determining trauma outcomes," *Journal of Trauma and Acute Care Surgery*, vol. 42, no. 1, pp. 62–66, 1997.

G. Holmes, B. Pfahringer, R. Kirkby, E. Frank, and M. Hall (eds.), "Multiclass alternating decision trees," In: *Machine Learning: ECML*, Springer, Berlin, Germany, pp. 161–172, 2002.

J. R. Quinlan, *C4.5: Programs for Machine Learning*, Morgan Kaufmann, San Mateo, CA, 1993.

The detail about the classification and regression trees is presented in:

L. Breiman, J. H., Friedman, R. A. Olshen, and C. J. Stone, *Classification and Regression Trees*, Wadsworth, Belmont, CA, 1984.

N. Speybroeck, "Classification and regression trees," *International Journal of Public Health*, vol. 57, no. 1, pp. 243–246, 2012.

For a detailed account of the statistics needed for model prediction using LR (notably how to compute maximum likelihood estimates, R^2, significance values), see the following text book and research paper:

V. Basili, L. Briand, and W. Melo, "A validation of object-oriented design metrics as quality indicators," *IEEE Transactions on Software Engineering*, vol. 22, no. 10, pp. 751–761, 1996.

D. Hosmer, and S. Lemeshow, *Applied Logistic Regression*, John Wiley & Sons, New York, 1989.

The following publications present a detailed analysis on Bayesian learners:

D. Heckerman, D. Geiger, and D. M. Chickering, "Learning Bayesian networks: The combination of knowledge and statistical data," *Machine Learning*, vol. 20, no. 3, pp. 197–243, 1995.

D. Heckerman, *A Tutorial on Learning with Bayesian Networks*, Springer, Berlin, Germany, 1998.

E. Keogh, and M. Pazzani, "Learning augmented Bayesian classifiers: A comparison of distribution-based and classification-based approaches," *Proceedings of the 7th International Workshop on Artificial Intelligence and Statistics*, FL, pp. 225–230, 1999.

X. Zhu, "Machine teaching for Bayesian learners in the exponential family," *Advances in Neural Information Processing Systems*, vol. 26, pp. 1905–1913, 2013.

The naïve Bayes classifier is proposed in:

G. Ridgeway, D. Madigan, T. Richardson, and J. O'Kane, "Interpretable Boosted Naïve Bayes Classification," *Proceedings of 9th International Conference on Knowledge Discovery and Data Mining*, pp. 101–104, 1998.

This research paper by Webb and Zheng presents the framework for multistrategy ensemble learning techniques:

G. I. Webb, and Z. Zheng, "Multistrategy ensemble learning: Reducing error by combining ensemble learning techniques," *IEEE Transactions on Knowledge and Data Engineering*, vol. 16, no. 8, pp. 980–991, 2004.

The Logitboost classifier is proposed in:

S. B. Kotsiantis, and P. E. Pintelas, "Logitboost of simple Bayesian classifier," *Informatica (Slovenia)*, vol. 29, no. 1, pp. 53–60, 2005.

The concept of AdaBoost classifier is presented in the following papers:

C. Domingo, and O. Watanabe, "MadaBoost: A modification of AdaBoost," *Proceedings of 13th Annual Conference on Computation Learning Theory*, San Francisco, pp. 180–189. 2000.

W. Fan, S. J. Stolfo, and J. Zhang, "The application of AdaBoost for distributed, scalable and on-line learning," *Proceedings of the 5th ACM SIGKDD International Conference on Knowledge Discovery and Data Mining*, pp. 362–366, ACM, New York, 1999.

G. Rätsch, T. Onoda, and K. R. Müller, "Regularizing adaboost," *Advances in Neural Processing Systems*, vol. 11, pp. 564–570, 1999.

The concept of random forest classifier is proposed in:

L. Breiman, "Random forests," *Machine Learning*, vol. 45, no. 1, pp. 5–32, 2001.

In this book, the authors explain the basic concepts of NN and then show how these models can be applied to applications:

J. A. Freeman, and D. M. Skapura, *Neural Network Algorithms, Applications, and Programming Techniques*, Addison-Wesley, Reading, MA, 1991.

The concept of multilayer perceptron is proposed in:

J. G. Attali, and G. Pagès, "Approximations of functions by a multilayer perceptron: a new approach," *Neural Networks*, vol. 10, no. 6, pp. 1069–1081, 1997.

L. M. Belue, and W. B. Kenneth, "Determining input features for multilayer perceptron," *Neuro Computing*, vol. 7, no. 2 pp. 111–121, 1995.

The concept of radial basis function is proposed in:

M. J. D. Powell, "The theory of radial basis function approximation in 1990," Department of Applied Mathematics and Theoretical Physics, University of Cambridge, Cambridge, 1990.

The concepts of probabilistic NNs are present in:

> D. F. Specht, "Probabilistic neural networks," *Neural Networks*, vol. 3, no. 1, pp. 109–118, 1990.

The overview of convolutional NNs can be obtained from:

> L. O. Chua, *Convolutional Neural Networks*, World Scientific, Singapore, 1998.

The authors present the basic ideas of SVM together with the latest developments and current research questions in:

> I. Steinwart, and A. Christmann, *Support Vector Machines*, Springer Science & Business Media, New York, 2008.

The rule-based approach is discussed in:

> T. J. Kowalski, and L. S. Levy, *Rule-Based Programming*, Softcover reprint of the original 1st edition, Springer, New York, 2011.

The concept of NNge is proposed in:

> M. Bouten, and C. van-den B, "Nearest-neighbour classifier for the perceptron," *EPL (Europhysics Letters)*, vol. 26, no. 1, pp. 69, 1994.

The concept of search-based techniques along with its practical applications and guidelines in software engineering are effectively presented by Harman et al. in:

> M. Harman, "The relationship between search based software engineering and predictive modeling," *Proceeding of 6th International Conference on Predictive Models in Software Engineering*, Timisoara, Romania, 2010.
>
> M. Harman, "The current state and future of search based software engineering," In: *Proceedings of Future of Software Engineering*, Washington, pp. 342–357, 2007.
>
> M. Harman, S. A. Mansouri, and Y. Zhang, "Search based software engineering: A comprehensive analysis and review of trends techniques and applications," Technical Report TR-09-03, Department of Computer Science, King's College London, London, 2009.
>
> M. Harman, P. McMinn, J. T. D. Souza, and S. Yoo, "Search based software engineering: Techniques, taxonomy, tutorial," In: *Empirical Software Engineering and Verification*, Springer, Berlin, Germany, pp. 1–59, 2012.
>
> M. Harman, and P. McMinn, "A theoretical and empirical study of search-based testing: Local, global, and hybrid search," *IEEE Transactions on Software Engineering*, vol. 36, no. 2, pp. 226–247, 2010.

The concept of voting feature intervals is summarized in:

> G. Demiröz, and H. A. Güvenir, "Classification by voting feature intervals," In: *Machine Learning: ECML-97*, Springer, Berlin, Germany, pp. 85–92, 1997.

Y. Li, "Selective voting for perceptron-like online learning," In: *Proceedings of the 17th International Conference on Machine Learning*, San Francisco, pp. 559–566, 2000.

The concept of AIS is proposed in:

J. E. Hunt, and D. E. Cooke, "Learning using an artificial immune system," *Journal of Network and Computer Applications*, vol. 19, no. 2, pp. 189–212, 1996.

The basic concept of *K*-nearest neighbor is presented in:

L. E. Peterson, "*K*-nearest neighbor," *Scholarpedia*, vol. 4, no. 2, 2009. http://www.scholarpedia.org/artice/k-nearest-neighbor.

The knowledge classification rules are presented in:

A. D. Pila, R. Giusti, R. C. Prati, and M. C. Monard, "A multi-objective evolutionary algorithm to build knowledge classification rules with specific properties," *Proceedings of 6th International Conference on Hybrid Intelligent Systems*, Rio de Janeiro, Brazil, pp. 41–45, 2006.

The information about genetic programming can be obtained from:

J. R. Koza, *Genetic Programming: On the Programming of Computers by Means of Natural Selection*, MIT Press, Cambridge, MA, vol. 1, 1992.

J. R. Koza, and J. P. Rice, *Genetic Programming II: Automatic Discovery of Reusable Programs*, MIT Press, Cambridge, MA, vol. 40, 1994.

This research paper provides knowledge about detecting multicollinearity and also discusses how to deal with the associated problems:

R. F. Gunst, and J. T. Webster, "Regression analysis and problems of multicollinearity," *Communications in Statistics-Theory and Methods*, vol. 4, no. 3, pp. 277–292, 1975.

The following research paper introduces and explains methods for comparing the performance of classification algorithms:

A. D. Forbes, "Classification-algorithm evaluation: Five performance measures based on confusion matrices," *Journal of Clinical Monitoring*, vol. 11, no. 3, pp. 189–206, 1995.

Yang addresses general issues in categorical data analysis with some advanced methods in:

K. Yang, *Categorical Data Analysis*, SAGE Benchmarks in Social Research Methods, Sage, United Kingdom, 2014.

Davis and Goadrich presents the deep connection between ROC space and PR space in the following paper:

J. Davis, and M. Goadrich, "The relationship between Precision-Recall and ROC curves," *Proceedings of the 23rd International Conference on Machine Learning*, pp. 233–240, ACM, New York, 2006.

This research paper presents the most powerful nonparametric statistical tests to carry out multiple comparisons using accuracy and interpretability:

S. García, A. Fernández, J. Luengo, and F. Herrera, "A study of statistical techniques and performance measures for genetics-based machine learning: accuracy and interpretability," *Soft Computing*, vol. 13, no. 10, pp. 959–977, 2009.

The concept of kappa coefficient is efficiently addressed in:

R. Hernández-Nieto, *Contributions to Statistical Analysis: The Coefficients of Proportional Variance, Content Validity and Kappa*, Universidade de Los Andes, Mérida, Venezuela, 2002.

This research paper serves as an introduction to ROC graphs and as a guide for using them in research:

T. Fawcett, "An introduction to ROC analysis," *Pattern Recognition Letters*, vol. 27, no. 8, pp. 861–874, 2006.

The root-mean-square error (RMSE) and the MAE to describe average model performance error are examined in:

C. J. Willmott, and K. Matsuura, "Advantages of the mean absolute error (MAE) over the root mean square error (RMSE) in assessing average model performance," *Climate Research*, vol. 30, no. 1, p. 79, 2005.

Browne provides a good introduction on the cross-validation methods in:

M. W. Browne, "Cross-validation methods," *Journal of Mathematical Psychology*, vol. 44, no. 1, pp. 108–132, 2000.

8

Validity Threats

The validity of the results is an important concern for any empirical study. The results of any empirical research must be valid for the population from which the samples are drawn. The samples are derived and generalized to the population that the researcher decides. Threats reduce the applicability of the research results. Hence, the researcher must identify the extent of validity of results in design stage and must also provide a list of threats to the validity of results after the results have been analyzed. This will provide the readers complete information about the limitations of the study.

In this chapter, we present categories of threat to validity, explain the threats with examples, and also list various threats identified from fault prediction studies. We also provide possible mitigation of these threats.

8.1 Categories of Threats to Validity

According to Campbell and Stanley (1963), the threats to validity of an experimental design can be broadly classified as internal or external. The internal validity concerns are related to all the issues that could introduce errors in research design and may threaten the conclusions of the study. Internal validity issues would explore the confidence in the conclusions of the study or may assess extraneous variables that could be responsible for the relationship between dependent and independent variables. However, the issues of external validity are related to the "generalization" and "representation" of the subjects and results of the study. To achieve high external validity, the researcher should make sure that the subjects and results of the study should be accurately represented and the results should also generalize to the subjects, which may not be investigated in the study. Cook and Campbell (1979) revised the categories of threats and extended the list to four types of threats namely: conclusion, internal, construct, and external. As shown in Figure 8.1, these threat categories can be considered as subcategories of internal and external validity threats and are described in Sections 8.1.1 through 8.1.4 with instances. It may be noted that a threat may belong to more than one category.

8.1.1 Conclusion Validity

Conclusion validity takes into account all the concerns that could affect the capability of concluding an accurate and legitimate association between the treatment (independent variables) provided by the experiment and the outcome (dependent variable) generated by it. All those concerns, which relate to the validity of the conclusion, should be verified to demonstrate this validity. For example, if the researcher has demonstrated the relationship between the dependent and the independent variable but has not assessed this relationship statistically with a specific confidence, then the threat to conclusion validity exists in the study. Similarly, if a researcher is statistically exploring the relationship of the independent and dependent variable, he should be very sure that the conditions of

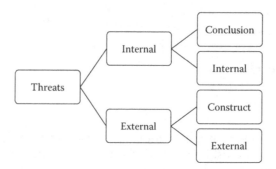

FIGURE 8.1
Categorization of threats.

the chosen statistical test (like the size of the sample, normality of the data, etc.) should be fulfilled. Researchers should apply nonparametric tests in cases where they are not completely sure that their data fulfills all the conditions of a parametric statistical test to eliminate conclusion validity. It may be noted that some researchers address this threat as "statistical conclusion validity." The various possible threats to conclusion validity are as follows (Cook and Campbell 1979; Wohlin et al. 2012):

- Inaccurate data: If data consists of erroneous observations, outliers, or noise, then it may lead to incorrect conclusions.

- Assumptions of statistical tests not satisfied: The statistical tests (specifically parametric tests) are based on primary assumptions, for example, t-test is based on the assumption that the samples should be normally distributed. These assumptions are required to be fulfilled before applying the test, and violating them may produce incorrect conclusions (see Section 6.4 for details on statistical tests).

- Lack of hypothesis formulation and analysis: If the researcher does not form appropriate hypothesis to analyze the research questions of the study and does not statistically analyze the results, the conclusions will not be valid.

- Biased results: If a researcher is looking for a particular outcome and influences the results to obtain that outcome it will produce biased results, as the results are no longer independent. Sometimes the researcher may intentionally or unintentionally produce results that satisfy the established research hypothesis.

- Low statistical test ability: The ability of statistical tests to reveal the pattern of underlying data is low because of inappropriate selection of significance level, which could lead to erroneous judgment.

- Validation bias: The results of the predicted models should be validated on data sets that are different than from which they are derived. The researcher may use cross-validation methods described in Section 7.7 to reduce this threat. The researcher may include multiple runs to validate the results.

- Inadequate number of samples: If the sample size is inadequate or very less, then the validity of the results is not assured.

- Inappropriate use of performance measures: The imbalanced nature of dependent variable requires the appropriate use of performance measures for evaluating the predictive ability of developed models.

- Use of immature subjects: If the experimental data is collected from groups that are not true representatives of industrial settings. For example, if the results are based on the samples collected from software developed by undergraduate students, then this may pose a serious threat to conclusion validity.
- Reliability of techniques applied: The settings of the algorithms or techniques should be standard and not be overtuned according to the data as this may produce overfitted results.
- Heterogeneity of validation samples or case studies: The samples should not be heterogeneous as then the variation in the results will be more influenced by the environment and nature of the samples rather than the techniques applied. However, this will pose a threat to generalizability and hence decrease external validity of results.
- Lack of expert evaluation: The conclusions or results should be evaluated by an expert to understand and interpret their true meaning and significance. Lack of expert judgment may lead to erroneous conclusions.
- Variety of data preprocessing or engineering activities not taken into account: An experiment involves a wide range of data preprocessing or other activities such as scaling, discretization, and so on. These activities can significantly influence the results if not properly taken into account.

8.1.2 Internal Validity

Internal validity is also known as causal validity, that is, showing that the changes made in the independent variable A cause changes in the dependent variable B. The researcher can conclude that variable A causes changes in variable B, if the following conditions hold:

1. Direction of relationship is known.
2. Variable A is related to variable B.
3. The relationship between variable A and variable B is not caused by some extraneous or "other" variable. For example, there is a correlation between coupling and defects. Size is related to both coupling and defects. Hence, the researchers should control for "size" to determine whether the relationship between coupling and defects hold.

Internal validity is the degree to which we can strongly conclude that the causes/changes in dependent variable B are because of only the independent variable A.

Internal validity concerns itself with all the possible factors except the independent variables of the study, which can cause the observed outcome (Neto and Conte 2013). Apart from the independent variables of a study, there could be other ("confounding") factors, which cannot be controlled by the researcher. Such extraneous variables are confounding variables, which may be correlated with the independent and/or the dependent variable. Therefore, such variables pose a threat to the results of the study as they could be responsible for the "causal" effect of the independent variables on the dependent variable (Wohlin et al. 2012). For example, a researcher who would like to investigate the relationship between object-oriented (OO) metrics and the probability that a particular class would be faulty or not should consider the confounding effect of class size on these associations. This is important, as a researcher cannot control the size of a class. Moreover, class size is correlated with the probability of faults, as larger classes tend to have more number of faults because of their size. Thus, size may affect the actual relationship between other OO metrics like coupling, cohesion, and so on and could be responsible for their "causal" relationship with the fault-prone nature of a class. Internal validity thus accounts for factors that are not controlled by the researcher and may falsely contribute to a relationship. The threats related to experimental settings also are part of the internal validity. The various possible internal validity threats that commonly occur in empirical studies are as follows:

- Confounding effects of variables: If the relationship between an independent variable and a dependent variable is affected/influenced by some other variable without the researcher's knowledge, then this may pose a threat to internal validity.
- Response of samples for a given technique: If the experiment is repeated, the response may be different at different times. For example, genetic algorithms may produce different results each time they are applied for predicting defects.
- Influence of human factors: If because of their own preferences or lack of capability of applying a given technique, the software group does not adapt to new techniques. For example, the software group may not prefer to adapt to inspection method of verification and would rather want to use walkthroughs as the preferred method of verification technique because of the more familiarity with this technique. Also, a programmer's capability, domain knowledge, and so on are certain human factors that can significantly influence the results and may have an effect on the relationship of dependent and independent variable.
- Use of poorly designed experimental artifacts: This threat is caused because of badly designed documents produced at various phases of software development. For example, in conducting systematic review, if the data extraction forms are not properly designed then the study may be affected negatively. Similarly, if a survey form does not contain clear queries, then the response given by the respondents may not be accurate.
- Selection of samples from different groups: The samples must be collected from different participants in order to reduce internal validity threat.
- Nondetermination of direction of relationship among variables: This threat is caused because of nondetermination of direction of correlation between two variables. For example, if there is a positive correlation between complexity and defect proneness, the question is that whether high complexity causes higher defect proneness in a class or whether high defect proneness causes higher complexity in a class.
- Ignoring relevant factors in experimental settings: If important factors are over looked in an empirical study, then the threat to internal validity may increase. For example, severity level of defects may not be taken into account in a study with intent to predict defect proneness in a module.
- Inappropriate experimental settings: The experimental settings may be improper while performing an experiment, which could lead to result bias.

Proper experimental design can lead to reduction in internal validity threats.

8.1.3 Construct Validity

Construct validity concerns itself with the gap, if any, between the theoretical concepts and the actual representation of the concepts (Barros and Neto 2011). It can be verified if the variables collected in an empirical study are correct and convey the same concept they intend to measure. Hence, this validity poses a threat if the researcher has not accurately and correctly represented the variables (independent and dependent) of the study. For example, the coupling attribute of a class (theoretical concept) in an OO software may be represented by a measure that correctly and precisely counts the number of other classes to which a particular class is interrelated. Similarly, a researcher who wishes to investigate the relationship between OO metrics and fault-prone nature of a class should primarily evaluate that all the selected metrics of the study are correct and effective indicators of concepts like coupling, cohesion, and so on. The bugs collected from a bug repository can represent the theoretical concept "fault." However, this bug data should be collected carefully and exhaustively with correct mapping to remove any unbiased representation of the faults in the classes. If the bugs are not properly collected, it may lead to an incorrect dependent variable. Thus, both the independent and the dependent variables should be carefully verified for use in experiments to eliminate the threats to construct validity. The various possible threats to construct validity are as follows:

- Misinterpretation of concepts and measures: If the concepts are misunderstood or are unclear, then it may lead to incorrect measurements. For example, if the basic concept of coupling is not well understood, then the metric that captures coupling may be inaccurate or incorrect.
- Reliability of measurement tools: The tools used to collect measures may be incorrect.
- Improper data-collection methods: This threat occurs if the data-collection methods are inappropriate. For example, where only fixed defects were to be taken into account, unfixed defects are also related to classes.
- Measurement bias: The classes of the variable may be subjective or based on human judgment. For example, fault severity is classified as high, medium, and low, and this involves subjective classification; hence, the experiment may produce biased results.
- Intentional misrepresentation of measures: The software professional may try to hide facts because of his/her personal benefits. For example, the software developer may not want to reveal actual number of defects encountered in the module developed by him/her.
- Unaccountability of related constructs: For example, if because of use of a new technique A, although maintainability increases, however, the testing effort may decrease. But as testing effort was not taken into account this important attribute is ignored.
- Guessing hypothesis: The people involved in the empirical study might try to prove the hypothesis and base their behavior on the hypothesis formed.
- Errors while combining data: During the process of data collection, there could be various errors. For example, there may be errors while collecting fault data such as erroneous mapping of faults to classes.

8.1.4 External Validity

A practitioner or researcher may wonder whether the results of the empirical study will be applicable to software systems with different purpose, programming language, or size? External validity concerns itself with the generalization of the results obtained by a study to the conditions and scenarios not accounted for in the study. Hence, it deals with effectiveness of results in those situations that are different from the subjects and settings of the study. For example, a study which evaluates the effectiveness of machine learning methods to identify fault-prone classes using data mined from open source software would have external validity concerns regarding whether the results computed using machine learning algorithms on a given data set hold true for industrial software data sets or for other open-source data sets. A study with high external validity is favorable as its conclusions can be broadly applied in different scenarios, which are valid across the study domain (Wright et al. 2010). Thus, the results obtained from a study with more number of data sets of different size and nature and recomputation of results using varied algorithms will help in establishing well-formed theories and generalized results. Such results will form widely acceptable and well-formed conclusions. The various possible threats to external validity are as follows:

- Inappropriate selection of subjects: The subjects may be incorrectly selected. For example, software programmers are given the questionnaire, where software testers could have more appropriately answered the questions in the questionnaire.

- Applicability of results across languages: The results of the study may not be generalized to the samples collected from software developed with different programming languages.

- Inadequate size and number of samples: The results of the study may not be generalized if they are evaluated on low number of data sets. The number of evaluated data sets should be high to increase the external validity. Moreover, the size of the evaluated software systems should be appropriate to allow generalizability of results across various industrial software data sets.

- Applicability of results across different variables: The results of the study may not be applicable with different and related outcome and independent variables.

- Applicability of results across different samples: The results obtained from a software developed in specific environment, with specific purpose, size, and other characteristics may not be applicable to the software developed in a different environment with dissimilar purpose, size, and other characteristics.

- Applicability of results across different environment: The results based on samples collected from open source software may not be generalizable to software developed in industrial environment or vice versa.

- Applicability of results when technique is varied: If the technique is slightly varied, will the obtained results be similar? This is important, as it is unlikely that the researchers apply the technique exactly as was applied in the original study.

- Results bias because of techniques or subjective classification of a variable: If there is result biasness because of use of specific technique, then the results produced may not be generalized. For example, if random forest technique is randomly selected, then it may not produce best results as compared to some other carefully selected technique. Similarly subjective classification of a variable such as severity of faults may result in specific and biased results.

- Nonspecification of experimental setting and relevant details: If the experimental setting or other important details are not clearly stated, then this may pose a threat to repeatability and replicability of the study.

- Data set not representative of industrial settings: If the software from which the data set is collected does not represent true industrial practices, this is a threat to external validity as the results cannot be generalized in real-world scenarios. For example, the software may be developed in the academic environment but the principles, standards, and practices for development should match the industrial practices. If this is not the case, it poses a serious threat to external validity of results.

The threats to external validity can be reduced by clearly describing the experimental settings and techniques used. The study must be carried out with the intent to enable researchers to repeat and replicate the study being carried out.

8.1.5 Essential Validity Threats

The threats to validity exhibit the practical importance of the produced results. The internal threats are the strongest form of validity threats as high internal validity proves that strong evidence regarding the causal relationship between variables is present. In software engineering-based empirical studies, the main aim is to show the generalizability of the results. Rather than showing a result based on data collected from software company X, it is more important to show that the results can be applied in practice to which software companies given the size and domain. The construct validity is on third priority followed by the conclusion validity.

8.2 Example—Threats to Validity in Fault Prediction System

Table 8.1 presents the summary of the case study characteristics discussed in Section 4.2, which assesses the relationship between OO metrics and different severity level of faults. We then present all the possible threats to the study, and how they can be reduced.

8.2.1 Conclusion Validity

The conclusion validity threats identified from example study presented in Table 8.1 are given below:

- The study uses public domain NASA data set KC1. Thus, the data set is verified and trustworthy, and does not contain erroneous observations as it is developed following best industrial practices in NASA.
- The study uses well-formed hypothesis to ascertain the relationship between OO metrics and fault proneness. Moreover, the values of statistical significance levels (0.01 and 0.05) used during correlation analysis as well as univariate and multivariate analysis increases the confidence in the conclusions of the study.
- The study uses tenfold cross-validation results that are widely acceptable methods in research (Pai and Bechta Dugan 2007; De Carvalho et al. 2010) for yielding conclusive results. Thus, reducing threat to conclusion validity.
- A data set is said to be imbalanced, if the class distribution of faulty and nonfaulty classes is nonuniform. A number of literature studies (Lessmann et al. 2008; Menzies et al. 2010) advocate the use of receiver operating characteristic (ROC) analysis as a competent measure for assessing unbalanced data sets. Thus, the use of ROC analysis avoids threats to conclusion validity.

TABLE 8.1

Details of Example Study

Data used	Description	NASA data set KC1 (public domain)
	Size	145 classes, 40K lines of code
	Language	C++
	Distribution	Faulty classes: 59
		Nonfaulty classes: 86
	Descriptive statistics stated	Min, max, mean, median, standard deviation, 25% quartile, and 75% quartile for each input metric
Independent variables	OO metrics	Chidamber and Kemerer metrics and LOC
Dependent variable	Fault proneness	Faults categorized into three severity levels: high, medium, and low. A model was also created with ungraded fault severity
	Distribution according to fault severity	High severity 23 classes
		Medium severity 58 classes
		Low severity 39 classes
Preprocessing performed	Outlier detection	Detected univariate and multivariate outliers (using Mahalanobis Jackknife distance)
	Input metrics normalization	Using min–max normalization
	Correlation analysis	Correlation coefficient values among different metrics analyzed. Significance level: 0.01
	Multicollinearity analysis	Conditional number using principal component method is <30
Algorithms used	LR	Univariate LR
		Multivariate LR
	Machine learning	DT
		ANN
Algorithm settings	DT	Chi-square automatic interaction detection (CHAID) algorithm
	ANN	Architecture 3 layers
		7 input units
		15 hidden units
		1 output unit
		Training Tansig transfer function
		Back propagation algorithm
		TrainBR function
		Learning rate 0.005
Model evaluation	Performance metrics	Sensitivity
		Specificity
		Completeness
		Precision
		ROC analysis
	Statistics reported for univariate LR and multivariate LR analyses	Coefficient (B), standard error (SE), statistical significance (Sig.), odds ratio (Exp [B]), R^2 statistic. Significance level: 0.01 and 0.05
Model development	Feature reduction	Univariate analysis
	Validation method	Tenfold cross-validation

- Outliers are unusual data points that may pose a threat to the conclusions of the study by producing bias. The study reduces this threat by performing outlier detection using univariate and multivariate analysis.

8.2.2 Internal Validity

The internal validity threats identified from example study presented in Table 8.1 are given below:

- The researchers do not have control over class size, and thus class size can act as a confounding variable in the relationship between OO metrics and fault proneness of a class. However, the study included the lines of code (LOC) metric (a measure of class size) as an independent variable in the analysis. But, evaluating the confounding effect of class was beyond the scope of the study. Thus, this threat to internal validity exists in the study.
- The study also examined correlation among different metrics, and it is seen that some independent variables are correlated among themselves. However, this threat to internal validity was reduced by performing multicollinearity analysis, where the conditional number was found to be <30 indicating effective interpretation of the predicted models as the individual effect of independent variables can be effectively assessed.
- A number of studies, which evaluate fault proneness of a class, do not take into account the severity of the faults. This is a possible threat to internal validity. However, this study accounts for three severity levels of faults.
- The study does not take into account and control the effect of programmer's capability/training and experience in model prediction at various severity levels of faults. Thus, this threat exists in the study.

8.2.3 Construct Validity

The construct validity threats identified from example study presented in Table 8.1 are given below:

- The association of defects with each class according to their severity was done very carefully to provide an accurate representation of fault-prone nature and fault severity. Moreover, the faults were divided into three severity levels: high, medium, and low, so that medium-severity level of faults can be given more attention and resources than a low-level fault. An earlier study by Zhou and Leung (2006) divided faults only into two severity levels: high and low. They combined both medium- and low-severity faults in the low category. This was a possible threat to construct validity, as medium-severity faults are more critical and should be prioritized over low-severity faults. However, this threat was removed in this study.
- The metrics used in the study are widely used and established metrics in the literature. Thus, they accurately represent the concepts they propose to measure. Moreover, the selected metrics are representative of all OO concepts like depth of inheritance tree (DIT) and number of children (NOC) metrics for inheritance, lack of cohesion in methods (LCOM) metric for cohesion, coupling between object (CBO) metric for coupling, weighted methods per class (WMC) metric for complexity, and response for

a class (RFC) and LOC metrics for size. Thus, the selected metric suite reduces the threat to construct validity by accurately and properly representing all OO concepts.

- The mapping of faults to their corresponding classes is done carefully. However, there could be an error in this mapping, which poses a threat to construct validity.
- The metrics and severity of faults for NASA data set KC1 are publically available. However, we are not aware as to how they were calculated. Thus, the accuracy of the metrics and severity levels of faults cannot be confirmed. This is a possible threat to construct validity.

8.2.4 External Validity

The external validity threats identified from example study presented in Table 8.1 are given below:

- The data set used is publically available KC1 data from NASA metrics data program. Since the data set is publically available, repeated and replicated studies are easy to perform increasing the generalizability of results. As discussed by Menzies et al. (2007), NASA uses contractors that are obliged by contract (ISO-9001) to demonstrate the understanding and use of current best industrial practices.
- The results of the study are limited to the investigated complexity metrics (Chidamber and Kemerer [CK] metrics and LOC) and modeling techniques (logistic regression [LR], decision tree [DT], and artificial neural network [ANN]). However, the selected metrics and techniques are widely used in literature and well established. Thus, the choice of such metrics and techniques does not limit the generalizability of the results.
- Fault severity rating in KC1 data set may be subjective. Thus, may limit the generalizability of study results.
- Data sets developed using other programming languages (e.g., Java) have not been explored. Thus, replicated and repeated studies with different data sets are important to establish widely acceptable results.
- The conclusions of the study are only specific to fault-proneness attribute of a class and the results of the study do not claim anything about the maintainability or effort attributes.
- The researchers have completely specified the parameter setting for each algorithm used in the study. This increases the generalizability of the results as researchers can easily perform replicated studies.
- The study uses ten fold cross-validation technique that uses ten iterations (the whole data set is partitioned into ten subsets, each iteration uses nine partitions for training and the tenth partition for validating the model and this process is repeated 10 times). Thus, the use of tenfold cross-validation increases the generalizability of our results.
- The study states the descriptive statistics of the data set used in the study. These descriptive statistics gives other researchers an insight into the properties of data sets. Researchers can thus effectively use the results of the study on similar types of data sets effectively.
- There is only one data set used in the study. This poses a threat to the generalizability of results. However, the data set used is an industrial data set developed by experienced developers. Thus, the results obtained may be applied for software industrial practices.

8.3 Threats and Their Countermeasures

Identification of threats is important to assess their impact on the study outcomes and conclusions. However, to increase the validity of the study, threats to the study should be mitigated. It is important to not only identify the threats to a study, but a researcher should perform a number of actions to address these threats and reduce their effect on the outcomes of the study. To give an overview of threats to fault prediction primary studies, we analyzed 56 fault prediction studies from the year 1999 to 2013. The aim of the study is to identify the threats to validity corresponding to each category of validity threats. The mitigation related to each threat is identified and reported. This study will guide the researchers and practitioners in identifying threats related to prediction studies and also will provide a possible solution to remove or minimize the identified threat.

Table 8.2 states all the fault prediction studies along with their unique study identifier. The systematic review was conducted by following the procedure given in Chapter 2. The "Threats to Validity" or "Limitations" section of all these studies was thoroughly analyzed. We examined and categorized all the threats in these studies into the four major threat categories, namely, conclusion, internal, construct, and external. The threats explicitly stated and dealt in these studies only are reported in this section. Thus, there may have been additional threats in these studies, however, not reported or addressed by the authors, and hence not included in this systematic review. Tables 8.3 through 8.6 state major threats encountered in fault prediction studies along with possible threat mitigation actions to address a specific threat. All the studies, which deal with a specific threat, are

TABLE 8.2

Fault Prediction Studies

Study No.	Reference	Study No.	Reference	Study No.	Reference
S1	El Emam et al. 1999	S20	Aggarwal et al. 2009	S40	Al Dallal 2012a
S2	Briand et al. 2000	S21	Catal and Deri 2009	S41	Al Dallal 2012b
S3	Glasberg et al. 2000	S22	Tosun et al. 2009	S42	Nair and Selverani 2012
S4	Briand et al. 2001	S23	Turhan and Bener 2009	S43	He et al. 2012
S5	El Emam et al. 2001	S24	Turhan et al. 2009	S44	Li et al. 2012
S6	El Emam et al. 2001a	S25	Zimmermann et al. 2009	S45	Ma et al. 2012
S7	Subramanyam and Krishnan 2003	S26	Afzal 2010	S46	Mausa et al. 2012
		S27	Ambros et al. 2010	S47	Okutan Vildiz 2012
S8	Zhou and Leung 2006	S28	Arisholm et al. 2010	S48	Pelayo and Dick 2012
S9	Aggarwal et al. 2007	S29	De Carvalho et al. 2010	S49	Rahman et al. 2012
S10	Kanmani et al. 2007	S30	Liu et al. 2010	S50	Rodriguez et al. 2012
S11	Menzies et al. 2007	S31	Menzies et al. 2010	S51	Canfora et al. 2013
S12	Oral and Bener 2007	S32	Singh et al. 2010	S52	Chen et al. 2013
S13	Pai and Becthaduran 2007	S33	Zhou et al. 2010	S53	Herbold 2013
S14	Lian Shatnawi et al. 2007	S34	Al Dallal 2011	S54	Menzies et al. 2013
S15	Lessmann et al. 2008	S35	Elish et al. 2011	S55	Nam et al. 2013
S16	Marcus et al. 2008	S36	Kpodjedo et al. 2011	S56	Peters et al. 2013
S17	Moser et al. 2008	S37	De Martino et al. 2011		
S18	Shatnawi and Li 2008	S38	Misirh et al. 2011		
S19	Turhan et al. 2008	S39	Ambros et al. 2012		

TABLE 8.3

Conclusion Validity Threats

Threat Type	Threat Description	Threat Mitigation	Studies that Encounter These Threats
Assumptions of statistical tests not satisfied	The data on which a particular statistical test or measure is applied may not be appropriate for fulfilling the conditions of the test. For example, application of analysis of variance (ANOVA) test requires the assumption of data normality and homogeneity of variance, which may not be fulfilled by the underlying data leading to erroneous conclusions.	A researcher should ensure that the conditions of a parametric statistical test are fulfilled or should use a nonparametric test where the conditions necessary for using parametric tests do not hold.	C: S26; S27; S28; S29; S36; S38; S39; S45; S48; S51.
Low statistical test ability	Use of inappropriate significance level while conducting statistical tests. Choice of an inadequate significance level will lead to incorrect conclusions.	The researcher should choose an appropriate significance level for statistical tests such as 0.01 or 0.05 to conclude significant results.	C: S26; S27.
Validation bias	The predicted models may use the same data for training as well as validation leading to the possibility of biased results. May be termed as sampling bias.	Use of cross-validation and multiple iterations to avoid sampling bias. Also, to yield unbiased results, the training data should be different from testing data.	C: S13; S15; S20; S21; S26; S29; S38; S46; S47.
Inappropriate use of performance measures	Inaccurate use of performance measures.	The performance measure used for analyzing the results should be appropriate to evaluate the studied data sets. For example, an unbalanced data set should be evaluated using Area Under the ROC curve (AUC) performance measure.	C: S21; S46.
	Inaccurate cutoffs for performance measure. For example, a recall value of greater than 70% may be considered appropriate for an effective model.	Appropriate cutoff values for performance measure should be justified and should be based on previous empirical studies and research experience.	N: S53.
Reliability of techniques applied	Parameter tuning of algorithms was not done appropriately. The parameters of a specific algorithm may be overtuned or undertuned leading to biased results.	Tuning parameters of a particular technique improves the results. However, on the other hand, use of default parameters avoids overfitting. Thus, the parameters of algorithms should be appropriately tuned.	C: S43.
Lack of expert evaluation	An expert should evaluate the generated classification rules to understand its true significance or meaning.	An expert to understand their true significance or meaning should evaluate the generated classification rules.	N: S28; S30; S50. N: S29.
Wide range of data preprocessing or engineering activities not taken into account	A number of preprocessing or engineering activities such as scaling, feature selection, and so on has not been taken into account for their effect on results and conclusions.	Different preprocessing techniques or engineering activities like scaling of data values, feature selection, and so on may improve the performance of the classifier and should be accounted for in the study. However, doing so is computationally infeasible. Thus, the selection of techniques and preprocessing activities should be such that the conclusions do not vary significantly.	P: S15.

TABLE 8.4

Internal Validity Threats

Threat Type	Threat Description	Threat Mitigation	Studies that Encounter These Threats
Confounding effects of variables	Does not account for the confounding effect of class size on the association between metrics and fault proneness.	The study should account for the confounding effect of class size on the relationship between metrics and fault proneness by controlling class size and its effect on the relationship.	C: S6. P: S34; S40; S41.
	Does not account for causality of association between specific metrics and dependent variable (fault proneness).	Controlled experiments where a specific measure such as cohesion or coupling can be varied in a controlled manner while keeping all other factors constant can demonstrate causality. Such experiments are difficult to perform in practice.	N: S1; S2; S4; S6; S10.
	Does not account for noise in the data collected from a specific repository. The noisy data can significantly affect the relationship between metrics and defects.	As noise in a repository such as Bugzilla could affect the relationship between bugs and metrics, it should be completely removed or the effect of noise should be accounted for in the relationship between faults (bugs) and metrics.	N: S27. P: S39.
Response of samples for a given technique	Certain techniques produce different results at different times because of randomness of a particular technique. For example, an algorithm like genetic algorithm randomly selects its initial population. Thus, a single run of a random algorithm may produce biased results.	Should execute multiple runs (>10 runs) of a random algorithm to obtain unbiased results.	N: S14; S18; S33; S34 S41.
Influence of human factors	Does not account for programmer's capability/training and other human factors on the explored relationship of metrics and fault proneness.	If possible, the study should evaluate the effect of programmer's capability/training and other human factors on the cause–effect relationship of metrics and fault proneness. However, data to evaluate such a relationship is difficult to collect in practice.	N: S6; S7; S12; S42; S45.
	Confounding effect of developer experience/domain knowledge on the cause–effect relationship between metrics and faults has not been accounted for.	If possible, the study should evaluate the effect of developer experience/domain knowledge on the cause–effect relationship between metrics and faults. However, data to evaluate such a relationship is difficult to collect in practice.	N: S3; S8; S12. P: S42.

(Continued)

TABLE 8.4 (*Continued*)
Internal Validity Threats

Threat Type	Threat Description	Threat Mitigation	Studies that Encounter These Threats
Ignoring relevant factors in experimental settings	The severity of faults was not taken into account while assessing the relationship of metrics and faults.	Severity of faults is important for prioritizing resources. Thus, studies should take into account the severity of faults.	N: S5; S9; S20; S24; S26; S35.
	Does not take into account attribute correlation effect on the relationship between dependent and independent variable.	The correlation of various attributes/independent variable may affect the cause–effect relationship. Thus, the study should take into account the attribute correlation.	N: S45.
	Does not account for the relationship between data set features and the algorithm used for developing the fault prediction model	The study should evaluate the suitability of algorithms/learners to data sets with specific characteristics.	N: S29.
	Does not account for extra aspects of software process, machine learning techniques, and other characteristics to get advanced solutions.	The study should account for other aspects of software process and machine learning techniques to get better and advanced solutions.	N: S38; S52; S53. P: S36.
Inappropriate experimental settings	Data may be inconsistent in nature. For example, in JM1 NASA data set, certain modules have identical values for all metrics but given different labels. Such type of data may significantly influence results.	It is important to remove inconsistent modules from a data set to yield appropriate results.	N: S21.
	Does not account for bias because of selection of classifiers in ensemble.	To account for bias while selecting classifiers in an ensemble, the researchers should use different voting schemes and analyze the overall performance of the ensemble.	C: S38.

TABLE 8.5

Construct Validity Threats

Threat Type	Threat Description	Threat Mitigation	Studies that Encounter These Threats
Misinterpretation of concepts and measures	Different OO metrics used as independent variables may not be well understood or represented.	The measures capturing the various attributes should be well understood.	N: S17. P: S2; S4. C: S13; S21.
	Inappropriate use of performance metrics for representing the theoretical concepts or improper computation of performance measure.	The performance metrics used for evaluating the results should be carefully chosen and should minimize the gap between theory and experimental concepts. Also, the computed performance measure should be verified by two independent researchers so that the results are not biased	N: S28; S51. C: S26, S37.
Unaccountability of generalizability across related constructs	To compare different algorithms, different settings, and internal details such as different number of rules and different quality measures are not accounted for. This leads to unfair comparison among different algorithms.	To conduct a fair comparison among algorithms, a baseline should be established to evaluate the number of rules, choose an appropriate quality measure, and so on. Only use of default parameters does not take into account better sets of rules.	N: S50.
	Combining results with different metrics (OO and procedural) for multiple data set combinations may produce inappropriate or biased results.	Use sensitivity analysis to check whether the results are biased.	C: S30; S48.
Reliability of measurement tools	Use of an automated tool for extraction of metrics (independent variables) or dependent variable.	Manual verification of the metrics/dependent variable generated by the tool.	N: S18; S19. P: S17. C: S4; S19; S25; S36; S41; S47.
Improper data-collection methods	The process or data-collection methods for collection of dependent variable may not be completely verified leading to improper data collection.	Researchers should specifically verify the collection process of the dependent variable. For example, acceptance testing activities for collection of faults.	N: S44. C: S2; S13.
	Mistakes in identification/assignment/classification of faults.	Manual verification of faults can minimize this threat.	N: S43; S53. C: S36; S52.
	Does not account for measurement accuracy of the software attributes such as coupling, faults, and so on.	A researcher should use appropriate data-collection methods and verify them. Public data sets whose characteristics have already been validated in previous research studies may also be used to instill confidence in software attributes data collection.	C: S15; S17; S30; S37; S45; S48; S49. P: S44; S51.

(Continued)

TABLE 8.5 (*Continued*)

Construct Validity Threats

Threat Type	Threat Description	Threat Mitigation	Studies that Encounter These Threats
Measurement bias	Incomplete fault data may create bias as only fixed faults are considered.	The study should account for bias because of incomplete faults.	N: S14; S18; S33; S34; S41.
	Appropriate threshold values for metrics are subjective in nature. Thus, results may not hold or change on other threshold values leading to bias.	As threshold values are subjective in nature, the study should validate other threshold values to increase the significance of the results.	N: S14, S50.
Errors while combining data	While collecting various faults (bugs) for a particular software, a researcher may incorrectly map bugs with classes.	Researchers should carefully map and verify bugs to corresponding classes.	N: S17; S27; S39. P: S33.

TABLE 8.6

External Validity Threats

Threat Types	Threat Description	Threat Mitigation	Studies that Encounter These Threats
Inappropriate selection of subjects	Investigated software systems may not be representative in terms of their size and number to industrial scenario.	The size and number of classes in the evaluated data set may not be representative of industrial systems. Thus, researchers should be careful while selecting data sets so that their results might be generalizable to industrial settings.	N: S34; S40; S41; S42.
	In software industry, a personnel may not be available for all software development phases. An industrial software undergoes a large organizational and personnel change. It is possible that the selected software system may not have undergone a large personnel and organization change as in an industrial scenario.	The selected software systems should have undergone a large number of organizational and personnel change to generalize results to real industrial settings.	C: S28.
Applicability of results across languages	Use of monoprogramming language software systems.	The software systems selected for evaluation should be developed in different programming languages to yield generalized results across languages.	N: S1; S2; S5; S8; S16; S19; S27; S32; S33; S34; S39; S40; S41; S49; S53; S56. C: S30.
Inadequate size and number of samples	The number of software systems evaluated is low affecting the generalizability of the results.	A large number of data sets should be evaluated to assure universal application of results.	N: S1; S4; S5; S7; S8; S13; S26; S35; S46; S53. P: S16; S19. C: S25; S47; S51; S56.
	The size of evaluated software systems may not be appropriate for industrial settings.	The evaluated data sets should be industry sized and/or should be of varied sizes (small/medium/large).	N: S2; S4; S8; S9; S10; S20; S29. P: S35. C: S21; S25; S28; S30; S36.
	The software systems evaluated may be extracted from a single versioning system/bug tracking system. Thus, application of results to other versioning/bug tracking systems is limited.	The evaluated software systems should be collected from different versioning systems/bug repositories to obtain generalized results.	C: S27; S39.

(Continued)

TABLE 8.6 (*Continued*)

External Validity Threats

Threat Types	Threat Description	Threat Mitigation	Studies that Encounter These Threats
Applicability of results across different variables	Results specific to a particular dependent variable like fault proneness.	The study should explore different dependent variables like maintainability and reliability to generalize results.	N: S1; S5; S6; S7; S20; S32.
	Results would be limited to the choice of independent variables.	Should evaluate a number of predictor variables—process metrics as well as product metrics.	N: S1; S7; S16; S25; S26; S30; S33; S51; S53.
Applicability of results across different samples	Evaluated software systems are of the same domain and application. Thus, limiting the generalizability of the results.	The evaluated software systems should be of different domain and applications, so that results can be widely used.	N: S30; S33; S42. P: S21; S55. C: S12; S36; S38.
	Evaluated software systems are developed by the same development teams/organizations leading to limitation of results to those specific software systems.	The data sets evaluated in the study should be developed by different development teams/organizations to yield generalizable results.	N: S30. P: S48.
Applicability of results across different environment	Evaluated software systems are not developed by experienced (average professional programmer) developers.	The study should explore software systems developed by experienced programmers so that the results could be generalized to industrial projects.	C: S12; S26; S27; S31; S36; S38; S39; S45; S49. C: S21; S30; S33.
	All evaluated software systems are open source in nature. Thus, results may not be applied to industrial data sets.	The software systems chosen should be representative of industrial as well as open source systems to improve the generalizability of results.	N: S16; S27; S34; S35; S37; S40; S41; S43; S44; S49; S53; S55; S56. P: S39.
	Evaluated software systems are academic in nature and are thus of limited conceptual complexity. Thus, their results may not be applicable to other open source or industrial systems.	The evaluated software systems should have high conceptual complexity so that the results may be generalized to other complex open source or industrial software.	C: S11; S21; S23; S26; S30; S50. N: S2; S10; S29.

(Continued)

TABLE 8.6 (*Continued*)

External Validity Threats

Threat Types	Threat Description	Threat Mitigation	Studies that Encounter These Threats
Results bias because of techniques or subjective classification of a variable	Model bias because of selection of specific learners.	A number of learners should be evaluated to increase the generalizability of results.	N: S46. P: S11; S12; S13; S17; S24; S33; S51; S56. C: S15.
	Classification of faults into severity levels could be subjective.	Researchers should clearly specify the criteria for labeling faults into different severity levels so that studies could be appropriately replicated. Inappropriate classification of faults may lead to biased results.	N: S8; S14; S18; S32.
Nonspecification of experimental setting and relevant details	Nonspecification of data characteristics and details for repeated studies.	The data sets should be publically available for repeated and replicated studies.	C: S11; S15; S30; S37; S44; S46.

categorized into three divisions, namely, N (not addressed the threat at all), P (partially addressed the threat), and C (completely addressed the threat). Thus, all the studies that encounter a particular threat and do not take any actions to mitigate its effect are categorized in N category. Similarly, all the studies that try to address a specific threat but have not been successful in completely removing its effect are categorized as P. Finally, all the studies that take appropriate actions to mitigate the existence of threat as effectively as possible are grouped into C.

Exercises

8.1 Identify the categories to which the following threats belong:

- Threat caused by not taking into account the effect of developer experience on the relationship between software metrics and fault proneness.
- Threat caused by only exploring systems developed using Java language.
- Threat caused by using the same data for testing and training.
- Threat caused by investigating a not publically available data set.
- Threat caused by exploring only open source systems.
- Threat caused by considering inappropriate level of significance.
- Threat caused by incomplete or imprecise data sets.

8.2 Consider a study where the lines of code is mapped to various levels of complexity such as high, medium, and low. What kinds of threats the mapping will impose?

8.3 Consider a systematic review where only journal papers are considered in the review. The review also uses an exclusion and inclusion protocol based on subjective judgment to select papers to be included. Identify the potential threats to validity.

8.4 Compare and contrast conclusion and external threats to validity.

8.5 What are validity threats? Why it is important to consider and report threats to validity in an empirical study?

8.6 Consider the systematic review given in Section 8.3, identify the threats of validity that exist in this study.

Further Readings

The concept of threats to validity is presented in:

C. Yu, and B. Ohlund, "Threats to validity of research design," 2010. http://www.creative-wisdom.com/teaching/WBI/threat.shtml.

This following research paper address external validity and raises the bar for empirical software engineering research that analyzes software artifacts:

H. K. Wright, M. Kim, and D. E. Perry, "Validity concerns in software engineering research," *Proceedings of the FSE/SDP Workshop on Future of Software Engineering Research*, ACM, New York, pp. 411–414, 2010.

This following research paper provides a tradeoff between internal and external validity:

J. Siegmund, N. Siegmund, and S. Apel, "Views on internal and external validity in empirical software engineering," 2014. http://www.infosun.fim.uni-passau.de/cl/publications/docs/SiSiAp15.pdf.

The impact of the assumptions made by an empirical study on the experimental design is given in:

J. Carver, J. V. Voorhis, and V. Basili, "Understanding the impact of assumptions on experimental validity," *IEEE Proceedings of the International Symposium on Empirical Software Engineering*, pp. 251–260, Redondo Beach, CA, 2004.

9

Reporting Results

The goal of the research is not just to discover or analyze something but the results of the study must be properly written in the form of research report or publication to enable the results to be available to the intended audience—software engineers, researchers, academicians, scientists, and sponsors. After the experiment has finished, the results of the experiment can be summarized for the intended audience. While reporting the results, the research misconduct, especially plagiarism, must be taken care of.

This chapter presents when and where to report the research results, provides guidelines for reporting research, and summarizes the principles of research ethics and misconduct.

9.1 Reporting and Presenting Results

When one decides to report the findings, the important questions to be considered are, To whom the results of the research should be addressed? How the results of the research should be presented? How to present the results of the research without being biased and influenced? The research report or publication will present the findings and their interpretation that reflects the goals and objectives of the research. It enables the intended audience to learn from the findings and also allow them to judge or assess the results of the findings. The content of the research report depends on the type of the report. Research report type may vary among the student theses, conference papers, technical reports, magazine articles, and journal papers. For example, in case of a doctoral student, initial findings of the study are presented in a conference, detailed findings are published in the journal for academic and research community, and finally, the research is organized in the form of doctoral degree thesis for external examination.

For a masters or doctoral student, reporting the results of the research is in the form of thesis that has to be examined as an essential part of the completion of degree. For a professional researcher, the quality of the publication in the form of research paper is most important as it may influence or build up his/her reputation or at least career growth. The technical report is produced as an outcome of the study carried out, which is a part of a grant funded by a body. The findings of the research may be produced in more than one form. Hence, several published forms may be produced of the same work. Despite different motives while reporting results, the published work has the following benefits:

- Allows presenting the methodology and the results to the outside world
- Allows software engineering organizations to apply the findings of the research in the industrial environment
- If the study is a systematic literature review, it will allow the researcher to have an idea of the current position of the research in the specific software engineering area

- Allows the researchers to replicate and repeat the study
- Provides guidelines to students and researchers for future work
- Allows the sponsor of the study to verify whether desired findings have been produced or not

9.1.1 When to Disseminate or Report Results?

The decision of reporting results of the research findings in the form of publication or research reports depends on many factors, including status of the research work, confidentiality, or ethical issues. For instance, when sufficient preliminary work has been done, it is a good idea to present the results in a conference or produce interim report to obtain feedback from the outside world, including sponsors. The researchers can then publish the detailed results. Consider an instance where research work is confidential, hence, decision about disclosing the work should be made at a right time. Another instance is when the researcher needs to obtain formal approval from the funding or professional body as an ethical policy before beginning to report the results. No matter what, the preparation of writing result reports should be started as early as possible.

9.1.2 Where to Disseminate or Report Results?

Publishing the results on the right platform and at a right time is a very important decision that needs to be made by a researcher. In the software engineering community, two most popular venues for publishing the findings of the research are journal papers or conference articles. Whether to publish in conference proceedings or as a journal paper depends on various factors:

1. Type of research work: The work carried out by the research may be survey/review, original work, or case study. The survey or systematic reviews are mostly considered by journals, as they mostly do not depict any new and innovative research or new findings. Hence, relevant journals may be considered for publishing them. The new or empirical work may be considered for publication in conference or journal depending on the status of the work.

2. Status of the work: As discussed in previous section, the status of the work is an important criterion for selection of the place for publication of the research. The new or initial idea or research may be considered for publication in a conference to obtain the initial feedback about the work. The detailed findings of the research may be considered to be published in journals.

3. Type of audience: The selection of publication venue also depends on the type of audience. The study can be considered for publication in journal, when the researcher wants to present the work to academic and research community. To make the work visible to software industry, the findings may be communicated to industry conferences, practitioners' magazines, or journals. The work may also be presented to sponsors or funders in the form of technical reports.

Thus, the initial findings of the scholarly articles may be communicated to the conferences and the detailed findings can be published as journal papers or a book chapter with atleast 30% of new material added. Some researchers also like to publish the results of the work on their website or home page.

9.1.3 Report Structure

The research findings can be organized in the form of a journal paper, technical report, conference article, or book chapter. The readers not only want to understand the findings of the research, but they also want to be able to repeat the analysis or replicate the study. Hence, in an empirical research, while writing the results, not only enough details of the techniques and methods used must be provided, the access to the original data must also be given. This will allow the readers and professionals to verify and replicate the results obtained from the existing study. Table 9.1 presents the structure of a journal paper.

The sections of the report format given in Table 9.1 are described below:

TABLE 9.1

Structure of Journal Paper

Title Author details with affiliation	
Abstract	What is the background of the study?
	What are the methods used in the study?
	What are the results and primary conclusions of the study?
Introduction	
Motivation	What is the purpose of the study? How does the proposed work relate to the previous work in the literature?
Research questions	What issues are to be addressed in the study?
Problem statement	What is the problem?
Study context	What are the experimental factors in the study?
Related work	How is the empirical study linked with literature?
Experimental design	
Variables description	What are the variables involved in the study?
Hypothesis formation	What are the assumptions of the study?
Empirical data collection	How is the data collected in the study?
	What are the details of data being used in the study?
Data analysis techniques	What are the data analysis techniques being used in the study?
	What are the reasons for selecting the specified data analysis techniques?
Analysis process	What are the steps to be followed during research analysis?
Research methodology	
Analysis techniques	What are the details related with the selected techniques identified in experimental design?
Performance measures	How will the performance of the models developed in the study analyzed?
Validation methods	Which validation methods will be used in the study?
Research results	
Descriptive statistics	What is the summary statistics of data?
Attribute selection	Which attributes (independent variables) are relevant in the study?
Model prediction	What is the accuracy of the predicted models? What are the model validation results?
Hypothesis testing	What are the results of hypothesis testing?
	Is the hypothesis accepted or rejected?

(Continued)

TABLE 9.1 (*Continued*)

Structure of Journal Paper

Discussion and interpretation	What are the interpretations of the findings?
	What are the answers to the research questions?
	What is the applications/practical significance of the results?
	Do the results support the findings in the literature?
	What are the differences between current findings and the related studies?
	How can the results be generalized?
Threats to validity	What are the limitations of the study?
Conclusion and future work	What are the main findings of the study?
	What are the future avenues in the area?
Acknowledgment	Who all contribute to the research?
References	What are the related citations to the work?
Appendix	What is the additional material that can help the readers?

9.1.3.1 Abstract

The abstract provides a short summary of the study, including the following components of the research:

1. Background
2. Methods used
3. Major findings/key results
4. Conclusion

The length of the abstract should vary between 200 and 300 words. The abstract should not provide long descriptions, abbreviations, figures, tables (or reference to figures and tables), and references.

The example abstract is shown below:

Background: Software fault prediction is the process of developing models that can be used by the software practitioners in the early phases of software development life cycle for detecting faulty constructs such as modules or classes. There are various machine learning techniques used in the past for predicting faults.

Method: In this study, we assess the predictive ability of 18 machine learning techniques for software fault prediction. We use object-oriented (OO) metrics to predict faulty or nonfaulty classes. The results are validated using data collected from two open source software. The results are obtained using area under the curve obtained from receiver operating characteristic curves.

Results: The results show the prediction capability of the machine learning techniques for predicting classes as fault prone or not fault prone. The naïve Bayes technique is best as compared to other 17 machine learning techniques.

Conclusion: Based on the results obtained, we conclude that the machine learning techniques can be efficiently used by researchers and software practitioners for predicting faulty classes.

9.1.3.2 Introduction

This section must answer the questions such as: What is the purpose of the study? Why the study is important? What is the context of the study? How the study enhances or adds to the current literature? Hence, the introduction section must include motivation behind the study, research aims or questions, and the problem statement. The motivation of the work provides the information about the need of the study to the readers. The purpose of the study is described in the form of research question, aim, or hypothesis. The relevant primary studies from the literature (with citations) are provided in this section to provide the summary of current work to the readers. The introduction section should also present the brief details about the approach of the empirical research or study being carried out. For example, this section should briefly state the techniques, data sets, and validation methods used in the study.

The introduction section should be organized in the following steps:

1. Stating the motivation of the work.
2. Establishment of the context by providing citation to the relevant research.
3. Stating the research purpose in the form of research questions.
4. Stating the approach of the research, including techniques and methods.

The extract from the introduction section describing the research aims is shown in Figure 9.1.

9.1.3.3 Related Work

It provides the relationship of current work with literature. This section must mention the related studies in the literature. The brief description of methods used in the work along with results may be specified for each related study (refer Chapter 4). If the current work is a replication or repetition of an existing work, then the past study that is being repeated or replicated must be described in detail. This section should also provide a brief description about the difference between the current study from the closest studies already carried out in the past.

9.1.3.4 Experimental Design

In this section, the details of experimental design phase given in Chapter 4 are reported. The purpose of this section is to provide details to the readers so that repeated and replicated studies can be carried out. It provides details about independent and dependent

---

In this work, the fault-prone classes are predicted using the object-oriented (OO) metrics design suite given by Chidamber and Kemerer (1994). The results are validated using latest version of Android data set containing 200 Java classes. Thus, this work addresses the following research questions:

RQ1: What is the overall predictive performance of the machine learning techniques on open source software?

RQ2: Which is the best machine learning technique for finding effect of OO metrics on fault prediction?

RQ3: Which machine learning techniques are significantly better than other techniques for fault prediction?

FIGURE 9.1
Portion from introduction section.

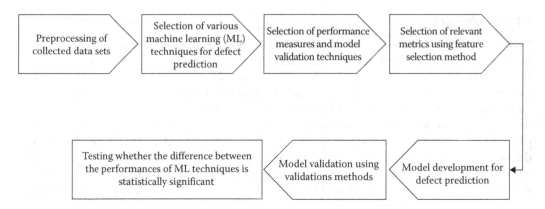

FIGURE 9.2
Example experimental process.

variable being used, data-collection procedure, and hypothesis for research. The size, nature, description about subjects, and the source of data should be provided here. The tools (if any) used to collect the data must also be provided in this section. The analysis process to be followed while conducting the research is presented. For detailed procedure of experimental design refer Chapter 4. The data-collection procedure is explained in Chapter 5. For example, the experimental process depicted in Figure 9.2 is followed for conducting research to find the answers to questions given in Figure 9.1.

9.1.3.5 Research Methods

This section provides insight about the procedures and methods used in the empirical work. The section describes the basic methods used to perform the study. The way in which the data is analyzed, including data preprocessing methods, performance measures, validation methods, statistical techniques, or any other techniques are included in this section. For example, if t-test is used in the study then a brief description of t-test along with significance levels should be stated. The methods used and the experimental settings with respect to techniques used to obtain the results must be clearly stated so that the researchers can repeat the work. For example, the number of hidden layers, number of neurons in the hidden layer, transfer function, learning rate, and so on must be specified if the study uses neural network technique for exploring relationships. Similarly data transformations and cleaning methods, techniques used for outlier analysis, feature subselection methods, and software packages must be described.

9.1.3.6 Research Results

The descriptive statistics summarizing the characteristics, the relevant attributes that are derived, and the results of the hypothesis are presented in this section. The model prediction and validation results are summarized using performance measures. In software engineering research, quantitative research is carried out more often as compared to qualitative research. For quantitative analysis, statistical analysis is usually done and then the hypotheses are accepted or rejected based on the statistical test results. The test statistic and significance level must be reported. For detailed information on hypothesis testing, refer Chapter 6.

9.1.3.7 Discussion and Interpretation

The results obtained to answer the established research questions or research hypothesis should be presented in this section. The results should be interpreted in light of the findings of the relevant studies from the literature. Questions such as What is the meaning of the findings of the study? and How do the results relate to the research questions formed in the study? must be addressed. The evidences that confirm any claims should be provided. The theoretical and practical implication of the study must also be provided (refer Chapter 7).

9.1.3.8 Threats to Validity

Threats to validity must be addressed throughout the experimental design and result analysis phase, but the complete threats can be stated only after result analysis and discussion phase. The limitations in terms of generalizations or biasness of the study must be reported in this section. For details on categories of validity threats, refer Chapter 8.

9.1.3.9 Conclusions

This section presents the main findings and contributions of the empirical study. The future directions are also presented in this section. It is important to focus on the commonalities and differences of the study from previous studies.

9.1.3.10 Acknowledgment

The persons involved in the research that do not satisfy authorship criterion should be acknowledged. These include funding or sponsoring agencies, data collectors, and reviewers.

9.1.3.11 References

References acknowledge the background work in the area and provide the readers a list of existing work in the literature. The references are presented at the end of the paper and should be cited in the text. There are software packages such as Mendeley available for maintaining the references.

9.1.3.12 Appendix

The appendix section presents the raw data or any related material that can help the reader or targeted audience to better understand the study.

The claims of contributions, novelty in the work, and difference of the work from the literature work are the main concerns that need to be addressed while writing a research paper.

9.2 Guidelines for Masters and Doctoral Students

The masters and doctoral students must conduct a research that makes original contribution to the existing state of the art of the software engineering discipline. Doctoral thesis is an essential outcome of the PhD work carried out by a student for completion of the

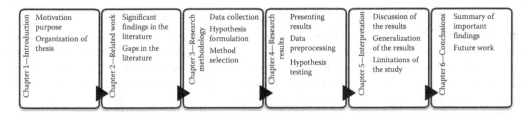

FIGURE 9.3
Format of thesis.

degree. It ensures that the student is capable of carrying out independent research at a later stage. The masters or doctoral work may involve conducting new research, carrying an empirical study, development of new tool, exploring an area, development of new technique or method, and so on.

The selection of right area and supervisor are the first and most important steps for the masters and doctoral students. Each thesis has common structure, although different topics. The general format of masters or doctoral thesis is presented in Figure 9.3, and the description of each section is given below. The thesis begins with abstract that provides a summary of the problem statement, purpose, data sources, research methods, and key findings of the work.

1. Chapter 1—Introduction

   The first chapter should clearly state the purpose, motivation, goals, and significance of the study by describing how the work adds to the existing body of knowledge in the specified area. This chapter should also describe the practical significance of the work to the researchers and software practitioners. The doctoral students should describe the original contribution of their work. This chapter provides the organization of the rest of the thesis. This part of the thesis is most critical and, hence, should be well written with strong theoretical background.

2. Chapter 2—Literature Review

   This chapter should not merely provide the summary of the literature, but rather should analyze and discuss the findings in the literature. It must also describe what is not found in the literature. This chapter provides the basis of the research questions and hypothesis of the study.

3. Chapter 3—Research Methodology

   The third chapter describes the research context. It describes the data-collection procedure, data analysis steps, and techniques description. The research settings and details of tools used are also provided.

4. Chapter 4—Research Findings

   The results of the tests of model prediction and/or hypothesis testing are presented in this section. The positive as well as the negative results should be reported. The results are summarized and presented in the form of tables and figures, respectively. This chapter can be divided into logical subsections.

5. Chapter 5—Results Interpretation

This chapter provides an in-depth discussion and interpretation of results. The aim of the chapter is to identify the emerging patterns, generalizations from the findings, and exception from these generalizations. The results should be discussed in light of the research questions formed in the introduction chapter.

6. Chapter 6—Conclusions and Future Work

This chapter provides the concluding remarks on the findings of the research area, and guidelines for future work are also proposed in this chapter.

9.3 Research Ethics and Misconduct

Empirical research in software engineering involves analyses of data obtained through software organizations or open source software. The data is required to perform empirical validations. However, there are a number of ethical issues involved while analyzing the data. These issues deal with data confidentiality, content disclosure, and dissemination of findings. For instance, whether the name of the organization should be revealed or not? How to obtain unbiased information from subjects/participants in the software organization? Whether it is ethical to use data from open source software? The researcher, while conducting the study, must consider these ethical issues. Section 1.5 provides the summary of ethics in research. The following ethical issues must be well thought out while reporting results:

1. The consent obtained from the participants must be honored.
2. The identification of the participants or the organization must not be revealed during the research.
3. The report should include description of any research bias that may have influenced the results of the research.
4. The empirical study must be written in such a way that the results can be reproduced.
5. The sponsoring agency must be acknowledged in the publication.

Research misconduct means fabrication, falsification, and plagiarism in conducting research or reporting results. According to the Federal Policy on Research Misconduct (www.aps. org/policy/statements/federa lpolicy.cfm), these three terms are defined as follows:

- Fabrication: Constructing up research data to produce results or making up false results.
- Falsification: Modifying or manipulating or omitting research data to produce results.
- Plagiarism: Copying or reusing someone else work without proper acknowledgment.

The research misconduct must be carefully examined to assess the validity of the suspected or reported incident. Plagiarism is a serious offence, and the issues involved in it are described in next subsection.

Guidelines for avoiding plagiarism

1. Provide proper citations and references to resources from where content or ideas are taken.
2. Any text/sentence copied from another source must be stated in quotations marks. For example, Emam et al. said "..."
3. Any paraphrased sentence should be properly cited.
4. Reference to the author's previous similar work must always be provided in the current work.
5. Author must always doubly check the citations.
6. If a figure or table is reproduced, the reference to the original work must be provided in the heading.
7. Plagiarism checking tools must be used to avoid honest or accidental mistakes.

FIGURE 9.4
Guidelines to researchers regarding plagiarism.

9.3.1 Plagiarism

IEEE defines plagiarism as "the reuse of someone else's prior ideas, processes, results, or words without explicitly acknowledging the original author and source." Plagiarism involves paraphrasing (rearranging the original sentence) or copying someone else's words without acknowledging the source. Reusing or copying one's own work is called self-plagiarism. IEEE/ACM guidelines define serious actions against authors committing plagiarism (ACM 2015). The institutions (universities and colleges) also define their individual policies to deal with plagiarism issues. The faculty and students should be well informed about the ethics and misconduct issues by providing them guidelines and policies, and imposing these guidelines on them. Figure 9.3 depicts guidelines for researchers and practitioners to avoid plagiarism. The researchers, publishers, employers, and agencies may use the plagiarism software for scanning the research paper. There are various open source softwares such as Viper and proprietary software such as Turnitin available for checking the documents for plagiarism.

Finally, it is the primary responsibility of research institutions to ensure, monitor, detect, and investigate research misconduct. Serious actions must be taken against individuals caught with plagiarism or misconduct.

Exercises

9.1 What is the importance of documenting the empirical study?

9.2 What is the importance of related work in an empirical study?

9.3 What is the importance of interpreting the results rather than simply stating them?

9.4 When reporting a replicated study, what are the most important things that a researcher must keep in mind?

9.5 What is research misconduct? Why plagiarism is considered a serious offence in research? How plagiarism can be avoided?

9.6 How can a researcher decide where to publish the research?

Further Readings

An excellent study that provides guidelines on empirical research in software engineering is given below:

B. Kitchenham, S. Pfleeger, L. Pickard, P. Jones, D. Hoaglin, K. El Emam, and J. Rosenberg, "Preliminary guidelines for empirical research in software engineering," *IEEE Transactions on Software Engineering*, vol. 28, no. 8, pp. 721–734, 2002.

The following paper provides a minitutorial on writing research in software engineering:

M. Shaw, "Writing good software engineering research papers," *Proceedings of the 25th International Conference on Software Engineering*, IEEE Computer Society, Portland, OR, pp. 726–736, 2003.

An article on reporting research results is presented in:

A. Jedlitschka, M. Ciolkowski, and D. Pfahl, "Reporting experiments in software engineering," In: F. Shull, J. Singer, and D. Sjøberg (eds.), *Guide to Advanced Empirical Software Engineering*, chapter 8. Springer, Berlin, Germany, pp 201–228, 2008.

Perry et al. provided a summary of various phases of empirical studies in software engineering in their article:

E. Perry, A. Porter, and L. Votta, "Empirical studies of software engineering: A roadmap," *Proceedings of the Conference on the Future of Software Engineering*, Limerick, Ireland, pp. 345–355, 2000.

The advisory notes on checking plagiarism are provided in:

C. Kaner, and R. Fiedler, "A cautionary note on checking software engineering papers for plagiarism," *IEEE Transactions on Education*, vol. 51, no. 2, pp. 184–188, 2008. doi:10.1109/TE.2007.909351.

ACM plagiarism policy is defined in:

R. F. Boisvert, and M. J. Irwin, "ACM plagiarism policy: Plagiarism on the rise," *Communications of the ACM*, vol. 49, no. 6, pp. 23–24, 2006.

10

Mining Unstructured Data

As seen in Chapter 5, software repositories can be mined to assess the data stored over a long period of time. Most of the previous chapters focused on techniques that can be applied on structured data. However, in addition to structured data, these repositories contain large amount of data present in unstructured form such as the natural language text in the form of bug reports, mailing list archives, requirements documents, source code comments, and a number of identifier names. Manually analyzing such large amount of data is very time consuming and practically impossible. Hence, text mining techniques are required to facilitate the automated assessment of these documents.

Mining unstructured data from software repositories allows analyzing the data related to software development and improving the software evolutionary processes. Text mining involves processing of thousands of words extracted from the textual descriptions. To obtain the data in usable form, a series of preprocessing tasks like tokenization, removal of stop words, and stemming must be applied to remove the irrelevant words from the document. Thereafter, a suitable feature selection method is applied to reduce the size of initial feature set leading to more accuracy and efficiency in classification.

There are various artifacts produced during the software development life cycle. The numerical and structured data mined using text mining could be effectively used to predict quality attributes such as fault severity. For example, consider the fault tracking systems of many open source software systems containing day-to-day fault-related reports that can be used for making strategic decisions such as properly allocating available testing resources. These repositories also contain unstructured data on vulnerabilities, which records all faults that are encountered during a project's life cycle. The Mozilla Firefox is one such instance of open source software that maintains fault records of vulnerabilities. While extensive testing can minimize these faults, it is not possible to completely remove these faults. Hence, it is essential to classify faults according to their severity and then address the faults that are more severe as compared to others. Text mining can be used to mine relevant attributes from the unstructured fault descriptions to predict fault severity.

In this chapter, we define unstructured data, describe techniques for text mining, and present an empirical study for predicting fault severity using fault description extracted from bug reports.

10.1 Introduction

According to the researchers, it has been reported that nearly 80%–85% of the data is unstructured in contrast to structured data like source code, execution traces, change logs, and so on (Thomas et al. 2014). The software library consists of fault descriptions, software requirement documents, and other related documents. For example, the software requirement document contains the requirement descriptions of the following type: "The software product must be able to distinguish between authorized and non-authorized access of a user. The software product must ensure that only authorized users access the product."

The researcher must be able to analyze the above requirement and predict its category such as security, usability, availability, and maintainability. Usually, it is difficult to analyze and assess the software documents, as they may be present in an unorganized manner. The document may also simply be a collection of words with no relationship between the words and with no semantic meaning. Further, the long text descriptions may contain data of almost all types, that is, it may be a linked data, a semistructured data, a numerical information, and so on. Thus, it is very essential to bring the data into a form, which is understandable and/or can be analyzed by the computer-aided tools.

Hence, a typical text mining problem is to extract relevant information from the software documents such as fault descriptions or software requirements specification document that are maintained in the software repositories, and thereby analyze these descriptions using suitable measures, tools, and techniques.

10.1.1 What Is Unstructured Data?

Structured data is organized and represented in a known format. For, example, bug repository metadata consists of bug severity, LOC added, LOC deleted, and priority. The LOC added and deleted are used to correct the bug.

Unstructured data can be defined as the data that does not have a clear semantics, that is, it is a natural language text that has no explicit data model. In other words, such kind of data is simply a collection of characters with no structure or format and there is no relationship between these words. This data cannot be assessed because of the size and nature of the data. It would be very expensive and time consuming to analyze and interpret vast number of documents and find relationship between them. This requires text mining algorithms and natural language processing techniques to transform into a structured form.

Unstructured data usually consists of data that is unlabeled, vague, noisy, and ambiguous. The data may be noisy because of misspellings, typographical errors, unconventional acronyms, and so on used in the document. Apart from this, there may be multiple phrases used for the same concept, thus resulting in presenting vague information. Unstructured data is difficult to mine and poses many challenges for software engineers, professionals, and practitioners to mine such data. Mining such kind of data requires the use of systematic and specialized text mining techniques along with the use of information retrieval modules.

10.1.2 Multiple Classifications

There are a number of different classification tasks used for classifying a given set of objects into some fixed categories. These classification tasks fall under the category of binary classifications or multiple classifications.

Binary classification is the term used when a set of objects could be categorized as either belonging to category "A" or category "Not A." In other words, each of the objects could associate themselves into one of the two categories available.

Multiple classifications in text mining problems differ from classification tasks previously explained in this book. In multiple classifications, there is a pool of a number of categories available and the task is to associate each of the objects to any one of the categories. There are multiple categories that a document may belong to, out of N categories available an object can belong to any of the categories. For example, in a requirement specification document, the requirement may belong to various quality attributes such as security, availability, usability, maintainability, and so on. There may be the case that a particular object does not fit into any of the available N categories, or an object is suitable to be fit into more than one category. Any kind of such combination is permissible in case of multiple classifications. The need to perform N separate classification tasks is time consuming and generally computationally expensive.

10.1.3 Importance of Text Mining

The amount of printed material is increasing at an incredible rate and a lot of such material is stored in libraries, stores, and so on in an unorganized manner. The printed material could be in any of the forms like textbooks, magazines, newspapers, articles, journals, research papers, and so on. The list is endless and it has been observed that with days passing by, the information in the form of printed material is only increasing. Apart from these printed materials, there are volumes of electronic text in the form of documents available in software libraries, open source repositories, social networking sites, medicines, educational institutes, and so on. As a result of this, users have to spend hours searching for a relevant material before they could find bits of desired information. This leads to unnecessary wastage of time and effort. In view of this, it has now become very essential to devise some mechanism that can categorize large amount of text present in online software libraries and repositories. In such a way, the information available can be organized, making it easier for both the users and the organization to use the material, thus leading to an effective utilization of the available resources.

After applying text mining techniques, the relevant information will be obtained from the vast set of unstructured data that can now be used by human experts. Automated tasks will also allow the practitioners to obtain answers to various queries in less time, and thus save a lot of resources and costs.

The field of text mining is gaining huge popularity in today's scenario, wherein the amount of information available online is increasing at an exponential rate.

10.1.4 Characteristics of Text Mining

Text mining involves a series of steps that can transform the unstructured data into a structured one. Text mining encapsulates the natural language processing techniques used by information retrieval modules. As the literal meaning of this term, text mining means the mining of relevant information from the text, which could be either semistructured or unstructured, to make it understandable by computer-aided machines, softwares, and tools. For example, text mining can be effectively used in predicting cost and effort by analyzing the maintenance requests. The information obtained can easily be classified to answer various queries.

Mining the data follows a series of steps, by which classification of the text into one of the predefined categories is done. This is referred to as text classification. In other words, a given document is categorized into one of the categories available by employing text mining techniques and using machine learning methods. The aim is to use a set of pre-classified documents to classify those that have not yet been done.

There are two approaches for text classification: global dictionary and local dictionary. For the global dictionary approach, all the words that occur at least once in the documents are included. Then, the classification is done for N categories. This approach works faster but the accuracy can be less. In the local dictionary approach, a separate dictionary considering the terms only related to a given category is created. Thus, the dictionary is small but the cost of computing N models to predict N categories is more than the global dictionary approach (Bramer 2007).

10.2 Steps in Text Mining

There are various methods that can be used to extract fixed number of attributes from the document. Text mining involves the processing of thousands of words that are extracted from the textual descriptions. The processing of the words is carried out by following a series of text mining steps, which are shown in Figure 10.1. The aim of text mining is to extract a set of potentially relevant attributes that could contribute to an efficient model prediction. These techniques include document representation, preprocessing, feature selection, and weighting. All these techniques are explained in the subsequent sections below.

10.2.1 Representation of Text Documents

The given documents need to be represented in a particular form so that they can be further explored and analyzed. The most common representation approach is referred to as the bag-of-words approach, in which the entire document is considered as a collection of words. The words in a textual document may refer to any kind punctuations, articles, nouns, or verbs. Thus, a document is assumed to be a collection of thousands and thousands of infinite words. It is very normal to visualize that all these words are not significant, and rather their usage can degrade the performance of the model so predicted. It is therefore necessary to eliminate the irrelevant words that add no meaning to the document, thus reducing the size of the total number of words. The words in the document are referred to as features in the context of text mining. The total number of features in a document make up a feature space or feature set. The aim is to reduce the size of the feature set so that the accuracy of the model predicted can be improved.

10.2.2 Preprocessing Techniques

Performing a series of preprocessing tasks can reduce an initial size of the feature space. These tasks remove all the types of punctuation characters and stop words. Also, the words are stemmed up to their original stem. The preprocessing includes tokenization, stop words removal, and stemming, as shown in Figure 10.2. Tokenization is concerned with the task of replacing the punctuation characters with blank spaces, removing all the nonprintable escape characters, and converting all the words to lowercases. Thereafter,

FIGURE 10.1
Steps in text mining.

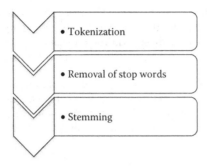

FIGURE 10.2
Preprocessing techniques.

all the stop words like prepositions, articles, conjunctions, verbs, pronouns, nouns, adjectives, and adverbs are removed from the document. Finally, the most important step of preprocessing is performed, which is referred to as stemming. It is defined as the process of removing the words that have the same stem, thereby retaining the stem as the selected feature. For instance, words like "move," "moves," "moved," and "moving" can be replaced with a single word as "move." After preprocessing steps, a set of features are obtained by reducing the initial size of the feature space.

Example 10.1:

Consider an example of software requirements given in Table 10.1. This example demonstrates the description of the various software requirements, which describe the nonfunctional requirement (NFR; quality attributes). Generally, it has been observed that these requirements are not properly defined in the software requirement document and are scattered throughout the document in an ad hoc fashion. As we know, these qualities play an important role for the development and behavior of the software

TABLE 10.1

Original Data Consisting of Twelve Requirements and Their Description

Req. No.	Requirement Description	NFR Type
RQ1	the product shall be available during normal business hours. As long as the user has access to the client pc the system will be available 99% of the time during the first six months of operation.	1
RQ2	the product shall have a consistent color scheme and fonts.	2
RQ3	the system shall be intuitive and self-explanatory.	3
RQ4	the user interface shall have standard menus buttons for navigation.	2
RQ5	out of 1000 accesses to the system, the system is available 999 times.	1
RQ6	the product shall be available for use 24 hours per day 365 days per year.	1
RQ7	the look and feel of the system shall conform to the user interface standards of the smart device.	2
RQ8	the system shall be available for use between the hours of 8 am and 6 pm.	1
RQ9	the system shall achieve 95% up time.	1
RQ10	the product shall be easy for a realtor to learn.	3
RQ11	the system shall have a professional appearance.	2
RQ12	the system shall be used by realtors with no training.	3

system and, therefore, these qualities should be incorporated in the architectural design as early as possible, which is not the case. Hence, in this example, we intend to employ text mining techniques to mine these requirements and convert them into a structured form that can then be used for the prediction of the unknown requirements into their respective nonfunctional qualities. The data presents few requirements extracted from promise data repository. Here, we categorize the requirements into three different type of NFRs, namely, availability (A), look-and-feel (LF), and usability (U). These three NFRs have been labeled as type 1, 2, and 3, respectively. The original data in its raw form containing the description of the NFRs along with the type of NFR is given in Table 10.1.

10.2.2.1 Tokenization

The main aim of text mining is to extract all the relevant words in a given set of documents. Tokenization is the first step in preprocessing. In tokenization, a document consisting of various characters is divided into a well-defined collection of tokens. The process involves removal of irrelevant numbers, punctuation marks, and replacement of special and non-text characters with blank spaces. After removing all the unwanted characters, the entire document is converted into lowercase. This tokenized representation forms the foundation of extracting words for sentences. Table 10.2 represents the tokenized data obtained after tokenizing the original data shown in Table 10.1.

10.2.2.2 Removal of Stop Words

Stop words are commonly used terms that contain no important information and, hence, are of no relevance in the text mining process. The English stop words include articles, nouns, punctuations, verbs, adjectives, adverbs, and so on. These are common words that are not useful for classification. For example, "a," "an," "is," "the," "for," and "of." The complete list of 724 stop words was obtained from the University of Glasgow and is shown in Table 10.3 (http://ir.dcs.gla.ac.uk/resources/linguistic_utils/).

TABLE 10.2

Tokenized Data Obtained after Tokenizing the Original Data

Req. No.	Requirement Description
RQ1	the product shall be available during normal business hours as long as the user has access to the client pc the system will be available of the time during the first six months of operation
RQ2	the product shall have a consistent color scheme and fonts
RQ3	the system shall be intuitive and self-explanatory
RQ4	the user interface shall have standard menus buttons for navigation
RQ5	out of accesses to the system the system is available times
RQ6	the product shall be available for use hours per day days per year
RQ7	the look and feel of the system shall conform to the user interface standards of the smart device
RQ8	the system shall be available for use between the hours of 8 am and 6 pm
RQ9	the system shall achieve up time
RQ10	the product shall be easy for a realtor to learn
RQ11	the system shall have a professional appearance
RQ12	the system shall be used by realtors with no training

TABLE 10.3

Top-100 Stop Words

a	Ah	Anybody	aside	be
able	Aint	Anyhow	ask	became
about	all	anymore	asking	because
above	allow	anyone	associated	become
abst	allows	anything	at	becomes
accordance	almost	anyway	auth	becoming
according	alone	anyways	available	been
accordingly	along	anywhere	away	before
across	already	apart	awfully	beforehand
act	also	apparently	back	begin
actually	although	appear	beginning	better
added	always	appreciate	beginnings	between
adj	am	appropriate	begins	beyond
affected	among	approximately	behind	biol
affecting	amongst	are	being	both
affects	an	aren	believe	brief
after	and	arent	below	briefly
afterwards	announce	arise	beside	but
again	another	around	besides	by
against	any	as	best	cmon

TABLE 10.4

Data Obtained after Removing the Stop Words from the Tokenized Data

Req. No.	Requirement Description
RQ1	product normal business hours long user access client pc system time months operation
RQ2	product consistent color scheme fonts
RQ3	system intuitive explanatory
RQ4	user interface standard menus buttons navigation
RQ5	accesses system times
RQ6	product hours day days year
RQ7	feel system conform user interface standards smart device
RQ8	system hours 8 am–6 pm
RQ9	system achieve time
RQ10	product easy realtor learn
RQ11	system professional appearance
RQ12	system realtors training

Table 10.4 represents the data obtained after removing the stop words from the tokenized data shown in Table 10.2.

10.2.2.3 Stemming Algorithm

Stemming recognizes a set of words and treats them equivalently. For example, "apply," "applies," and "applying" are treated as equivalent. Hence, by using stemming algorithm, the derived words are reduced. Porter's Stemming Algorithm (Porter 1980), developed

TABLE 10.5

Data Obtained after Performing Stemming

Req. No.	Requirement Description
RQ1	product normal busi hour long user access client pc system time month oper
RQ2	product consist color scheme font
RQ3	system intuit explanatori
RQ4	user interfac standard menus button navig
RQ5	access system system time
RQ6	product hour day day year
RQ7	feel system conform user interfac standard smart devic
RQ8	system hour 8 am 6 pm
RQ9	system achiev time
RQ10	product easi realtor learn
RQ11	system profession appear
RQ12	system realtor train

in 1980, is the most widely used. The algorithm is imported from NuGet Package Manager for .NET Framework (www.nuget.org). Porter's Stemming Algorithm provides a set of rules that iteratively reduce English words by replacing them with their stems. Table 10.5 represents the stemmed data obtained after performing stemming on data given in Table 10.4.

10.2.3 Feature Selection

After removing the stop words and replacing them by stems, the number of words in the document is still very large. Hence, feature-selection method is applied to further reduce the dimensionality of the feature set.

The aim of feature-selection methods is to reduce the size of the feature set by removing the words that are considered irrelevant for the classification. This will result in smaller size of the data set, thereby leading to lesser amount of computation requirement. Such kind of data set is highly beneficial for the classification algorithms that do not scale well with the large-size feature space. One of the major advantages of feature selection is the reduction of the curse of dimensionality, thereby leading to better classification accuracy.

The feature-selection methods are based on the concept of using an evaluation function being applied to a single word. There are a number of such methods that are being used in the literature, for instance, document frequency, term frequency, mutual information, information gain, odds ratio, χ^2 statistic, and term strength. All of these feature-scoring methods rank the features (selected after the preprocessing step) by their independently determined scores, and then select the top scoring features. There is another approach that is used to reduce the size of the initial feature set. This approach is known as feature transformation. This approach does not weigh terms to discard the lower weighted terms, but compacts the vocabulary based on feature concurrencies. Principal component analysis is a well-known method for feature transformation.

Infogain measure is the most commonly used feature-selection method. This method is used to rank all the features obtained after preprocessing, and then the top N scoring features are selected based on the rank. Infogain measure aims to identify those words from the document that aim to simplify the target concept.

Now, the total number of bits required to code any particular class distribution C is H(C0). It is given by the following formula:

$$N = \sum_{c \in C} n(c)$$

$$p(c) = \frac{n(c)}{N}$$

$$H(C) = -\sum_{c \in C} p(c) \log_2 p(c)$$

Now, suppose A is a group of attributes, then the total number of bits needed to code a class once an attribute has been observed is given by the following formula:

$$H(C|A) = -\sum_{a \in A} p(a) \sum_{c \in C} p(c|a) \log_2 \left[p(c|a) \right]$$

Now, the attribute that obtains the highest information gain is considered to be the highest ranked attribute, which is denoted by the symbol A_i.

$$\text{Infogain}(A_i) = H(C) - H(C|A_i)$$

Table 10.6 shows the list of words that are sorted on the basis of Infogain measure. These words are the unique words that are obtained after stemming the data. As we can see from the stemmed data, there are a total of 39 unique words. The Infogain of all these words was calculated and then they were given the rank, which is shown in Table 10.6. On the basis of this table, top-5 words, top-25 words, and so on can also be obtained by using Infogain measure.

To understand the concept of Infogain measure, the calculation of Infogain corresponding to two words "realtor" and "system" has been shown below. As presented in Table 10.6, the word "realtor" has been ranked 1 and the word "system" has been given the rank 38. This will become clear from their respective Infogain values. Table 10.7 represents the matrix of 0's and 1's corresponding to the two words, namely, "system" and "realtor." This

TABLE 10.6

List of Words Sorted on the Basis of Infogain Measure

Rank	Words	Rank	Words	Rank	Words	Rank	Words
1	realtor	11	train	21	conform	31	month
2	hour	12	user	22	smart	32	oper
3	time	13	consist	23	devic	33	day
4	interfac	14	color	24	profession	34	year
5	standard	15	scheme	25	appear	35	8am
6	access	16	font	26	normal	36	6pm
7	intuit	17	menus	27	busi	37	achiev
8	selfexplanatori	18	button	28	long	38	system
9	easi	19	navig	29	client	39	product
10	learn	20	feel	30	pc		

TABLE 10.7

Matrix Representing Occurrence of a Word in
a Document

Doc#	System	Realtor	NFRType
1	1	0	1
2	0	0	2
3	1	0	3
4	0	0	2
5	1	0	1
6	0	0	1
7	1	0	2
8	1	0	1
9	1	0	1
10	0	1	3
11	1	0	2
12	1	1	3

table represents whether a particular word is present in a document or not. If a word is present in a document, then it is assigned a value of "1," otherwise it is assigned a value of "0."

Now, first the entropy of the entire data set is found out. In our example 10.1, the data set consists of 12 documents, out of which five documents belong to type 1 NFR, four documents belong to type 2 NFR, and the remaining three documents belong to type 3 NFR. Thus, the entropy of the data set is as below:

$$\text{Entropy}(S) = -\left(\frac{5}{12}\right)\left(\log_2\frac{5}{12}\right) - \left(\frac{4}{12}\right)\left(\log_2\frac{4}{12}\right) - \left(\frac{3}{12}\right)\left(\log_2\frac{3}{12}\right)$$

$$= -0.42(-1.25) - 0.33(-1.6) - 0.25(-2)$$

$$= 0.525 + 0.528 + 0.5$$

$$= 1.553$$

The Infogain measure of the word "realtor" is as below:

$$\text{Infogain}(S, \text{realtor}) = \text{Entropy}(S) - \frac{2}{12}\text{Entropy}(1) - \frac{10}{12}\text{Entropy}(0)$$

$$= 1.553 - 0.17\left[-\left(\frac{2}{2}\right)\left(\log_2\frac{2}{2}\right)\right] - 0.83\left[-\left(\frac{5}{10}\right)\left(\log_2\frac{5}{10}\right) - \left(\frac{4}{10}\right)\left(\log_2\frac{4}{10}\right) - \left(\frac{1}{10}\right)\left(\log_2\frac{1}{10}\right)\right]$$

$$= 1.553 - 0 - 0.83\left[-0.5(-1) - 0.4(-1.32) - 0.1(-3.32)\right]$$

$$= 1.553 - 0.83\left[0.5 + 0.528 + 0.332\right]$$

$$= 0.424$$

Similarly, the Infogain measure of the word "system" is as below:

$$\text{Infogain}(S, \text{System}) = \text{Entropy}(S) - \left(\frac{8}{12}\right)\text{Entropy}(1) - \left(\frac{4}{12}\right)\text{Entropy}(0)$$

$$= 1.553 - \frac{8}{12}\left\{-\left[\left(\frac{4}{8}\right)\left(\log_2\frac{4}{8}\right)\right] - \left[\left(\frac{2}{8}\right)\left(\log_2\frac{2}{8}\right)\right] - \left[\left(\frac{2}{8}\right)\left(\log_2\frac{2}{8}\right)\right]\right\}$$

$$- \frac{4}{12}\left\{-\left[\left(\frac{1}{4}\right)\left(\log_2\frac{1}{4}\right)\right] - \left[\left(\frac{2}{4}\right)\left(\log_2\frac{2}{4}\right)\right] - \left[\left(\frac{1}{4}\right)\left(\log_2\frac{1}{4}\right)\right]\right\}$$

$$= 1.553 - 0.67\left[-0.5(-1) - 0.25(-2) - 0.25(-2)\right]$$

$$- 0.33\left[-0.25(-2) - 0.5(-1) - 0.25(-2)\right]$$

$$= 1.553 - 0.67[0.5 + 0.5 + 0.5] - 0.33[0.5 + 0.5 + 0.5]$$

$$= 1.553 - 1.005 - 0.495$$

$$= 0.053$$

Thus, we can see, the Infogain value of the two words, namely, "realtor" and "system" is 0.424 and 0.053, respectively, which is calculated by using the above formulae. Similarly, the Infogain measure of all the words obtained after stemming was calculated, and then these words were ranked based on their value. The top few words were then used for the developing the prediction model.

10.2.4 Constructing a Vector Space Model

Once a series of preprocessing tasks have been completed (removal of stop words, stemming) and relevant features have been extracted using a particular feature-selection method (Infogain), we will have the total number of attributes or terms as N, which can be represented as $t_1, t_2, \ldots, t_N$. The ith document is then represented as a N-dimensional vector consisting of n values, which are written as $(X_{i1}, X_{i2}, \ldots X_{iN})$. Here, X_{ij} is a weight measuring the importance of the jth term t_j in the ith document. The complete set of vectors for all documents under consideration is called a vector space model.

Term frequency (TF) is the count of total occurrences of a term in a given document. There are various methods that can be used for weighting the terms. Term frequency inverse document frequency (TFIDF) is the most popularly used method for calculating the weights. The term frequency can be represented in many ways as listed below:

1. Simple count of frequency of occurrences of terms.
2. Binary indication for the presence or absence of a term.
3. Normalizing the term frequency counts.

The weighted frequency count is 0 if the term is not present in the document, and a nonzero value otherwise. There are many ways to represent the normalized or weighted term frequency in a document. The following formula can be used to compute the normalized frequency count:

$$TF = \begin{cases} 0, \text{if frequency count is zero} \\ 1+\log\{1+\log[\text{frequency}(t)]\} \end{cases}$$

Apart from term frequency, it is important to define inverse document frequency that represents the importance of a term. The importance value is decreased if the term is present in many documents as it reduces the discriminative power (Bramer 2007).

$$IDF = \log_2 \frac{n}{n_j}$$

where:

n_j is the total number of documents containing j^{th} term

n is the total number of documents

This value is a combination of the terms that occur frequently in a particular document with the terms, which occur rarely among a group of documents.

TFIDF method is considered to be the most efficient method for weighting the terms. The TFIDF value of a term given in a document (X_{ij}) is defined as the product of two values, which correspond to the term frequency and the inverse document frequency, respectively. It is given as:

$$\text{TFIDF}(X_{ij}) = t_{ij} \times \log_2\left(\frac{n}{n_j}\right)$$

where:

t_{ij} is the frequency of the jth term in document i

n_j is the total number of documents containing j^{th} term

n is the total number of documents

Term frequency takes the terms that are frequent in the given document to be more important than the others. Inverse document frequency takes the terms that are rare across a group of documents to be more important than the others.

Example 10.2

Consider Table 10.8, the number of occurrences of each term in the corresponding document is shown. The row represents occurrence of each term in a document and

TABLE 10.8

Term Frequency Matrix Depicting the Frequency Count of Each Term in the Corresponding Document

Document/Term	t_1	t_2	t_3	t_4	t_5
d_1	0	2	8	0	0
d_2	5	20	8	15	0
d_3	14	0	0	5	0
d_4	20	4	13	0	5
d_5	0	0	9	7	4

column represent occurrences of a given term in each document. Based on the table, the TFIDF value can be calculated. For example, TFIDF value of term *t4* in document *d2* is calculated as below:

$$\text{TF} = t_4 = 1 + \log\left[1 + \log(15)\right]$$

$$= 1.337$$

$$\text{IDF}(t_4) = \log_2\left(\frac{5}{3}\right) = 0.736$$

$$\text{TFIDF} = t_4 \times \log_2\left(\frac{n}{n_4}\right) = 1.337 \times 0.736 = 0.985$$

Now, before using the set of *N*-dimensional vectors, we first need to normalize the values of the weights. It has been observed that "normalizing" the feature vectors before submitting them to the learning technique is the most necessary and important condition.

Table 10.9 shows the TFIDF matrix corresponding to the top-5 words. Now, this matrix represents the structured form of the original raw data that can now be used for the development of the prediction model.

10.2.5 Predicting and Validating the Text Classifier

Once the training documents have been converted into numerical form, we can use the techniques described in Chapter 7 for model prediction, and the accuracy of the model can be measured using performance measures such as recall, precision, *F*-measure, and receiver operating characteristics (ROC) analysis given in Section 7.5.

The data can be collected from software repositories (e.g., SVN, CVS, GIT), and the details are presented in Chapter 5.

TABLE 10.9

TFIDF Matrix of Top-5 Words of NFR Example

Doc	Realtor	Hour	Time	Interfac	Standard	NFR Type
1	0	2.115477	2.115477	0	0	1
2	0	0	0	0	0	2
3	0	0	0	0	0	3
4	0	0	0	2.70044	2.70044	2
5	0	0	2.115477	0	0	1
6	0	2.115477	0	0	0	1
7	0	0	0	2.70044	2.70044	2
8	0	2.115477	0	0	0	1
9	0	0	2.115477	0	0	1
10	2.70044	0	0	0	0	3
11	0	0	0	0	0	2
12	2.70044	0	0	0	0	3

10.3 Applications of Text Mining in Software Engineering

Text mining is being widely used in software engineering, where the large amount of information available in the form of texts or online is of unstructured form. Thus, it becomes very difficult to interpret the data and make suitable analysis resulting in inappropriate conclusions. In the field of software engineering, there are various artifacts produced during the software development life cycle such as software requirements specification (SRS) document, fault reports, and design documents. Text mining is gaining a huge attention that is being a great help to engineers, practitioners, and researchers working in this field. Some of the widely applicable areas of software engineering where text mining techniques are employed are given below.

10.3.1 Mining the Fault Reports to Predict the Severity of the Faults

One of the most popular applications of text mining is to mine the fault descriptions available in software repositories and extract relevant information from the description, which is in the form of some relevant words extracted by employing text mining techniques. Thus, the data is reduced to a structured format, which can now be applied for the development of prediction models. With the help of this structured data, the severities of the document could be predicted, which is one of the most important aspect of the fault reports. The prediction of fault severity is very important as the faults with higher severity could be dealt first on a priority basis, thus leading to an efficient utilization of the available resources and manpower. Menzies and Marcus (2008) mined fault description using text mining and machine learning techniques using rule-based learning.

10.3.2 Mining the Change Logs to Predict the Effort

Another very important application of text mining could be in the analysis of change logs of the different versions of the software and mining the fault reports. These fault reports contain the fault description, which could be mined in a similar way, and the amount of effort and time required to correct the fault could be predicted. On the basis of predicted effort, the change management board can take the required action as to whether the fault should be corrected or not. This appropriate decision by the change management board could lead to an effective utilization of the resources and the staff required in correcting the fault.

10.3.3 Analyzing the SRS Document to Classify Requirements into NFRs

Yet another very popular application of text mining is to classify the requirements stated in the SRS into their respective NFRs (quality attributes). We are familiar with the fact that the stakeholders are not able to state the NFRs as clearly as it is required, and these requirements are scattered throughout the document in an ad hoc and unorganized form. This can prove to be very harmful as NFRs contribute to the overall development of the software. Thus, these requirements should be incorporated at an early stage of design architecture. Hence, there is a need to mine the description of the requirements that are stated improperly in the SRS. These requirements could then be analyzed and thereby classified into their respective nonfunctional qualities like availability, usability, look-and-feel, maintainability, security, performance, and so on. As a result of this, critical quality constraints could be taken into account and development of an efficient software product meeting the stakeholder's real needs could be achieved.

Apart from these, there are various other applications of text mining that are restricted not only to the area of software engineering, but also to other fields of the literature like medicine, networking, chemicals, and so on.

10.4 Example—Automated Severity Assessment of Software Defect Reports

An example study is presented in this section to illustrate the techniques of text mining and to demonstrate their applicability in solving real-world fault severity prediction problems. As already specified, there are a number of areas wherein the concept of text mining can be put into use and the unstructured data can be converted into an organized and structured data. The intent of this example study is to highlight one of these applications and the use of model prediction in predicting the unknown instances in text mining problem.

10.4.1 Introduction

As the complexity and size of the software is increasing, the introduction of faults into the software has become an implicit part of the development, which cannot be avoided whatsoever the circumstances may be. This causes the faults to enter the software, thereby leading to functional failures. There are a number of bug tracking systems such as Bugzilla and CVS that are widely used to track the faults present in various open source software repositories. The faults, which are introduced in the software, are of varying severity levels. As a result, these bug tracking systems contain the fault reports that include detailed information about the faults along with their IDs and associated severity level. A severity level is used by many organizations to measure the effect of fault on the software. This impact may range from mild to catastrophic, wherein catastrophic faults are most severe faults that may lead the entire system to go to a crash state. The faults that have a severe impact on the functioning of the software and may adversely affect the software development are required to be handled on priority basis. Faults having high-severity level must be dealt with on a priority basis as their presence may lead to a major loss like human life loss, crash of an airplane, and so on. However, the data present in such systems is generally in unstructured form. Hence, text mining techniques in combination with machine learning techniques are required to analyze the data present in the defect tracking system.

In this study the information from the NASA's database called project and issue tracking system (PITS) is mined, by developing a tool that will first extract the relevant information from PITS-A using text mining techniques. After extraction, the tool will then predict the fault severities using machine learning techniques. The faults are classified into five categories of severity by NASA's engineers as very high, high, medium, low, and very low.

In this study multilayer perceptron technique is used to predict the faults at various levels of severity. The prediction of fault severity will help the researchers and software practitioners to allocate their testing resources on more severe areas of the software. The performance of the predicted model will be analyzed using area under the ROC curve (AUC) obtained from ROC analysis.

Testing is an expensive activity that consumes maximum resources and cost in the software development life cycle. This study is particularly useful for the testing professionals for quickly predicting severe faults under time and resource constraints. For example, if only 25% of testing resources are available, the knowledge of the severe faults will help the testers to focus the available resources on fixing the severe faults. The faults with higher severity level should be tested and fixed before the faults with lower severity level are tested and fixed (Menzies and Marcus 2008). The testing professionals can select from the list of prioritized faults according to the available testing resources using models predicted at various severity levels of faults. The models developed in this study will also guide the testing professional in deciding when to stop testing—when an acceptable number (perhaps decided using past experiences) of faults have been corrected and fixed, then the testing team may decide to stop testing and allow the release of the software.

10.4.2 Data Source

The PITS-A data set supplied by NASA's software Independent Verification and Validation (IV & V) Program was used to validate the results. The data in this data set include issues related to robotic satellite missions and human rated systems, which have been collected for more than 10 years. The data sets comprise of the fault reports, wherein each fault report contains the description of that corresponding fault, ID of the fault, and associated severity level of the fault. According to NASA's engineers, each fault can be categorized into one of the five severity levels, which are very high, high, medium, low, and very low.

Faults, which fall into the category of very high severity level are the faults that may threaten the safety and security. The recovery from such faults is impossible and failures, which happen because of the occurrence of such faults, may result in cascading system failures. Because of these reasons, such faults are extremely rare in nature. PITS-B does not have any fault at very high severity level. Hence, in the empirical study, the severity level 1 is not taken into account and only the last four severity levels, namely, severity 2 (high), severity 3 (medium), severity 4 (low), and severity 5 (very low) are considered.

10.4.3 Experimental Design

The focus here is to elaborate on the design that will be used for the overall model prediction. Initially, the fault reports were analyzed and the textual description corresponding to each fault was extracted. The textual descriptions were then analyzed using a tool that is named as "Text Analyzer and Miner." This tool was developed at Delhi Technological University. This tool has been developed using C# Language in the Windows Form Template on Visual Studio 2012 with Windows 7 as the operating platform. The tool has been developed using the .NET Framework.

This tool incorporates a series of text mining techniques—tokenization, stop words removal, stemming, feature selection using Infogain measure, and finally TFIDF weighting, as explained in the sections above. The objective of these techniques was to remove irrelevant words from the document and retain only those words that contribute to an effective model prediction.

Multilayer perceptron (MLP) has been used as the machine learning technique for the development of the prediction model. The evaluation measures used, depict how well the model has performed in predicting the dependent variable, which is "severity" in this case. The independent variables were the top few words that were found out on the basis of Infogain measure. In this study, top-5, 25, 50, and 100 words were considered and the performance of the model was evaluated with respect to these words corresponding to each of the severity level, namely, high, medium, low, and very low. Table 10.10 demonstrates the top-100 words obtained after the ranking done by Infogain measure. From this table, top-5, 25, and 50 words can also be obtained.

MLP is one of the most popular algorithm that is used for supervised classification. It is responsible for mapping a set of input values onto a set of appropriate output values. MLP technique is based on the concept of back propagation. Back propagation is a type of learning procedure that is used to train the network. It comprises of two passes—forward pass and backward pass. In the forward pass, the inputs are fed to the network and then the effect is propagated layer by layer by keeping all the weights of the network fixed. In the backward pass, the updation of weights takes place according to the error computed. The process is repeated over and over again until the desired performance is achieved.

To evaluate the performance of the predicted model, there were different performance measures that were used. These were sensitivity, AUC, and the cutoff point. All these measures determine how well the model has predicted what it was intended to predict. The explanation for these measures has been provided in Chapter 7. The validation method

TABLE 10.10

Top-100 Terms in PITS-A Data Set, Sorted by Infogain

Rank	Words	Rank	Words	Rank	Words	Rank	Words
1	requir	26	rvm	51	includ	76	trace
2	command	27	obc	52	specifi	77	symbol
3	softwar	28	lsrobc	53	set	78	differ
4	srs	29	system	54	projecta	79	ground
5	lsfs	30	telemetri	55	text	80	interrupt
6	test	31	code	56	time	81	reset
7	flight	32	execut	57	task	82	design
8	line	33	number	58	column	83	document
9	referenc	34	variabl	59	access	84	question
10	engcntrl	35	initi	60	refer	85	safe
11	mode	36	issu	61	uplink	86	sourc
12	file	37	provid	62	monitor	87	attitud
13	messag	38	locat	63	perform	88	door
14	script	39	section	64	paramet	89	lead
15	error	40	control	65	fp	90	contain
16	data	41	memori	66	note	91	event
17	oper	42	verifi	67	capabl	92	appear
18	spacecraft	43	indic	68	fsw	93	process
19	link	44	address	69	ivv	94	current
20	ac	45	sequenc	70	vm	95	lsrvml
21	defin	46	lsrup	71	rate	96	checksum
22	function	47	point	72	list	97	engin
23	state	48	cdh	73	case	98	load
24	valu	49	fault	74	check	99	support
25	tabl	50	specif	75	flexelint	100	transit

used in the study is holdout validation (70:30 ratio) in which the entire data set is divided into 70% training data and the remaining 30% as test data. A partitioning variable is used that splits the given data set into training and testing samples in 70:30 ratio. This variable can have the value either 1 or 0. All the cases with the value of 1 for the variable are assigned to the training samples, and all the other cases are assigned to the testing samples. To get more generalized and accurate results, validation has been performed using 10 separate partitioning variables. Each time, MLP method is used for training, and the testing samples are used to validate the results.

10.4.4 Result Analysis

The results of applying preprocessing steps such as tokenization, removal of stop words, and stemming is shown Figure 10.3. Figure 10.3 depicts that initially 156,499 words were found for the PITS-A data set. However, after applying a set of preprocessing steps, these words were reduced to 89,119. Further, after calculating Infogain, we validated the results on top-100 words. The list of top-100 words for PITS-A data set is given in Table 10.10. A few words, in this table, may not be understandable as these are stemmed. For example, *requir* in Table 10.10 is the stemmed form of the original word *requirement*.

The words can be used to predict severity at different level. Each document is a vector with 100 values with corresponding TFIDF scores for each word. The model prediction results were obtained using MLP method corresponding to top-5, 25, 50, and 100 words. All the four severity levels are taken into consideration with regard to high, medium, low, and very low. The results are depicted in Tables 10.11 through 10.18. The AUC values for 10 runs for top-5, 25, 50, and 100 words are shown in these tables. The use of multiple runs will reduce the threats to conclusion validity and produce generalized results.

As it is clear from Tables 10.11 and 10.12, MLP has performed very well in predicting high-severity faults as compared to medium, low, and very low severity faults. This is because the average value of AUC for high-severity defects is 0.86 approximately. On the other hand, the performance of the MLP model with respect to medium, low, and very low severity defects can be considered as nominal because the average AUC values are 0.78, 0.72, and 0.74, respectively. Thus, it can be concluded that MLP has performed

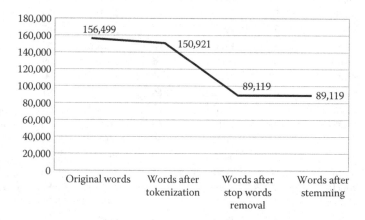

FIGURE 10.3
Results of applying preprocessing to the PITS-A data set.

TABLE 10.11

Results for Top-5 Words Corresponding to High and Medium Severity Faults

Runs	High Severity Defects			Medium Severity Defects		
	AUC	Sensitivity	Cutoff	AUC	Sensitivity	Cutoff
1	0.873	0.778	0.1818	0.785	0.545	0.4159
2	0.824	0.689	0.1632	0.786	0.519	0.4363
3	0.853	0.781	0.1655	0.759	0.553	0.432
4	0.852	0.733	0.2107	0.785	0.581	0.3892
5	0.862	0.78	0.1698	0.798	0.848	0.4652
6	0.872	0.772	0.1095	0.727	0.52	0.5539
7	0.83	0.2	0.1561	0.778	0.543	0.4477
8	0.84	0.753	0.1784	0.777	0.557	0.4361
9	0.897	0.833	0.1775	0.829	0.81	0.4311
10	0.868	0.765	0.1811	0.782	0.514	0.438
Average	0.857	0.708	–	0.781	0.599	–

TABLE 10.12

Results for Top-5 Words Corresponding to Low and Very Low Severity Faults

Runs	Low Severity Defects			Very Low Severity Defects		
	AUC	Sensitivity	Cutoff	AUC	Sensitivity	Cutoff
1	0.744	0.683	0.3562	0.726	0.6	0.0481
2	0.706	0.685	0.3402	0.733	0.714	0.0625
3	0.719	0.654	0.3049	0.701	0.571	0.0309
4	0.718	0.635	0.2837	0.783	0.714	0.0546
5	0.751	0.708	0.3277	0.789	1	0.0443
6	0.673	0.583	0.2995	0.599	0.75	0.0325
7	0.742	0.657	0.3316	0.732	0.667	0.0482
8	0.62	0.57	0.3058	0.844	0.857	0.0617
9	0.753	0.754	0.3175	0.794	0.6	0.0343
10	0.743	0.69	0.3363	0.732	0.667	0.0506
Average	0.716	0.662	–	0.743	0.714	–

exceptionally well in predicting high-severity faults than in predicting medium, low, and very low severity faults when top-5 words were considered for classification.

On such similar lines, conclusion can be drawn regarding the performance of MLP when taking into account top-25 words. It can be seen from Tables 10.13 and 10.14 that MLP has predicted high-severity faults with much correctness, as the maximum value of AUC is 0.903 with approximately 83% value of sensitivity. The performance of MLP is also good in predicting faults at other severity levels, namely, medium, low, and very low. This is because the average values of AUC at these severity levels are 0.80, 0.77, and 0.76 approximately. Thus, it can be said that MLP model is recommended for predicting the faults as the number of words considered for classification increases.

TABLE 10.13

Results for Top-25 Words Corresponding to High and Medium Severity Faults

Runs	High Severity Defects			Medium Severity Defects		
	AUC	Sensitivity	Cutoff	AUC	Sensitivity	Cutoff
1	0.891	0.804	0.2413	0.838	0.786	0.5085
2	0.856	0.781	0.2427	0.785	0.705	0.4721
3	0.903	0.829	0.2463	0.81	0.726	0.4763
4	0.87	0.793	0.2432	0.769	0.714	0.4796
5	0.902	0.833	0.2361	0.822	0.756	0.3956
6	0.898	0.804	0.2402	0.808	0.713	0.4209
7	0.863	0.812	0.2838	0.778	0.685	0.4291
8	0.855	0.789	0.2951	0.756	0.703	0.3382
9	0.859	0.787	0.2316	0.768	0.688	0.4621
10	0.873	0.783	0.2121	0.822	0.754	0.4023
Average	0.877	0.801	–	0.795	0.723	–

TABLE 10.14

Results for Top-25 Words Corresponding to Low and Very Low Severity Faults

Runs	Low Severity Defects			Very Low Severity Defects		
	AUC	Sensitivity	Cutoff	AUC	Sensitivity	Cutoff
1	0.813	0.741	0.2058	0.697	0.667	0.0216
2	0.754	0.627	0.2063	0.841	0.75	0.0262
3	0.794	0.690	0.2218	0.811	0.714	0.0343
4	0.753	0.636	0.3234	0.717	0.600	0.0244
5	0.800	0.735	0.3027	0.696	0.667	0.0343
6	0.774	0.672	0.2168	0.81	0.800	0.0384
7	0.706	0.652	0.2834	0.804	0.778	0.0367
8	0.71	0.627	0.297	0.515	0.500	0.0213
9	0.745	0.671	0.2568	0.855	0.800	0.0373
10	0.845	0.742	0.3051	0.807	0.625	0.019
Average	0.769	0.679	–	0.755	0.690	–

From Tables 10.15 and 10.16, it can be seen that the performance of MLP with respect to to top-50 words is exceptionally good for all types of faults with the average value of AUC being 0.91, 0.82, 0.80, and 0.81 at high, medium, low, and very low severities, respectively. So, from the discussion, it can be concluded that MLP method has worked very well for predicting the faults when taking into account top-50 words for classification.

Even when top-100 words are considered for classification (Tables 10.17 and 10.18), it is seen that the performance of MLP is exceptionally good in predicting high-severity faults as the maximum value of AUC is 0.928 with 85.3% sensitivity value. The performance of MLP model is nominal for other severity faults.

TABLE 10.15

Results for Top-50 Words Corresponding to High and Medium Severity Faults

Runs	High Severity Defects			Medium Severity Defects		
	AUC	Sensitivity	Cutoff	AUC	Sensitivity	Cutoff
1	0.916	0.865	0.2292	0.812	0.733	0.4776
2	0.905	0.817	0.1824	0.83	0.742	0.402
3	0.919	0.841	0.2642	0.831	0.739	0.4214
4	0.915	0.84	0.2209	0.83	0.736	0.5072
5	0.937	0.885	0.3282	0.829	0.75	0.4168
6	0.943	0.876	0.3365	0.846	0.757	0.4338
7	0.892	0.81	0.1858	0.824	0.748	0.4764
8	0.906	0.826	0.1963	0.824	0.748	0.4531
9	0.906	0.828	0.2702	0.792	0.71	0.43
10	0.875	0.793	0.3138	0.771	0.704	0.3972
Average	0.911	0.838	–	0.819	0.736	–

TABLE 10.16

Results for Top-50 Words Corresponding to Low and Very Low Severity Faults

Runs	Low Severity Defects			Very Low Severity Defects		
	AUC	Sensitivity	Cutoff	AUC	Sensitivity	Cutoff
1	0.801	0.692	0.224	0.754	0.667	0.026
2	0.848	0.743	0.273	0.848	0.6	0.020
3	0.808	0.704	0.246	0.816	0.75	0.032
4	0.773	0.703	0.185	0.896	0.75	0.028
5	0.806	0.712	0.191	0.848	0.778	0.027
6	0.812	0.721	0.226	0.673	0.6	0.017
7	0.795	0.681	0.230	0.799	0.714	0.040
8	0.792	0.708	0.256	0.846	0.833	0.056
9	0.823	0.768	0.240	0.801	0.6	0.022
10	0.755	0.653	0.260	0.820	0.778	0.031
Average	0.801	0.708	–	0.810	0.707	–

10.4.5 Discussion of Results

Although we have used few words to predict models, the AUC in many cases is high (see Tables 10.13 and 10.15). The AUC values of models predicted using the PITS-A data set are very high. The best results obtained for predicting faults at various severity levels are shown below:

- For severity = high, average AUC = 0.824–0.943
- For severity = medium, average AUC = 0.727–0.846

TABLE 10.17

Results for Top-100 Words Corresponding to High and Medium Severity Faults

	High Severity Defects			Medium Severity Defects		
Runs	AUC	Sensitivity	Cutoff	AUC	Sensitivity	Cutoff
1	0.923	0.863	0.2566	0.782	0.689	0.3952
2	0.912	0.82	0.3002	0.754	0.687	0.3787
3	0.918	0.833	0.2244	0.78	0.732	0.3807
4	0.928	0.853	0.2777	0.802	0.725	0.4713
5	0.899	0.816	0.2476	0.792	0.694	0.4344
6	0.904	0.816	0.2944	0.827	0.735	0.4387
7	0.918	0.848	0.2967	0.83	0.752	0.3464
8	0.889	0.809	0.2033	0.811	0.73	0.528
9	0.917	0.821	0.2636	0.803	0.708	0.4026
10	0.901	0.819	0.1994	0.807	0.742	0.4584
Average	0.910	0.829	–	0.798	0.719	–

TABLE 10.18

Results for Top-100 Words Corresponding to Low and Very Low Severity Faults

	Low Severity Defects			Very Low Severity Defects		
Runs	AUC	Sensitivity	Cutoff	AUC	Sensitivity	Cutoff
1	0.756	0.693	0.306	0.758	0.6	0.0231
2	0.785	0.711	0.3199	0.837	0.778	0.0215
3	0.826	0.741	0.2844	0.881	0.75	0.0421
4	0.823	0.753	0.2327	0.807	0.75	0.0197
5	0.766	0.688	0.2922	0.85	0.8	0.0502
6	0.819	0.75	0.233	0.687	0.5	0.0209
7	0.835	0.743	0.2711	0.642	0.6	0.0212
8	0.809	0.719	0.2061	0.663	0.571	0.0142
9	0.819	0.69	0.2195	0.659	0.571	0.021
10	0.805	0.729	0.2982	0.669	0.571	0.0261
Average	0.804	0.722	–	0.745	0.649	–

- For severity = low, average AUC = 0.62–0.848
- For severity = very low, average AUC = 0.515–0.896

The results show that there is marginal difference between the AUC of models predicted using top-50 words and the AUC of models predicted using only top-5 words in most of the cases. Hence, it is notable that using only very few words (only 5) the models perform nearly as well as the models predicted in most of the cases using large number of words. For example, individual results of model using PITS-A data set for top-50 words at high-severity level are 0.943. However, after reducing the words by 90% (i.e., from 50 words to 5 words), the maximum value of AUC obtained for top-5 words at high-severity level is 0.897.

10.4.6 Threats to Validity

In the PITS-A data set, the severity rating assigned to various faults are subjective and may be inaccurate. Thus, the generalizability of the results is possibly limited. Using holdout method at 10 runs for each model predicted reduces the threat to conclusion validity.

The conclusions are pertinent to only dependent variable, fault severity, as it seems to be most popular dependent variable in empirical studies. The validity of the models predicted in this study is not claimed when the dependent variable changes, like maintainability or effort.

While these results provide guidance for future research on the attributes extracted from fault descriptions at different severity levels, further validations are needed with different systems to draw stronger conclusions.

10.4.7 Conclusion

In today's scenario, there is an emerging need for the development of the defect prediction models, which are capable of detecting the fault introduced in the software. Not only this, the most important faults to consider in terms of the faults introduced in the software is its severity level that may range from mild to catastrophic. Catastrophic faults are the most severe faults that must be identified and then dealt with as early as possible to prevent any kind of damage to the software much further. With this intent, development of the fault prediction model has been carried out using MLP as the classification method. The data set employed for validation is the PITS data set, which is being popularly used by NASA's engineers. The data set was mined using text mining steps and the relevant information was extracted in terms of top few words (top-5, 25, 50, and 100 words). These words were used to predict the model. The predicted model was then used to assign a severity level to each of the fault found during testing.

It was observed from the results that MLP model has performed exceptionally well in predicting the faults at high-severity level irrespective of the number of words considered for classification. This observation is evident from the values of AUC lying in the range of 0.824–0.943. The performance of the model is even good for predicting medium severity faults as the maximum value of AUC is as high as 0.846. On the other hand, with respect to the faults at low and very low severity levels, the performance of the model is considered to be nominal. Thus, it can be concluded that the performance of the model is best when top-50 words are taken into account. Also, with very few words (only 5), the model has performed nearly as well as the models predicted in most of the cases using large number of words. From this analysis, it can be said that the model is suitable for predicting the severity levels of the faults even with very few words. This would be highly beneficial for an overall development of the organization in terms of proper allocation of testing resources and the available manpower.

Exercises

10.1 What is text mining? What are the applications of text mining in software engineering?

10.2 Explain the steps in text mining. Why mining relevant attributes is important before applying data analysis techniques?

10.3 Differentiate between unstructured and structured data.

10.4 Explain the different preprocessing steps in text mining.

10.5 What is the process and purpose of stemming in text mining?

10.6 How can important attributes be selected in text mining?

10.7 Explain information gain measure.

10.8 What is the purpose of weighting the terms in a document? What are the steps followed to compute TFIDF.

10.9 Consider the following data collected from a list of 500 documents, the top-5 words are given in the table, calculate the TDIDF value for each record.

Term	Number of Documents Containing the Term
Train	200
User	100
Learn	4
Display	50
System	20

10.10 Consider the following example given below, calculate the Infogain for each term.

Document/Term	T1	T2	T3	T4
D1	1	0	1	0
D2	0	0	0	1
D3	1	1	0	1
D4	0	0	1	1
D5	1	1	0	0

Further Readings

The following books provides techniques and procedures for information retrieval:

W. B. Frakes, and R. Baeza-Yates, *Information Retrieval: Data Structures & Algorithms*, Prentice Hall, Upper Saddle River, NJ, 1992.

G. Salton, and M. J. McGill, *Introduction to Modern Information Retrieval*, McGraw-Hill, New York, 1983.

The following papers provides term weighting approaches in text mining:

G. Salton, and C. Buckley, "Term weighting approaches in automatic text retrieval," *Information Processing and Management*, vol. 24, no. 5, pp. 513–523, 1998.

The information gain method is presented in:

D. McSherry, and C. Stretch, "Information Gain," Technical Notes, University of Ulster, Coleraine, 2003.

An excellent survey of text mining techniques is presented in:

M. W. Berry, *Survey of Text Mining: Clustering, Classification, and Retrieval*, Springer, New York, 2003.

In the following paper a novel approach that applies frequent item for text clustering is presented:

F. Beil, M. Ester, and X. Xu, "Frequent term-based text clustering," *Proceedings of the ACM SIGKDD International Conference on Knowledge Discovery in Databases*, pp. 436–442, Edmonton, Canada, July 2002.

The application of machine learning techniques in machine learning is given in:

M. Ikonomakis, S. Kotsiantis, and V. Tampakas, "Text classification using machine learning techniques," *WSEAS Transactions on Computers*, vol. 4, no. 8, pp. 966–974, 2005.

F. Sebastiani, "Machine learning in automated text categorization," *ACM Computing Surveys*, vol. 34, no. 1, pp. 1–47, 2002.

11

Demonstrating Empirical Procedures

The objective of this chapter is to demonstrate and present the practical application of the empirical concepts and procedures presented in previous chapters. This chapter also follows the report structure given in Chapter 9. The work presented in this chapter is based on change prediction.

The three important criteria for comparing results across various studies are (1) the study size, (2) the way in which the performance of the developed models is measured, and (3) statistical tests used. Also, the availability of data sets has always been a constraint in the software engineering research. The use of stable performance measures is another factor to be considered. The statistical tests for comparing the actual significance of results are not much used in change prediction models. Moreover, the models should be validated on the different data from which they are actually derived to increase the confidence in the conclusions of the study. To resolve these issues, in this chapter, we compare one statistical technique and 17 machine learning (ML) techniques for investigating the effect of object-oriented (OO) metrics on change-prone classes. The hypothesis is based on the fact that there is a statistical difference between the performance of the compared techniques.

11.1 Abstract

Software maintenance is a predominant and crucial phase in the life cycle of a software product as evolution of a software is important to keep it functional and profitable. Planning of the maintenance activities and distribution of resources is a significant step towards developing a software within the specified budget and time. Change prediction models help in identification of classes/modules that are prone to change in the future releases of a software product. These classes represent the weak parts of a software. Thus, change prediction models help software industry in proper planning of maintenance activities as change-prone classes should be allocated greater attention and resources for developing a good quality software.

11.1.1 Basic

Change proneness is an important software quality attribute as it signifies the probability that a specific class of a software would change in the forthcoming release of a software. A number of techniques are available in literature for development of change prediction models. This study aims to compare and assess one statistical and 17 ML techniques for effective development of change prediction models. The issues addressed are (1) comparing of the ML techniques and the statistical technique over popular data sets, (2) use of various performance measures for evaluating the performance of change prediction models,

(3) use of statistical tests for comparing and assessing the performance of ML techniques, and (4) validation of models from different data sets from which they are trained.

11.1.2 Method

To perform comparative analysis of one statistical and 17 ML techniques, the study developed change prediction models on six open source data sets. The data sets are application packages of the widely used Android OS. The developed models are statistically assessed using statistical tests on a number of performance measures.

11.1.3 Results

The results of the study indicate logistic regression (LR), multilayer perceptron (MLP), and bagging (BG) techniques as good techniques for developing change prediction models. The results of the study can be effectively used by software practitioners and researchers in choosing an appropriate technique for developing change prediction models.

11.2 Introduction

11.2.1 Motivation

Recently, there has been a surge in the number of studies that develop models for predicting various software quality attributes such as fault proneness, change proneness, and maintainability. These studies help researchers and software practitioners in efficient resource usage and developing cost-effective, highly maintainable good quality software products. Change proneness is a critical software quality attribute that can be assessed by developing change prediction models. Identification of change-prone classes is crucial for software developers as it helps in better planning of constraint project resources like time, cost, and effort. It would also help developers in taking preventive measures such as better designs and restructuring for these classes in the earlier phases of software development life cycle so that minimum defects and changes are introduced in such classes. Moreover, such classes should be meticulously tested with stringent verification processes like inspections and reviews. This would help in early detection of errors in the classes so that developers can take timely corrective actions. Although a number of techniques have been exploited and assessed to develop models for ascertaining change-proneness attribute, the search for the best technique still continues. The academia as well as the industry is tirelessly exploring the capabilities of different techniques to evaluate their effectiveness in developing efficient prediction models. Thus, there is an urgent need for comparative assessment of various techniques that can help the industry and researchers in choosing a practical and useful technique for model development.

11.2.2 Objectives

To develop software quality prediction models, various software metrics are used as the independent variables and a particular software quality attribute as the dependent variable. The model is basically a set of classification rules that can predict the dependent variable on the basis of the independent variables. The classification model can be created by a number of techniques such as the statistical technique (LR) or ML techniques. The capability of a technique can be assessed by evaluating the model developed by the particular technique. The various performance evaluators that are used in the study are classification accuracy, precision, specificity, sensitivity, *F*-measure, and area under the receiver operating characteristic (ROC) curve (AUC). According to Afzal and Torkar (2008), the use of a number of performance measures strengthens the conclusions of the study. This empirical study ascertains the comparative performance of statistical and ML techniques for the prediction of change-proneness attribute of a class in an OO software. Moreover, this study assesses models developed using tenfold cross-validation that are widely acceptable models in research (Pai and Bechta Dugan 2007; De Carvalho et al. 2010). Use of tenfold cross-validation reduces validation bias and increases the conclusion validity of the study. The study also strengthens its conclusions by statistically comparing the models developed by various techniques using Friedman and Nemenyi post hoc test. Furthermore, this study analyzes the application packages of a widely used mobile operating system named Android, which is open source in nature. Selection of such subjects for developing model increases the generalizability of the study and increases the applicability of the study's results.

11.2.3 Method

This study empirically validates six open source data sets to evaluate the performance of 18 different techniques for change-proneness prediction. The comparative assessment of various techniques is evaluated with the help of Friedman statistical test and Nemenyi post hoc test. The Friedman test assigns a mean rank to all the techniques on the basis of a specific performance measure and tests whether the predictive performance of all the techniques is equivalent. In case the predictive performance of various techniques is found to be statistically significantly different, there is a need to perform Nemenyi post hoc test. The Nemenyi test compares the results of each pair of techniques to ascertain the better performing technique among the two compared techniques. It computes the critical distance for the performance of two techniques and checks whether a pair of techniques are significantly different from each other or not. The study evaluates the change prediction models using six performance measures. The data sets used in the study are collected from the GIT repository using the defect collection and reporting system (DCRS) tool.

11.2.4 Technique Selection

LR, a statistical technique, is well-established for developing prediction models for ascertaining various software quality attributes. A number of studies in literature have used it for predicting fault proneness (Briand et al. 2000; El Emam et al. 2001; Aggarwal et al. 2009), maintainability (Li and Henry 1993; Alshayeb and Li 2003; Bandi et al. 2003), or change proneness (Zhou et al. 2009; Lu et al. 2012; Elish and Al-Khiaty 2013) attribute of a class in an OO software.

ML techniques have recently gained importance in the domain of software quality prediction as they can easily learn from data and past examples, and then efficiently classify new instances. ML techniques generalize based on examples and help in establishing cost-effective models. They have been extensively used for classification tasks in the last decade (Koru and Liu 2005; Thwin and Quah 2005; Koten and Gray 2006; Singh et al. 2009;

Dejager et al. 2013; Malhotra and Khanna 2013). This study evaluates the capabilities of 17 ML techniques for developing change prediction models and compares their predictive capability with LR. The ML techniques explored in this case study include adaptive boosting (AB), alternating decision tree (ADT), BG, Bayesian network (BN), decision table (DTab), J48 decision tree, repeated incremental pruning to produce error reduction (RIPPER), LogitBoost (LB), MLP, naïve Bayes (NB), non-nested generalized exemplars (NNge), random forests (RF), radial basis function (RBF) network, REP tree (REP), support vector using sequential minimal optimization (SVM-SMO), voted perceptron (VP), and ZeroR techniques implemented in the WEKA tool.

11.2.5 Subject Selection

The study analyzes a widely used mobile operating system named Android, which was developed by Google and is based on Linux kernel. It is dominating the mobile market with around 70% of smartphones using it as a preloaded OS. The Android OS uses GIT as the version control system. The source code of Android OS is available under the open source license. This study analyzes six application packages of the Android OS and the data is collected by comparing two Android versions, namely, Ice Cream Sandwich and Jelly Beans.

11.3 Related Work

Evaluation of change-proneness attribute of a class in an OO software helps software practitioners in proper allocation of resources to the identified change-prone classes. As change-prone classes are source of defects and changes, these classes require more attention and resources than a class that is not change-prone. Such a practice helps in efficient management of time, cost, and effort, and helps in producing a good quality software product within rigid deadlines and limited budgets.

Few researchers (Zhou et al. 2009; Lu et al. 2012) evaluated the change-proneness attribute in OO software data sets and established a relationship between OO metrics and change-prone nature of a class. A study by Elish and Al-Khiaty (2013) also investigated the use of evolution metrics along with Chidamber and Kemerer (1994) metrics suite for determination of change-prone classes. Malhotra and Khanna (2014a) devised a new metrics using gene expression programming for identifying change prone classes. Certain other researchers (Malhotra and Khanna 2013; Malhotra and Bansal 2014) explored the effectiveness of a few ML algorithms for developing change prediction models. Koru and Liu (2007) suggest tree-based models for ascertaining change-prone classes. Koru and Tian (2005) investigate the relationship between modules exhibiting high structural values and modules exhibiting high change values. A study by Han et al. (2010) advocates behavioral dependency measure as an important indicator for change-prone classes.

This study extensively compares the performance of 17 ML techniques with a traditional technique, LR, for developing an effective change prediction model. It will provide guidelines to researchers and practitioners for efficient selection of a classification technique. Furthermore, the use of Android data set for performing such comparisons favors wide applicability of the results of the study considering the popularity of Android applications.

11.4 Experimental Design

11.4.1 Problem Definition

The study investigates the effectiveness of OO metrics for predicting change-prone classes, and compares the performance of statistical and ML techniques for developing models to ascertain change-prone classes using various performance measures. The objective of the study is to statistically assess the performance of a number of techniques to provide future guidance for effective use of these techniques. The study also investigates the relationship of various techniques with various data sets. This study is quite effective as it is designed to minimize a number of validity threats and to increase the generalizability of its results. The main design considerations of the study are discussed as follows:

- Use of tenfold cross-validation method to build models using various techniques reduces validation bias and increases the conclusion validity of the study.
- Use of a number of performance measures (accuracy, sensitivity, specificity, precision, F-measure, and AUC) to analyze the predicted models strengthens the conclusions of the study.
- Assessment of results statistically using Friedman and Nemenyi test substantiates the effectiveness of the results.
- Use of widely prevalent open source software Android OS increases the generalizability of the results of the study. Thus, reducing external validity of the study.
- Finally, a comprehensive comparison of a number of techniques provides a strong research evidence on the applicability and use of these techniques for change prediction tasks.

11.4.2 Research Questions

The study explores the following research questions:

RQ1: Are OO metrics related to change?

It is important to ascertain the relationship between various metrics that are representative of different OO attributes like size, abstraction, inheritance, coupling, and so on and change-prone nature of a class. A change prediction model can only be developed using this relationship, where different OO metrics are independent variables and the change-prone attribute of a class is the dependent variable of the class.

RQ2: What is the capability of various techniques on data sets with varying characteristics?

There are a number of techniques available that can be used for developing change prediction models. Different techniques may show contrasting results on different data sets. Thus, it is important to evaluate the capability of different techniques using varied data sets as they may have varying characteristics.

RQ3: What is the comparative performance of different techniques when we take into account different performance measures?

A number of performance measures are available in literature to assess a prediction model. It is important to analyze the capability of a technique by evaluating the change prediction model developed by the technique using various performance measures. The comparative performance of techniques could be different when we take into account different performance measures. For example, model developed by technique A may give good accuracy values but very low *F*-measure values.

RQ4: Which pairs of techniques are statistically significantly different from each other for developing change prediction models?

Apart from evaluating all the techniques together, the study compares the performance of pairs of techniques using different performance measures. Pairwise comparisons gives us an insight into actual comparative performance of the two techniques forming the pair, and whether the techniques are significantly different in their performance from each other.

RQ5: What is the comparative performance of ML techniques with the statistical technique LR?

The basic functioning of statistical technique involves strict data assumptions and formulation of hypothesis. However, ML techniques do not require any initial hypothesis and the model developed using these techniques is flexible and adaptable to changing data (Malhotra and Khanna 2014b). Thus, there is a need to ascertain the comparative performance of the traditional statistical technique LR with ML techniques to evaluate which category of techniques is better for development of change prediction model.

RQ6: Which ML technique gives the best performance for developing change prediction models?

Different ML techniques work differently and have different characteristics such as speed, model accuracy, interpretability, and simplicity. This study evaluates the best ML technique for developing an effective and efficient change prediction model by evaluating change prediction models developed by 17 ML techniques.

11.4.3 Variables Selection

Over the years, a number of researchers have successfully used OO metrics for developing models that efficiently predict various software quality attributes. Continuous measurement of various attributes of a software like its size, cohesive capabilities, use of inheritance, and so on is important to gain an understanding of the project. Metrics can help in effectively managing the software project and developing a high-quality product.

This study uses a set of 18 commonly used OO metrics for developing change prediction models that include all the metrics proposed by Chidamber and Kemerer (1994), namely, weighted methods of a class (WMC), depth of inheritance tree (DIT), number of children (NOC), coupling between objects (CBO), response for a class (RFC), and lack of cohesion among methods (LCOM). The study also analyzed the quality model for object-oriented design metric suite that consists of data access metric (DAM), measure of aggression (MOA), method of functional abstraction (MFA), cohesion among methods of a class (CAM), and number of public methods (NPM). Afferent coupling (Ca) and efferent coupling (Ce) metrics proposed by Martin (2002) were also included. Some other miscellaneous metrics included in the study were inheritance coupling (IC) metric, coupling between methods of a class (CBM), average method complexity (AMC), lines of code (LOC), and LCOM3 (Henderson version of LCOM) metric. The detailed definition of each metric can be referred from Chapter 3. These metrics are the independent variables of the study and measure various OO properties of a software like cohesion, size, coupling, reusability, and so on.

The objective of the study is to ascertain change-prone classes. Thus, change proneness, that is, the likelihood of change in a class after the software goes into operation phase is the dependent variable of the study. To comprehend change in a class, LOC inserted or deleted from a class is taken into account.

11.4.4 Hypothesis Formulation

To answer RQ3, we developed the following set of hypothesis. The hypothesis evaluates the change prediction models developed using various techniques on different performance measures. Each hypothesis is based on a different performance measure.

11.4.4.1 Hypothesis Set

- Hypothesis for accuracy measure
  - H_0 *null hypothesis:* Change prediction models developed using all the techniques (LR, AB, ADT, BG, BN, DTab, J48, RIPPER, LB, MLP, NB, NNge, RF, RBF, REP, SVM-SMO, VP, and ZeroR) do not show significant differences when evaluated using accuracy measure.
  - H_a *alternate hypothesis:* Change prediction models developed using all the techniques (LR, AB, ADT, BG, BN, DTab, J48, RIPPER, LB, MLP, NB, NNge, RF, RBF, REP, SVM-SMO, VP, and ZeroR) show significant differences when evaluated using accuracy measure.
- Hypothesis for sensitivity measure
  - H_0 *null hypothesis:* Change prediction models developed using all the techniques (LR, AB, ADT, BG, BN, DTab, J48, RIPPER, LB, MLP, NB, NNge, RF, RBF, REP, SVM-SMO, VP, and ZeroR) do not show significant differences when evaluated using sensitivity measure.
  - H_a *alternate hypothesis:* Change prediction models developed using all the techniques (LR, AB, ADT, BG, BN, DTab, J48, RIPPER, LB, MLP, NB, NNge, RF, RBF, REP, SVM-SMO, VP, and ZeroR) show significant differences when evaluated using sensitivity measure.
- Hypothesis for specificity measure
  - H_0 *null hypothesis:* Change prediction models developed using all the techniques (LR, AB, ADT, BG, BN, DTab, J48, RIPPER, LB, MLP, NB, NNge, RF,

RBF, REP, SVM-SMO, VP, and ZeroR) do not show significant differences when evaluated using specificity measure.

- H_a *alternate hypothesis:* Change prediction models developed using all the techniques (LR, AB, ADT, BG, BN, DTab, J48, RIPPER, LB, MLP, NB, NNge, RF, RBF, REP, SVM-SMO, VP, and ZeroR) show significant differences when evaluated using specificity measure.

- Hypothesis for precision measure
  - H_0 *null hypothesis:* Change prediction models developed using all the techniques (LR, AB, ADT, BG, BN, DTab, J48, RIPPER, LB, MLP, NB, NNge, RF, RBF, REP, SVM-SMO, VP, and ZeroR) do not show significant differences when evaluated using precision measure.
  - H_a *alternate hypothesis:* Change prediction models developed using all the techniques (LR, AB, ADT, BG, BN, DTab, J48, RIPPER, LB, MLP, NB, NNge, RF, RBF, REP, SVM-SMO, VP, and ZeroR) show significant differences when evaluated using precision measure.

- Hypothesis for *F*-measure
  - H_0 *null hypothesis:* Change prediction models developed using all the techniques (LR, AB, ADT, BG, BN, DTab, J48, RIPPER, LB, MLP, NB, NNge, RF, RBF, REP, SVM-SMO, VP, and ZeroR) do not show significant differences when evaluated using *F*-measure.
  - H_a *alternate hypothesis:* Change prediction models developed using all the techniques (LR, AB, ADT, BG, BN, DTab, J48, RIPPER, LB, MLP, NB, NNge, RF, RBF, REP, SVM-SMO, VP, and ZeroR) show significant differences when evaluated using *F*-measure.

- Hypothesis for AUC measure
  - H_0 *null hypothesis:* Change prediction models developed using all the techniques (LR, AB, ADT, BG, BN, DTab, J48, RIPPER, LB, MLP, NB, NNge, RF, RBF, REP, SVM-SMO, VP, and ZeroR) do not show significant differences when evaluated using AUC performance measure.
  - H_a *alternate hypothesis:* Change prediction models developed using all the techniques (LR, AB, ADT, BG, BN, DTab, J48, RIPPER, LB, MLP, NB, NNge, RF, RBF, REP, SVM-SMO, VP, and ZeroR) show significant differences when evaluated using AUC performance measure.

11.4.5 Statistical Tests

To statistically compare the performance of different techniques for developing change prediction models, the study performs Friedman's statistical test, which is followed by a post hoc test named Nemenyi test. If the null hypothesis of Friedman test is rejected, we will perform post hoc analysis using Nemenyi test. These nonparametric tests suggested by Demšar (2006) are used for comparison of various techniques. Lessmann et al. (2008) also used these tests to compare a number of fault prediction models, which were developed using 22 classification techniques. The tests are nonparametric and hence are not based on data normality assumptions. The details on these tests can be found in Chapter 6.

11.4.6 Empirical Data Collection

This study uses the DCRS tool for data collection. The tool was developed by undergraduate students of Delhi Technological University and is used for collecting data from open source repositories that use GIT as the version control system. The tool is developed in Java language (Malhotra et al. 2014).

To collect change data, the DCRS tool analyzes the source code of an open source software from two specific consecutive versions to extract change logs. A change log records all the changes made in the source code of a file. Changes could be because of defect correction, change in requirements, technological upgrade, or any other reason. The change record stores information such as timestamp of the commit, a change identifier, a defect identifier if the change is because of defect correction, change description, and a listing of all modified files of source code along with all changed LOC.

Each data point consists of OO metrics and change statistics. All the metrics generated by the tool are obtained with the aid of another open source tool Chidamber and Kemerer Java Metrics (http://gromit.iiar.pwr.wroc.pl/p_inf/ckjm/metric.html). The change statistics account for the LOC changes in a source code file. The tool generates a change report that states the name of the Java source code file along with total LOC added for all the changes, total LOC deleted for all the changes, and the total LOC changes in a class. Each data point also states a binary variable named ALTER. The variable is given a value of "yes" if the total LOC changes of a particular class is greater than zero, otherwise it is assigned a "no."

This study analyzes the source code of Android OS, which uses Git for version control management. To collect change data, the two versions of Android OS selected are Ice Cream Sandwich and Jelly Bean, whose source code can be obtained from Git repository hosted by Google (https://android.googlesource.com). The code for Android OS was distributed in a number of application packages, namely, Contacts, Calendar, Bluetooth, and so on. Only Java source code or class files were analyzed in these packages ignoring all the other files like media files, layout files, and so on. This study analyzes six packages of Android OS: Bluetooth, Contacts, Calendar, Gallery2, MMS, and Telephony. The details of classes in each of the packages are given in Table 11.1. The table also shows the percentage of classes that changed from one version to the next and the number of data points, that is, common classes in each package. Data collection was done by the DCRS tool by "Cloning," that is, copying and transferring the entire repository from a server (local or remote) to an end user machine. A detailed description of the DCRS tool can be referred from Chapter 5. Figure 11.1 shows the change statistics of each data set. For example, Telephony data set has 19,313 inserted LOC; 22,228 deleted LOC; and 41,541 total changed lines.

TABLE 11.1

Software Data Set Details

Software Name	Versions Analyzed	No. of Data Points	% of Changed Classes
Bluetooth	4.3.1–4.4.2	72	19%
Contacts	4.3.1–4.4.2	210	48%
Calendar	4.3.1–4.4.2	106	19%
Gallery2	4.3.1–4.4.2	374	41%
MMS	2.3.7–4.0.2	195	30%
Telephony	4.2.2–4.3.1	249	63%

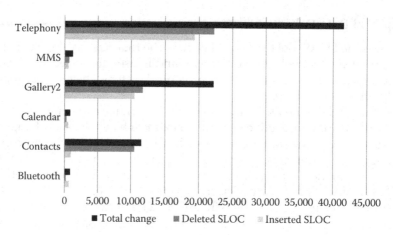

FIGURE 11.1
Change statistics of all data sets.

11.4.7 Technique Selection

The study compares and analyzes the capability of 18 different techniques for developing change prediction models. The various techniques are selected on the basis of various aspects, which are discussed as follows:

- LR: The technique is a traditional technique which is widely used in literature (Li and Henry 1993; Briand et al. 2000; El Emam et al. 2001a; Alshayeb and Li 2003; Bandi et al. 2003; Aggarwal et al. 2009; Zhou et al. 2009; Lu et al. 2012; Elish and Al-Khiaty 2013) for developing a number of software quality models. It is a well-recognized technique that can be easily used in a probabilistic framework.

- AB: The technique works well even with noisy data or outliers. It is fast, simple, and flexible.

- ADT: The technique is easy to interpret and combines the simplicity of decision trees and good performance of a boosting algorithm (Freund and Mason 1999; De Comite et al. 2003).

- BG: This technique is quite effective in terms of classification accuracy over a number of regression techniques. Moreover, it reduces variance and avoids overfitting to training data (Malhotra and Khanna 2013).

- BN: This technique is simple and easy to understand.

- J48: This technique builds models that are easy to interpret, are accurate, and are efficient in terms of speed (Zhao and Zhang 2008).

- RIPPER: This technique is a rule-based classifier, which is easy to interpret and generate. It is a fast algorithm (Uzun and Tezel 2012).

- LB: This technique produces prediction models with high accuracy.

- MLP: This technique is adaptive in nature and supports parallel architecture. It can easily handle nonlinear data (Malhotra and Khanna 2015; Malhotra 2014).

- RF: This technique is robust to noise and performs well even with outliers. It is simple and fast (Malhotra and Khanna 2015).

- REP Tree: It is a fast decision tree technique that reduces variance (Mohamed et al. 2012).
- SVM: This technique effectively handles high dimensionality, redundant features, and complex functions. It is robust in nature (Malhotra 2014).
- VP: This technique is comparable in terms of accuracy to SVM. However, literature claims that the technique has better learning and prediction speed as compared to the traditional SVM technique (Freund and Schapire 1999; Sassano 2008).

Some other techniques such as DTab, NNge, RBF, and ZeroR were also selected for analyzing their performances.

11.4.8 Analysis Process

The various steps performed for analysis are as follows:

- The descriptive statistics of all the data sets are collected and analyzed.
- Next, identify all the outliers in a particular data set and remove them. The change prediction models were developed with the remaining data points using 18 techniques.
- Next, to reduce the dimensionality of the input data set, use correlation-based feature selection (CFS) method and identify the important metrics for each corresponding data set. This step eliminates the noisy and redundant features of each data set.
- Now develop models using all the 18 techniques on the six selected data sets using tenfold cross-validation method. The change prediction models developed by all the techniques are evaluated using six performance measures.
- Analyze the developed models using Friedman statistical test and evaluate the developed hypothesis.
- Finally, perform Nemenyi post hoc test to find the pairs of techniques that are statistically significantly different from each other.

11.5 Research Methodology

This section briefly states the description of the techniques and the various performance measures used in the study. It also briefly describes the validation method.

11.5.1 Description of Techniques

This case study evaluates the performance of 18 techniques. Out of these 18 techniques, LR is a statistical technique and all other techniques are ML techniques. A brief description of each technique is given below. To develop effective change prediction models, we first need to reduce the dimensionality of our input features. This is done by applying CFS method proposed by Hall (2000). The method identifies a set of all noisy and unwanted features and eliminates them before model development. This helps in improving the results of the model.

- LR: It is a technique to predict a dependent variable such as change proneness from a set of independent variables, that is, various OO metrics to ascertain the variance in the dependent variable caused by the independent variables (Hosmer and Lemeshow 1989).

- AB: It is an important algorithm for boosting methods. The technique develops an efficient classifier by combining a number of weak performing classifiers (Witten and Frank 2005). The technique allots weights to weak learners according to their performance. It gains knowledge by analyzing the incorrect predictions of the previous models. The default settings for this technique in WEKA tool are 10 iterations and 100% of weight mass percentage.

- ADT: It is a generalized version of decision tree and works with a combination of boosting technique. It has two types of nodes: decision nodes and prediction nodes. The decision nodes represent a predicate decision and the prediction nodes contain a number (Freund and Mason 1999). To classify an instance, a path is followed in which all decision nodes evaluate to true and all prediction nodes occurring in between are summed up. The technique uses 10 boosting iterations and no saving of instance data as its parameter settings in the WEKA tool.

- BG: It is based on the concept of constant improvization by using a number of similar training sets (Breiman 1996). Training sets are produced by creating bootstrap duplicates of the original training set. A minor disturbance in the training set might cause significant changes in the predictor. Each training set is used to train a function and the output class of BG is the result output by the majority of functions. The parameter setting for BG in WEKA tool were a bag size percent of 100, REP tree classifier, and 10 iterations.

- BN: It is a network with a set of nodes that are connected by directed edges. It helps in ascertaining relationship between probabilistic values of relationship dependency and random variables. The strength of connections between the random variables is assessed quantitatively and the end result is a joint probability distribution from the network. WEKA uses a simple estimator and K2 algorithm as default settings for this algorithm.

- DTab: It represents a data structure that contains complex data entries in the upper levels of the table (Witten and Frank 2005). It is a hierarchical representation where the complex entries are simplified with the aid of additional attributes. The technique selects various attributes for simplifying the table structure by analyzing all possible combinations of the existent attributes. The technique selects the combination of attributes that gives the best result. The parameter settings were accuracy as evaluation measure and best first technique for searching in WEKA tool.

- J48: It is an implementation of the C4.5 algorithm in Java for the WEKA tool. The parameter settings for the technique in WEKA tool were 0.25 as the confidence interval for pruning and a minimum number of two instances in the leaf.

- RIPPER: It is a rule-based learner which formulates a set of rules that minimize the error in output predictions (Cohen 1995). The technique has four stages: build, grow, prune, and optimize. The parameter settings for the technique in the WEKA tool were three folds, two optimizations, seed as one, and use of pruning as true.

- LB: It is a boosting technique that uses additive LR (Friedman et al. 2000). The technique uses a likelihood threshold of -1.79, weight threshold of 100, 10 iterations, a shrinkage parameter of 1, and reweighing as the parameter settings in the WEKA tool.

- MLP: It is based on the functioning of nervous system and is capable of modeling complex relationships with ease. Apart from an input and an output layer, they have a number of intermediate hidden layers. Synaptic weights are assigned and adjusted in these intermediate layers with back propagation training algorithm. The technique comprises of two passes: forward as well backward. The backward pass produces an error signal that helps in reducing the difference between actual and desired output (Haykin 1998). WEKA uses a learning rate of 0.005 and sigmoid function for transfer. The number of hidden layers was set as 1.

- NB: This ML algorithm is based on Bayes theorem and creates a probabilistic model for prediction. All the features of the technique are treated independently, and it uses only a small training set for developing classification models. The default settings of WEKA uses kernel estimator and supervised discretization as false for NB.

- NNge: This technique uses NNge that are hyperrectangles. These can be viewed as if rules. The parameter settings were five attempts for generalization and five folders for mutual information in WEKA tool.

- RF: It is composed of a number of tree predictors. The tree predictors are based on a random vector, which is sampled independently with the same distribution. The output class of RF is the mode of all the individual trees in the forest (Breiman 2001). RF is advantageous because of its noise robustness, parallel nature, and fast learning. The RF were used with 10 trees as parameter settings in the WEKA tool.

- RBF network: This technique is an implementation of normalized Gaussian RBF network. To derive the centers and widths of the hidden layer, the algorithm uses m-means. The LR technique is used to combine the outputs from the hidden units. The technique uses a parameter settings of two clusters and one clustering seed in the WEKA tool.

- REP: It is a fast decision tree learning technique that uses information gain or variance reduction to build a decision tree. The technique performs reduced error pruning with backfitting. The default parameter settings for this technique in WEKA tool were as seed of 1, maximum depth of -1, minimum variance of 0.001, and 3 number of folds.

- Support vector machine (SVM): It aims to construct an optimal hyperplane that can efficiently separate the new instances into two separate categories (Cortes and Vapnik 1995). WEKA uses sequential minimal optimization algorithm to train the SVM. The parameter settings used by the technique in WEKA tool were random seed of 1, tolerance parameter of 0.01, a c value of 1.0, and an epsilon value of 1.0E-12 and a polykernel.

- VP: It is a technique that is based on the Rosenblatt Frank's (Frank 1958) perceptron technique. The technique can be effectively used in high-dimensional spaces with the use of kernel functions. The parameter settings for the technique in WEKA tool were an exponent value of 1 and a seed of 1.

- ZeroR: It is a technique that uses 0-R classifier. The technique predicts the mean if the class is numeric or the mode if the class is nominal.

11.5.2 Performance Measures and Validation Method

The performance measures used for analyzing the change prediction models were accuracy, sensitivity, specificity, precision, *F*-measure, and AUC. A brief description of the performance measures are as follows (for detailed description, refer to Chapter 7):

- Accuracy: It represents the percentage of correct predictions. It takes into account both correctly classified change-prone classes as well as correctly classified non-change-prone classes. The higher the accuracy of the model, the better is the model.
- Sensitivity: It represents the percentage of correctly classified change-prone classes.
- Specificity: It represents the percentage of correctly classified non-change-prone classes.
- Precision: It represents the percentage of predicted change-prone classes that were correct, that is, it depicts how many predicted change-prone classes are actually change prone.
- *F*-measure: This measure couples both precision and sensitivity, and represents the harmonic mean of precision and sensitivity.
- AUC: It is a plot of sensitivity on the vertical axis and a measure of 1-specificity on the horizontal axis. It is a cumulative measure of both sensitivity and specificity. The higher the AUC value, the better is the model.

The change prediction models were developed using tenfold cross-validation method. The tenfold cross-validation method involves division of data points into 10 partitions, where nine subsets are used for training the model while the tenth partition is used for model validation. A total of 10 iterations are performed, each time using a different set as the validation set (Stone 1974).

11.6 Analysis Results

This section describes the various steps performed for analyzing the models. It presents the results of the study.

11.6.1 Descriptive Statistics

This section gives a brief description of descriptive statistics of each data set and also performs correlation analysis. Tables 11.2 through 11.7 state the descriptive statistics, that is, the minimum (Min.), maximum (Max.), mean (Mean), standard deviation (SD), 25% percentile, and 75% percentile for all the metrics used in the study for each data set.

After analyzing Tables 11.2 through 11.7, we present a brief description of various characteristics of all the data sets.

TABLE 11.2

Descriptive Statistics for Android Bluetooth Data Set

Metric Name	Min.	Max.	Mean	SD	Percentile (25%)	Percentile (75%)
WMC	1	118	15	20.24	4	17.25
DIT	1	1	1	0	1	1
NOC	0	0	0	0	0	0
CBO	0	10	1.94	2.32	0	3
RFC	2	119	16	20.24	5	18.25
LCOM	0	6903	307	1042.07	6	140.25
Ca	0	5	0.76	1.15	0	1
Ce	0	10	1.40	2.09	0	2
NPM	0	16	4.51	3.89	1.75	7
LCOM3	1.01	2	1.28	0.34	1.06	1.33
LOC	9	718	104.86	127.65	31	116.5
DAM	0	1	0.67	0.39	0.36	1
MOA	0	6	0.72	1.26	0	1
MFA	0	0	0	0	0	0
CAM	0.13	1	0.42	0.24	0.25	0.542
IC	0	0	0	0	0	0
CBM	0	0	0	0	0	0
AMC	1	5	4.93	0.47	5	5

TABLE 11.3

Descriptive Statistics for Android Contacts Data Set

Metric Name	Min.	Max.	Mean	SD	Percentile (25%)	Percentile (75%)
WMC	0	60	10.86	10.13	4	14
DIT	0	4	0.9	0.42	1	1
NOC	0	15	0.11	1.06	0	0
CBO	0	15	1.27	2.07	0	2
RFC	0	61	11.86	10.14	5	15
LCOM	0	1770	104.74	229.76	6	91
Ca	0	15	0.43	1.44	0	0.75
Ce	0	12	0.83	1.49	0	1
NPM	0	43	7.10	6.85	2.25	9
LCOM3	0	2	1.39	0.40	1.07	2
LOC	0	422	70.50	69.02	24	91.75
DAM	0	1	0.65	0.45	0	1
MOA	0	5	0.19	0.59	0	0
MFA	0	0.96	0.01	0.08	0	0
CAM	0	1	0.39	0.22	0.21	0.5
IC	0	1	0	0.06	0	0
CBM	0	1	0	0.06	0	0
AMC	0	5	4.15	1.78	5	5

TABLE 11.4

Descriptive Statistics for Android Calendar Data Set

Metric Name	Min.	Max.	Mean	SD	Percentile (25%)	Percentile (75%)
WMC	1	110	14.47	17.20	5	17.75
DIT	0	1	0.95	0.21	1	1
NOC	0	1	0.37	0.19	0	0
CBO	0	12	1.70	2.12	0	2
RFC	2	111	15.47	17.20	6	18.75
LCOM	0	5995	244.12	755.04	10	148.75
Ca	0	4	0.56	0.89	0	1
Ce	0	11	1.19	1.84	0	2
NPM	1	65	9.39	10.43	3	11.75
LCOM3	1	2	1.26	0.32	1.05	1.31
LOC	7	936	107.43	134.61	31	133.5
DAM	0	1	0.70	0.37	0.44	1
MOA	0	10	0.55	1.29	0	1
MFA	0	0	0	0	0	0
CAM	0.05	1	0.38	0.24	0.20	0.5
IC	0	1	0.01	0.09	0	0
CBM	0	1	0.01	0.09	0	0
AMC	0	5	4.65	1.19	5	5

TABLE 11.5

Descriptive Statistics for Android Gallery2 Data Set

Metric Name	Min.	Max.	Mean	SD	Percentile (25%)	Percentile (75%)
WMC	0	251	11.55	16.74	4	14
DIT	0	3	0.64	0.54	0	1
NOC	0	23	0.35	1.70	0	0
CBO	0	42	3.06	4.66	1	4
RFC	0	252	12.55	16.74	5	15
LCOM	0	31,375	200.85	1,646.30	6	91
Ca	0	42	1.35	4.13	0	1
Ce	0	16	1.76	2.17	0	3
NPM	0	251	8.62	14.94	3	10
LCOM3	0	2	1.36	0.38	1.07	1.5
LOC	8	1,507	75.46	106.99	24	93
DAM	0	1	0.70	0.42	0.15	1
MOA	0	8	0.49	1.08	0	1
MFA	0	1	0.01	0.07	0	0
CAM	0	1	0.41	0.23	0.25	0.5
IC	0	2	0.01	0.14	0	0
CBM	0	4	0.02	0.23	0	0
AMC	0	5	2.87	2.42	0	5

TABLE 11.6

Descriptive Statistics for Android MMS Data Set

Metric Name	Min.	Max.	Mean	SD	Percentile (25%)	Percentile (75%)
WMC	0	128	10.87	14.14	3	14
DIT	0	4	0.85	0.54	1	1
NOC	0	5	0.19	0.72	0	0
CBO	0	12	2.00	2.09	0	3
RFC	0	129	11.84	14.17	4	15
LCOM	0	8128	153.33	612.94	3	91
Ca	0	11	0.89	1.65	0	1
Ce	0	10	1.65	1.39	0	2
NPM	0	53	7.70	9.63	2	8.25
LCOM3	1	2	1.45	0.40	1.08	2
LOC	0	882	66.58	94.75	13	83
DAM	0	1	0.57	0.45	0	1
MOA	0	6	0.23	0.73	0	0
MFA	0	1	0.01	0.12	0	0
CAM	0	1	0.47	0.27	0.25	0.66
IC	0	3	0.01	0.21	0	0
CBM	0	5	0.02	0.34	0	0
AMC	0	5	3.15	2.23	0.35	5

TABLE 11.7

Descriptive Statistics for Android Telephony Data Set

Metric Name	Min.	Max.	Mean	SD	Percentile (25%)	Percentile (75%)
WMC	1	213	20.00	32.28	4	20
DIT	0	4	0.97	0.47	1	1
NOC	0	4	0.05	0.33	0	0
CBO	0	20	2.84	3.5	1	4
RFC	2	214	21.00	32.28	5	21
LCOM	0	22,578	709.05	2,569.41	6	190
Ca	0	16	1.29	2.10	0	2
Ce	0	17	1.81	2.72	0	3
NPM	0	212	14.43	28.91	2	14
LCOM3	1	2	1.33	0.37	1.05	1.5
LOC	6	1,100	121.78	175.41	26	129
DAM	6	1	0.41	0.43	0	0.95
MOA	0	37	1.22	3.63	0	1
MFA	0	1	0.04	0.20	0	0
CAM	0	1	0.45	0.28	0.21	0.58
IC	0	1	0.01	0.06	0	0
CBM	0	1	0.01	0.06	0	0
AMC	0	5	4.28	1.61	5	5

11.6.2 Outlier Analysis

A critical step in data preprocessing involves outlier analysis. All data points that vary significantly from other data points are considered as outliers (Barnett and Lewis 1994). All outliers in a data set should be removed before data analysis so that the results are not biased by the values of the outliers. This study identified outliers with the help of inter-quartile range filter of the WEKA tool. Table 11.8 shows the number of outliers identified in each data set. All the identified outliers were removed before analyzing data and using it for model development.

11.6.3 CFS Results

This study uses 18 metrics as independent variables. It is essential to remove noisy and redundant features, and reduce the feature set. CFS method, proposed by Hall (2000), was used to extract appropriate features from each data set to get a reduced feature set that can efficiently predict change. The features extracted after application of CFS on each data set are shown in Table 11.9. According to the table, the LOC metric and the CAM metric are highly correlated with change as they are selected by five and four data sets, respectively. The WMC, MOA, CBO, and Ce are also good indicators of change as they are selected by two data sets each for model development. Other selected features include RFC, DIT, NPM, AMC, and MFA.

11.6.4 Tenfold Cross-Validation Results

This section states the results of change prediction models developed using all the 18 techniques using different performance measures. Tables 11.10 through 11.15 present the results specific to accuracy, sensitivity, specificity, precision, F-measure, and AUC performance measures for change prediction models developed using the 18 techniques analyzed in the study. Each row of the tables represent the change prediction models developed by a specific technique on all the data sets, whereas each column of the table represents the change prediction models developed on a particular data set. To highlight the best performing results of a given technique for all the data sets, specific values in the table are shown with superscript 'a' in each row. The table that are shown in bold formatting depict the best performing technique of a particular data set. All the cells with values ND represent that the specific measure is not defined for that particular technique in that particular data set.

1. Validation results using accuracy measure

   Table 11.10 represents the accuracy measure of change prediction models developed using all the 18 techniques on all the six Android data sets used in the study. According to the table, the most number of best accuracy values of various

TABLE 11.8

Outlier Details

Data Set Name	Number of Outliers
Bluetooth	7
Contacts	12
Calendar	6
Gallery2	43
MMS	22
Telephony	37

TABLE 11.9

Metrics Selected by CFS Method

Data Set Name	Metrics Selected
Bluetooth	WMC, RFC, LOC, CAM
Contacts	DIT, CBO, NPM, LOC
Calendar	CBO, Ce
Gallery2	Ce, LCOM3, LOC, MOA, CAM
MMS	LCOM3, LOC, DAM, MOA, CAM, AMC
Telephony	WMC, LCOM3, LOC, MFA, CAM

TABLE 11.10

Validation Results Using Accuracy Performance Measure

Technique	Bluetooth	Contacts	Calendar	Gallery2	MMS	Telephony
AB	66.15	73.23[a]	43.00	63.14	72.25	64.62
ADT	66.15	72.22[a]	44.00	61.93	69.94	63.67
BG	73.84	73.73	54.00	62.53	73.98[a]	63.20
BN	69.23	74.24[a]	66.00	61.93	72.25	63.20
DTab	56.92	74.74[a]	48.00	59.81	72.25	65.09
J48	56.92	70.20[a]	46.00	56.79	68.78	60.37
RIPPER	69.23	71.71[a]	41.00	57.70	56.06	57.07
LR	75.38[a]	72.72	54.00	62.53	74.56	**68.86**
LB	66.15	74.24	46.00	63.44	**75.72**	67.45
MLP	75.38[a]	**75.25**	55.00	59.81	73.98	66.03
NB	75.38[a]	70.20	53.00	60.42	72.83	66.98
NNge	72.30	71.71	77.00[a]	**64.04**	73.41	58.49
RF	64.61	72.72[a]	52.00	59.51	68.78	58.01
RBF	80.00[a]	70.20	36.00	61.63	72.83	66.98
REP	58.46	74.74[a]	36.00	59.81	67.05	64.15
SVM	**83.07[a]**	66.66	**83.00**	61.93	74.56	72.16
VP	56.92	45.45	82.00[a]	57.40	67.63	40.09
ZeroR	52.30	45.45	59.00[a]	53.47	43.35	48.11

techniques were achieved over the Android Contacts data set. However, the technique that developed the model exhibiting best accuracy values differed in all the data sets. The change prediction model developed using the SVM technique achieved the best accuracy value of 83.07% on Android Bluetooth data set and 83.00% on Android Calendar data set. The MLP technique gave the best accuracy value (75.25%) on Android Contacts data set and the LB technique (accuracy value: 75.72%) on the Android MMS data set. The best accuracy measures on Gallery2 and Telephony data sets were given by NNge and LR technique with an accuracy value of 64.04% and 68.86%, respectively.

2. Validation results using sensitivity measure

Table 11.11 is a representation of sensitivity values of change prediction models developed by all the techniques of the study on six Android data sets. We can

TABLE 11.11

Validation Results Using Sensitivity Performance Measure

Technique	Bluetooth	Contacts	Calendar	Gallery2	MMS	Telephony
AB	63.63	75.28	41.17	61.71	76.59[a]	63.77
ADT	72.72	74.15[a]	47.05	60.15	70.21	64.56
BG	72.72	75.28[a]	58.82	60.15	74.46	62.20
BN	63.63	75.28[a]	11.76	60.93	72.34	64.56
DTab	63.63	75.28[a]	47.05	59.37	74.46	63.77
J48	63.63	69.66	52.94	60.15	74.46[a]	62.20
RIPPER	63.63	69.66[a]	52.94	60.15	51.06	54.33
LR	72.72	73.03	52.94	64.06	76.59[a]	68.50
LB	63.63	75.28[a]	47.05	61.71	74.46	67.71
MLP	72.72	76.40[a]	52.94	60.93	74.46	64.56
NB	72.72[a]	71.91	52.94	62.50	72.34	66.92
NNge	18.18	69.66[a]	35.29	50.00	42.55	65.35
RF	63.63	74.15[a]	52.94	**64.84**	65.95	58.26
RBF	**81.81**[a]	69.66	35.29	59.37	74.46	66.92
REP	27.27[a]	73.03[a]	**70.58**	54.68	61.70	65.35
SVM	0.00	42.69	0.00	10.15	14.89	**88.18**[a]
VP	36.36	**100.00**[a]	0.00	25.78	19.14	0.00
ZeroR	36.36	89.88[a]	17.64	28.90	57.44	48.81

again see that the most number of best sensitivity values of different techniques are achieved over Android Contacts data set. The RBF technique (81.81%) and the VP technique (100%) achieved the best sensitivity values on Bluetooth and Contacts data sets, respectively. The REP technique and the RF technique gave best sensitivity values over the Calendar and Gallery2 data sets with sensitivity values of 70.58% and 64.84%, respectively. The LR technique and the AB technique both gave a sensitivity value of 76.59% on MMS data set. The best performing sensitivity value on Telephony data set was given by the SVM technique (88.18%).

3. Validation results using specificity measure

The specificity values of all the change prediction models of all the six android data sets is depicted in Table 11.12. The Bluetooth and Contacts data set showed the most number of techniques with best performing change prediction models when evaluated using specificity measure. The SVM model gave the best specificity values in all the data sets except Telephony data set. The VP technique gave the best specificity value on Telephony data set.

4. Validation results using precision measure

Table 11.13 shows the precision values of all the change prediction models developed on six Android data sets using 18 techniques. According to the table, the Telephony data set showed the best precision values on majority of techniques. The best precision value was exhibited by the RBF, SVM, NNge, VP, SVM, and LR techniques on Android Bluetooth, Contacts, Calendar, Gallery2, MMS, and Telephony data sets, respectively (precision measures—Bluetooth: 45%, Contacts: 71.69%, Calendar: 33.33%, Gallery2: 71.77%, MMS: 63.63%, Telephony: 76.99%).

TABLE 11.12

Validation Results Using Specificity Performance Measure

Technique	Bluetooth	Contacts	Calendar	Gallery2	MMS	Telephony
AB	66.66	71.55[a]	43.37	64.03	70.63	65.88
ADT	64.81	70.64	43.37	63.05	69.84[a]	62.35
BG	74.07[a]	72.47	53.01	64.03	73.80	64.70
BN	70.37	73.39	77.10[a]	62.56	72.22	61.17
DTab	55.55	74.31[a]	48.19	60.09	71.42	67.05
J48	55.55	70.64[a]	44.57	54.67	66.66	57.64
RIPPER	70.37	73.39[a]	38.55	56.15	57.93	61.17
LR	75.92[a]	72.47	54.21	61.57	73.80	69.41
LB	66.66	73.39	45.78	64.53	76.19[a]	67.05
MLP	75.92[a]	74.31	55.42	59.11	73.80	68.23
NB	75.92[a]	68.88	53.01	59.11	73.01	67.05
NNge	83.33	73.39	85.54[a]	72.90	84.92	48.23
RF	64.81	71.55[a]	51.80	56.15	69.84	57.64
RBF	79.62[a]	70.64	36.14	63.05	72.22	67.05
REP	64.81	76.14[a]	28.91	63.05	69.04	62.35
SVM	**100.00**[a]	**86.23**	**100.00**[a]	**94.58**	**96.82**	48.23
VP	61.11	0.91	98.79	77.33	85.71	**100.00**[a]
ZeroR	55.55	9.17	67.46	68.96	38.09	47.05

TABLE 11.13

Validation Results Using Precision Performance Measure

Technique	Bluetooth	Contacts	Calendar	Gallery2	MMS	Telephony
AB	28.00	68.36	12.96	51.97	49.31	73.63[a]
ADT	29.62	67.34	14.54	50.65	46.47	71.92[a]
BG	36.36	69.07	20.40	51.33	51.47	72.47[a]
BN	30.43	69.79	9.52	50.64	49.27	71.30[a]
DTab	22.58	70.52	15.68	48.40	49.29	74.31[a]
J48	22.58	65.95	16.36	45.56	45.45	68.69[a]
RIPPER	30.34	68.13[a]	15.00	46.38	31.16	67.64
LR	38.09	68.42	19.41	51.25	52.17	**76.99**[a]
LB	28.00	69.79	15.09	52.31	53.84	75.43[a]
MLP	38.09	70.83	19.56	48.44	51.47	75.22[a]
NB	38.09	65.30	18.75	49.07	50.00	75.22[a]
NNge	18.18	68.13[a]	**33.33**	53.78	51.28	65.35
RF	26.92	68.04[a]	18.36	48.25	44.92	67.27
RBF	**45.00**	65.95	10.16	50.33	50.00	75.22[a]
REP	13.63	71.42	16.90	48.27	42.64	72.17[a]
SVM	ND	**71.69**	ND	54.16	**63.63**	71.79[a]
VP	16.00	45.17	0.00	**71.77**[a]	33.33	ND
ZeroR	14.28	44.69	10.00	37.00	25.71	57.94[a]

5. Validation results using *F*-measure

The *F*-measure values of all the change prediction models developed using 18 techniques on six Android data sets are presented in Table 11.14. The data set with the most number of best performing technique results when evaluated using *F*-measure values is Android Contacts data set. The RBF technique and the MLP technique showed the best *F*-measure value of 58.06% and 73.51%, respectively, on Android Bluetooth and Android Contacts data set. The best performing technique on Calendar and Gallery2 data sets were NNge and LR techniques, respectively, with an *F*-measure value of 34.28% and 56.94%. The LB technique and the SVM technique showed an *F*-measure value of 62.50% and 79.15%, respectively, on MMS and Telephony data sets. These values were the best *F*-measure values for change prediction models on corresponding data sets.

6. Validation results using AUC measure

Table 11.15 displays the AUC values of all the change prediction models on each data set using all the 18 techniques of the study. The Android MMS data set showed the most number of best performing AUC results for various techniques. However, the best performing technique on Gallery2 and Telephony data set was LB with AUC values of 0.685 and 0.744, respectively. The NB technique, the MLP technique, the NNge technique, and the LR technique showed best AUC results on Bluetooth, Contacts, Calendar, and MMS data sets, respectively (AUC results—Bluetooth: 0.829, Contacts: 0809, Calendar: 0.604 and 0.811).

TABLE 11.14

Validation Results Using *F*-Measure Performance Measure

Technique	Bluetooth	Contacts	Calendar	Gallery2	MMS	Telephony
AB	38.88	71.65[a]	19.71	56.42	60.00	68.33
ADT	42.10	70.58[a]	22.22	55.00	55.93	68.04
BG	48.48	72.04[a]	30.30	55.39	60.86	66.94
BN	41.17	72.43[a]	10.52	55.31	58.62	67.76
DTab	33.33	72.82[a]	23.52	53.33	59.32	68.64
J48	33.33	67.75[a]	25.00	51.85	56.45	65.28
RIPPER	41.17	68.88[a]	23.37	52.38	38.70	60.26
LR	50.00	70.65	28.12	**56.94**	62.06	72.50[a]
LB	38.88	72.43[a]	22.85	56.63	**62.50**	71.36
MLP	50.00	**73.51[a]**	28.57	53.97	60.86	69.49
NB	50.00	68.44	27.69	54.98	59.13	70.83[a]
NNge	18.18	68.88[a]	**34.28**	51.82	46.51	65.35
RF	37.83	70.96[a]	27.27	55.33	53.44	62.44
RBF	**58.06**	67.75	15.78	54.48	59.82	70.83[a]
REP	18.18	72.22[a]	27.27	51.28	50.43	68.59
SVM	ND	53.52	ND	17.10	24.13	**79.15[a]**
VP	22.22	62.23[a]	ND	31.88	24.32	ND
ZeroR	20.51	59.70[a]	12.76	32.45	35.52	52.99

TABLE 11.15

Validation Results Using AUC Performance Measure

Technique	Bluetooth	Contacts	Calendar	Gallery2	MMS	Telephony
AB	0.660	0.749	0.428	0.672	0.765[a]	0.719
ADT	0.649	0.76	0.486	0.676	0.768[a]	0.704
BG	0.764	0.803[a]	0.602	0.655	0.797[a]	0.709
BN	0.664	0.754	0.463	0.648	0.772[a]	0.673
DTab	0.637	0.729	0.483	0.630	0.789[a]	0.713
J48	0.628	0.721	0.522	0.618	0.740[a]	0.634
RIPPER	0.627	0.695[a]	0.504	0.614	0.601	0.623
LR	0.800	0.768	0.514	0.667	**0.811[a]**	0.733
LB	0.668	0.79	0.493	**0.685**	0.795[a]	**0.744**
MLP	0.786	**0.809[a]**	0.535	0.652	0.805	0.719
NB	**0.829[a]**	0.769	0.577	0.644	0.797	0.732
NNge	0.507	0.715[a]	**0.604**	0.614	0.637	0.567
RF	0.680	0.768[a]	0.551	0.654	0.746	0.644
RBF	0.760	0.766	0.363	0.629	0.796[a]	0.716
REP	0.441	0.738[a]	0.452	0.618	0.696	0.667
SVM	0.500	0.644	0.500	0.523	0.558	0.682[a]
VP	0.489	0.504	0.487	0.515	0.522[a]	0.500
ZeroR	0.438	0.490[a]	0.425	0.488	0.467	0.476

7. Performance of various techniques on different data sets

Different techniques perform differently on various data sets. This study uses six Android data sets. Figures 11.2 through 11.7 depict graphs of the top six performing techniques on each data set. A technique is a good performer if the change prediction model developed by the model gives high accuracy values, high precision values, high *F*-measure values, and high AUC values. Since AUC is defined as a plot between specificity and sensitivity, we are not taking individual values

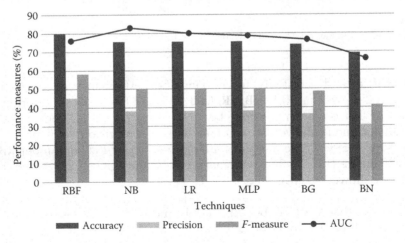

FIGURE 11.2
Top six performing techniques on Android Bluetooth data set.

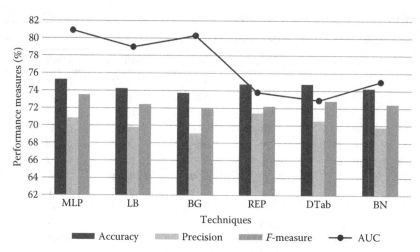

FIGURE 11.3
Top six performing techniques on Android Contacts data set.

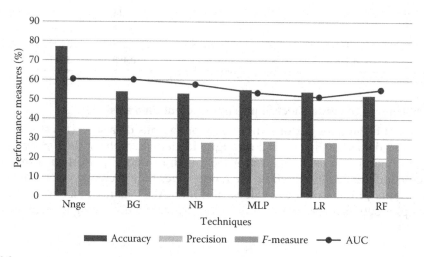

FIGURE 11.4
Top six performing techniques on Android Calendar data set.

of specificity and sensitivity for ranking the performance of the techniques. Figure 11.2 shows that RBF, NB, LR, MLP, BG, and BN are best performing techniques in that order on Android Bluetooth data set. According to the figure, the RBF technique showed an accuracy of 80%, a precision value of 45%, and a F-measure of 58%. These measures are represented by the bars under RBF technique in Figure 11.2. The line with marker show the AUC in percentage. The RBF technique showed an AUC value of 78%.

Figure 11.3 shows that MLP, LB, BG, REP, DTab, and BN techniques develop the top six performing models on Android Contacts data set. The AUC value of the model developed by the MLP technique was as high as 80.9%. The accuracy, precision, and F-measure values for the MLP model were 75%, 73%, and 70%, respectively.

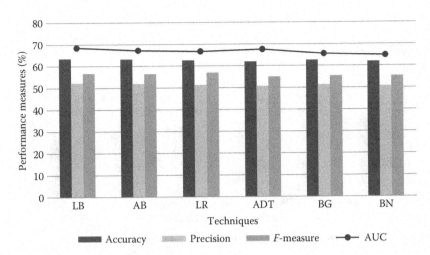

FIGURE 11.5
Top six performing techniques on Android Gallery2 data set.

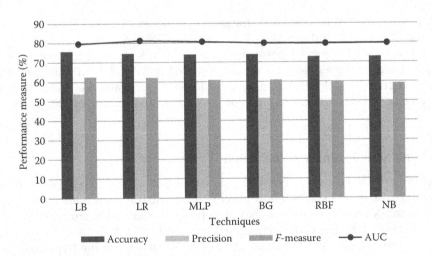

FIGURE 11.6
Top six performing techniques on Android MMS data set.

According to Figure 11.4, the techniques developing the best performing change prediction models on the Android Calendar data set were NNge, BG, NB, MLP, LR, and RF. The highest accuracy value was shown by the NNge model as 77%, while all other models showed accuracy values between 52% and 55%. The AUC % of all the models ranged from 51% to 60%.

Figure 11.5 depicts the top six techniques that gave the best change prediction models on the Android Gallery2 data set as LB, AB, LR, ADT, BG, and BN. However, as shown in the figure, there was not much difference in the values of different techniques. While the accuracy values ranged from 61% to 63%, the precision values ranged from 50% to 52%, the *F*-measure values ranged from 55% to 57%, and the AUC % values ranged from 62% to 69%. Similar results were shown by Android MMS data set, as shown in Figure 11.6. The top six best ranking techniques were LB, LR, MLP, BG, RBF, and NB. The AUC values for the

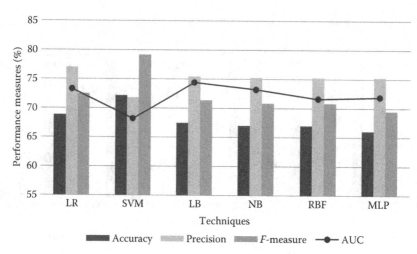

FIGURE 11.7
Top six performing techniques on Android Telephony data set.

MMS data set ranged from 72% to 76%, the precision values from 50% to 54%, the *F*-measure values from 59% to 61%, and the AUC % values from 79% to 81%.

Figure 11.7 shows the top six performing models on Android Telephony data set. The techniques used for developing these models were LR, SVM, LB, NB, RBF, and MLP. The change prediction model developed by the LR technique gave the best results with an accuracy value of 69%, a precision value of 77%, a *F*-measure value of 72%, and an AUC % of 73%.

11.6.5 Hypothesis Testing and Evaluation

This section states the results of Friedman statistical test using different performance measure. The Friedman statistic is based on chi square distribution with $N - 1$ degrees of freedom. Here, N represents the number of techniques used in the study. Thus, this study has 17 degrees of freedom as 18 techniques are used in the study. A detailed description of Friedman test can be referred from Chapter 6. While testing the hypothesis, we state the Friedman statistic value and *p*-value. The hypothesis was checked at $\alpha = 0.05$. The Friedman test is applied by evaluating a specific performance measure of all the techniques on all the six data sets used in the study.

1. Testing hypothesis for accuracy measure

    Table 11.16 represents the Friedman mean ranks of all the data sets using accuracy measure. According to the table, the best technique was SVM as it achieved the lowest rank of 4.75. The performance of the SVM technique was closely followed by the LR technique. The worst performing techniques were VP and ZeroR. The results of Friedman test was significant at $\alpha = 0.05$ as the *p*-value obtained is <0.05. The Friedman statistic value was 44.683. A low *p*-value indicates that we reject the null hypothesis H_0. Thus, change prediction models of different techniques when evaluated using accuracy measure show significant differences, which means that the performance of all the techniques is behaviorally different when evaluated using accuracy measure.

TABLE 11.16

Friedman Mean Ranks Using Accuracy Measure

Technique	Mean Rank	Technique	Mean Rank
SVM	4.75	RBF	9.08
LR	4.83	DTab	9.75
MLP	5.58	ADT	10.67
LB	5.67	REP	11.83
BG	6.67	RF	12.33
NNge	7.08	RIPPER	14.00
BN	7.58	J48	14.33
NB	8.17	VP	14.08
AB	9.00	ZeroR	15.58

2. Testing hypothesis for sensitivity measure

The Friedman mean ranks obtained using the sensitivity measure are depicted in Table 11.17. As shown in the table, the best performing technique was the LR technique with a mean rank of 4.42. The second rank was given to the MLP technique. The Friedman test results were significant at $\alpha = 0.05$ as the p-value obtained was 0.005. The Friedman statistic was computed as 35.68. These results show that we accept the alternate hypothesis. The models developed using sensitivity measure show significantly different results.

3. Testing hypothesis for specificity measure

The Friedman mean ranks using the specificity measure are not shown, as specificity is not a good measure for ranking different techniques. However, the p-value was computed as 0.005, which means that we reject the null hypothesis. Thus, the change prediction models developed using specificity measure are statistically different when evaluated using the specificity measure. The Friedman statistic value using specificity measure was computed as 36.01.

4. Testing hypothesis for precision measure

Table 11.18 shows the Friedman mean ranks of various techniques when evaluated using the precision measure. According to the table, the top two performing

TABLE 11.17

Friedman Mean Ranks Using Sensitivity Measure

Technique	Mean Rank	Technique	Mean Rank
LR	4.42	DTab	9.25
MLP	5.58	BN	9.50
LB	6.50	J48	9.92
NB	6.58	REP	10.33
BG	6.75	RIPPER	11.92
AB	7.58	ZeroR	12.92
RBF	8.75	VP	14.00
ADT	8.83	NNge	14.42
RF	8.83	SVM	14.92

TABLE 11.18

Friedman Mean Ranks Using Precision Measure

Technique	Mean Rank	Technique	Mean Rank
LR	4.33	DTab	9.08
MLP	4.92	BN	9.92
BG	5.42	ADT	10.50
LB	5.50	REP	10.67
NB	7.75	RF	12.17
NNge	8.42	RIPPER	12.58
SVM	8.42	J48	13.00
RBF	8.50	VP	14.08
AB	8.75	ZeroR	17.00

techniques were LR and MLP. The worst performing techniques were VP and ZeroR. The p-value of 0.001 is much less than 0.05, so we reject the null hypothesis. The Friedman statistic was computed as 41.41. Hence, the precision values of change prediction models developed using different techniques differ significantly. Thus, the results of all the techniques are behaviorally different when evaluated using precision measure.

5. Testing hypothesis for F-measure

To evaluate null hypothesis, we performed Friedman test using F-measure values. Table 11.19 presents the mean ranks obtained by all the techniques when we used F-measure for evaluating the various techniques. The LR technique and the MLP technique gave the best results with mean ranks of 3.50 and 4.42, respectively. The least effective techniques in terms of F-measure values were ZeroR and VP techniques. The p-value for the Friedman test was <0.05, indicating acceptance of alternate hypothesis. The Friedman statistic value was obtained as 53.661. The results show that change prediction models developed using all the techniques show significant differences when evaluated using F-measure values.

6. Testing hypothesis for AUC measure

Table 11.20 presents the mean ranks of all the techniques when the change prediction models developed by them on all the six data sets are evaluated using the

TABLE 11.19

Friedman Mean Ranks Using F-Measure

Technique	Mean Rank	Technique	Mean Rank
LR	3.50	ADT	9.33
MLP	4.42	RF	9.58
LB	5.00	REP	10.67
BG	5.42	NNge	11.67
NB	6.92	J48	11.92
AB	7.75	RIPPER	12.00
RBF	8.17	SVM	15.08
DTab	8.17	ZeroR	16.00
BN	8.83	VP	16.58

TABLE 11.20

Friedman Mean Ranks Using AUC Measure

Technique	Mean Rank	Technique	Mean Rank
LR	3.58	BN	9.67
MLP	3.75	DTab	10.00
NB	3.92	J48	11.58
BG	4.08	NNge	12.25
LB	4.67	RIPPER	13.42
RF	7.58	REP	13.42
ADT	8.33	SVM	13.67
RBF	8.67	VP	15.83
AB	8.75	ZeroR	17.83

AUC measure. The LR and the MLP techniques again achieved the top two ranks, while the VP and the ZeroR technique gave the worst results. The Friedman statistic value was calculated as 69.115 with a *p*-value much less than 0.05. Thus, we accept the alternate hypothesis, which indicates statistically significant differences in the performance of change prediction models developed by all the techniques.

11.6.6 Nemenyi Results

After performing Friedman tests using various performance measures, we need to ascertain the pair of techniques that differ significantly from each other in terms of performance of change prediction models using various measures. To find such pairs, we applied Nemenyi post hoc test on each pair of techniques using five performance measures (accuracy, sensitivity, precision, *F*-measure, and AUC). We do not include specificity as a performance measure for evaluating the pairwise performance of different techniques. The hypothesis for Nemenyi test is as follows:

- Null hypothesis: The performance of change prediction models developed using a specific pair of techniques (technique A and technique B) do not differ significantly when analyzed over multiple performance measures (accuracy, sensitivity, precision, *F*-measure, and AUC).
- Alternate hypothesis: The performance of change prediction models developed using a specific pair of techniques (technique A and technique B) differs significantly when analyzed over multiple performance measures (accuracy, sensitivity, precision, *F*-measure, and AUC).

The critical distance computed for Nemenyi test is as follows (Demšar 2006):

$$CD = q_\alpha \sqrt{\frac{k(k+1)}{6n}}$$

Here, k corresponds to the number of techniques, which is 18 in this study, and n corresponds to the number of data sets, which is six. The critical values (q_α) are studentized range statistic divided by $\sqrt{2}$. We evaluate the significance at $\alpha = 0.05$. The computed critical distance value is 10.75. The computed critical distance value is compared with the difference between average ranks allocated to two techniques. If the difference is at least equal to or greater than the critical distance value, the two techniques differ significantly at the chosen significance level $\alpha = 0.05$. Table 11A.1 in Appendix shows the critical distances after application of Nemenyi test on all the 153 possible pairs of techniques using the five specified performance measures. All pairs with significant critical difference, that is, the difference between individual ranks greater than the critical distance are bold formatted. There are two technique pairs (LR–ZeroR and SVM–ZeroR) that are significantly different from each other when accuracy is used as a performance measure. However, there is no pair with statistically significant difference when sensitivity is used as a performance measure. When precision, F-measure, and AUC are used as performance measures, four, eight, and ten technique pairs, respectively, are found to have statistically significant difference. On analyzing the results, it can be seen that ZeroR and VP techniques are different from a number of other techniques when using different performance measures. Only LR–ZeroR technique pair is statistically significant using four performance measures, namely, accuracy, precision, F-measure, and AUC.

11.7 Discussion and Interpretation of Results

This section analyzes the results of the study with respect to each research question.

RQ1: Are OO metrics related to change?

The study analyzes 18 techniques for their capability to develop efficient change prediction models. The developed models in the study use 18 OO metrics, which represent various OO characteristics like abstraction, coupling, inheritance, and so on. These metrics have been effective in developing change prediction models that have been evaluated using various performance measures. The metrics are used to ascertain the dependent variable change proneness (ALTER variable) and have yielded good results with accuracy up to 83%, sensitivity and specificity up to 100%, precision value up to 77%, F-measure value up to 73%, and AUC value up to 0.8. This indicates a relationship between various OO metrics and change proneness. Moreover, Table 11.9 shows highly correlated metrics with change on each data set using the CFS technique. Thus, we can summarize that OO metrics are related with change in a software data set. These metrics can be used to develop quality benchmarks that can be used by software practitioners for developing good quality products.

RQ2: What is the capability of various techniques on data sets with varying characteristics?

The capability of different techniques varies on different data sets, while developing change prediction models. Tables 11.10 through 11.15 show that various techniques work differently on different data sets as the best performance using a specific measure is given by a different technique for each data set.

For example, the RBF technique gave the best *F*-measure value for change prediction model on Android Bluetooth data set, and the MLP technique gave the best *F*-measure value for change prediction model on Android Contacts data set. However, the NNge technique, the LR technique, the MLP technique, and the SVM technique gave the best *F*-measure values for Android Calendar, Gallery2, MMS, and Telephony data sets, respectively. Moreover, Figures 11.2 through 11.7 clearly show that the top-performing techniques on each data set differ if we take into account their cumulative performance. An analysis of Figures 11.2 through 11.7 indicate that LR, MLP, and BG are high-performing techniques as they rank among top six techniques in five of the six data sets used in the study. Other good performing techniques include NB and LB, which rank among top six technique in four out of the six data sets of the study. Certain techniques (SVM, ADT, AB, RF, DTab, REP, NNge) gave good results in only one of the data sets. These techniques may be influenced by certain characteristics of a particular data set. However, we need to perform more such studies to actually evaluate which type of techniques get influenced by the characteristics of a data set.

RQ3: What is the comparative performance of different techniques when we take into account different performance measures?

To answer this question, we formulated research hypotheses given in Section 11.4.4.1. The hypothesis testing was done with the help of Friedman test. The results of the study indicate that we reject the null hypothesis for all the selected performance measures. Thus, the results of change prediction models developed using different techniques were significantly different from each other when evaluated using accuracy, sensitivity, specificity, precision, *F*-measure, and AUC performance measures. Tables 11.16 through 11.20 show Friedman mean ranks of various techniques using different performance measures. As can be seen, the LR technique gave the best results using all the performance measure except accuracy. Thus, we conclude that the LR technique is an effective technique for developing high-performing change prediction models. Also, the MLP technique is a good ML technique for developing models that predict change-prone classes as it achieves good Friedman ranks in all performance measures.

The results show that the best performing ML technique for the development of change prediction models is MLP. As can be seen from Tables 11.16 through 11.20, the MLP technique received the best ranks after LR technique except in the case of accuracy measure. The results show that although the accuracy measure has predicted all outcome classes for model predicted using SVM technique as not change prone (no predictive ability), but the Friedman test ranks the SVM technique as the best in terms of measuring accuracy. This is because of the presence of imbalance values of the outcome variable in the data sets. In imbalanced data sets, there are less change-prone classes as compared to non-change-prone classes. For example, for Bluetooth and Calendar data sets the change-prone classes are only 19%, and for MMS data set, the change-prone classes are 30%. The SVM technique predicted all classes as not change prone, hence, the specificity and accuracy values were very high specifically for Bluetooth and Calendar data sets (more than 80%) and thus contributed toward high ranking of SVM in terms of accuracy. The accuracy measure gives false results when the data is imbalanced and the technique

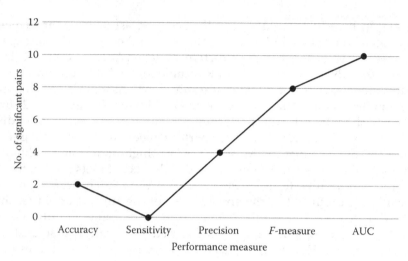

FIGURE 11.8
Summary of Nemenyi test.

classifies the classes into one single outcome category. Hence, this study does not base the results interpretation in light of accuracy measure.

RQ4: Which pairs of techniques are statistically significantly different from each other for developing change prediction models?

To answer this research question, we performed Nemenyi post hoc test among all possible pairs of techniques in the study. The test was performed using accuracy, sensitivity, precision, F-measure, and AUC values among 153 pairs of techniques. Figure 11.8 depicts the number of pairs of techniques that showed significantly different results using different performance measures. According to the figure, two pairs of techniques showed significant results using accuracy measure, but no pair of technique showed significant results using the sensitivity measure. On evaluation of precision, F-measure, and AUC measures, four, eight, and ten pairs of techniques, respectively, were significantly different from each other. Only one pair of technique showed significant Nemenyi results on four performance measures, namely, accuracy, precision, F-measure, and AUC, which was LR–ZeroR. On analyzing the pairs with significant differences, it can be seen that ZeroR and VP techniques are statistically significantly different from a number of other techniques like LR, BG, LB, MLP, and NB.

RQ5: What is the comparative performance of ML techniques with the statistical technique LR?

The LR technique has been ranked higher in most of the performance measures followed by MLP. The performance of model predicted using the ML techniques was comparable to the model predicted using the LR technique.

RQ6: Which ML technique gives the best performance for developing change prediction models?

The MLP technique has showed high performance capability in predicting change-prone classes. The MLP technique has the ability to model complex relationships and works well with reduced set of input variables. We applied CFS before using the MLP technique.

11.8 Validity Evaluation

This section presents the threats to validity.

11.8.1 Conclusion Validity

The statistical tests have been carefully selected and we have applied post hoc analysis using Nemenyi test to further validate the results. The significance values have also been carefully selected, thus reducing the conclusion validity threats. The results have been validated using tenfold cross-validation method, increasing the confidence on results. Moreover, the study analyzes change prediction models using six performance measures (Afzal and Torkar 2008). Thus, strengthening the conclusions of the study.

11.8.2 Internal Validity

The internal validity explores the "causal effect" of the independent variables on the dependent variable (Zhou et al. 2009). The causal effect can be validated by performing controlled experiments that keep a particular characteristic like coupling varying and the other characteristics like cohesion, inheritance, and so on constant (Briand et al. 2001). The focus of the study was not to study the causal effect of the metrics. Thus, the threat to internal validity exists in the study.

11.8.3 Construct Validity

Construct validity ensures correct representation of the various concepts depicted by the dependent and independent variables of the study (Zhou et al. 2009). Few researchers (Briand et al. 1998, 1999a, 2000) have already investigated OO metrics (independent variables) to ascertain their accuracy. Thus, this threat is reduced. The dependent variable, that is, change proneness is counted by analyzing the number of lines inserted or deleted in a class. The DCRS tool helps in efficient calculation of change by analyzing change logs. Thus, the dependent variable is measured accurately.

11.8.4 External Validity

The threat to external validity represents the generalizability of the results of the study (Malhotra and Khanna 2013). This threat explores whether the results of the study are universally applicable or not. Threat to external validity can only be minimized by performing replicated and repeated studies on a number of data sets and then by comparing the results. The threat to external validity is minimized in this study by analyzing six open source data sets, which are related to Android applications. However, all the data sets were developed in Java language. In future, researchers may perform

similar studies to explore different programming languages and other project characteristics to yield generalizable conclusions.

11.9 Conclusions and Future Work

This study analyzes the performance of 18 different techniques (both statistical and ML)
to develop effective change prediction models on the widely used Android data set. The
metrics and change information from Android data set was collected using DCRS tool.
The performance of various techniques was assessed using six different performance
measures, namely, accuracy, sensitivity, specificity, precision, F-measure, and AUC. The
aim of the study is to compare the capability of various techniques to rank them according
to their effectiveness based on different performance measures. The study also explores
pairs of techniques that are statistically different from each other using Nemenyi post hoc
test. The results of the study can be summarized as below:

- The study uses 18 OO metrics to ascertain change proneness of classes in a software. The change prediction models developed using these metrics were effective
  and yielded good results. Thus, various OO metrics are related to change in a class.
  The results of the study after application of the CFS technique indicate that LOC
  and CAM metrics are good indicators for change followed by WMC, MOA, CBO,
  and Ce metrics.

- Various techniques perform differently on different data sets. The results of the
  study indicate different top-performing techniques on each data set. Three techniques that gave good results on almost all the data sets were LR, MLP, and BG.
  Apart from these techniques, NB and LB techniques also performed well on a
  majority of data sets. The top-performing techniques were evaluated using the
  cumulative values of accuracy, precision, F-measure, and AUC performance
  measures.

- The performance of different techniques evaluated in the study varies when we
  assess them using different performance measures. The results of the study show that
  the performance of change prediction models was significantly different from each
  other when evaluated using accuracy, sensitivity, specificity, precision, F-measure,
  and AUC values. Also, the LR technique was ranked as a top technique when change
  prediction models were evaluated using sensitivity, precision, F-measure, and AUC
  values. Another well-performing technique was MLP technique.

- The study compared the performance of all the techniques pairwise using Nemenyi
  test. The pairwise comparison was performed using accuracy, sensitivity, precision, F-measure, and AUC values among 153 pairs of techniques. There was only
  one technique pair (LR–ZeroR) that showed significant difference on four performance measures, namely, accuracy, precision, F-measure, and AUC.

More studies in the future should be conducted to evaluate different statistical and ML
algorithms using other performance measures such as H-measure, G-measure, and so on.
Also, future studies can incorporate evolutionary computation techniques such as artificial immune systems and genetic algorithms for developing change prediction models.

Appendix

TABLE 11A.1

Nemenyi Test Results

Pair No.	Algorithm Pair	Accuracy	Sensitivity	Precision	*F*-Measure	AUC
1.	AB–ADT	1.67	1.25	1.75	1.58	0.42
2.	AB–BG	2.33	0.83	3.33	2.33	4.67
3.	AB–BN	1.42	1.92	1.17	1.08	0.92
4.	AB–DTab	0.75	1.67	0.33	0.42	1.25
5.	AB–J48	5.33	2.34	4.25	4.17	2.83
6.	AB–RIPER	5.00	4.34	3.83	4.25	4.67
7.	AB–LR	4.17	3.16	4.42	4.25	5.17
8.	AB–LB	3.33	1.08	3.25	2.75	4.08
9.	AB–MLP	3.42	2.00	3.83	3.33	5.00
10.	AB–NB	0.83	1.00	1.00	0.83	4.83
11.	AB–Nnge	1.92	6.84	0.33	3.92	3.50
12.	AB–RF	3.33	1.25	3.42	1.83	1.17
13.	AB–RBF	0.08	1.17	0.25	0.42	0.08
14.	AB–REP	2.83	2.75	1.92	2.92	4.67
15.	AB–SVM	4.25	7.34	0.33	7.33	4.92
16.	AB–VP	5.08	6.42	5.33	8.83	7.08
17.	AB–ZeroR	6.58	5.34	8.25	8.25	9.08
18.	ADT–BG	4.00	2.08	5.08	3.91	4.25
19.	ADT–BN	3.09	0.67	0.58	0.50	1.34
20.	ADT–DTab	0.92	0.42	1.42	1.16	1.67
21.	ADT–J48	3.66	1.09	2.50	2.59	3.25
22.	ADT–RIPER	3.33	3.09	2.08	2.67	5.09
23.	ADT–LR	5.84	4.41	6.17	5.83	4.75
24.	ADT–LB	5.00	2.33	5.00	4.33	3.66
25.	ADT–MLP	5.09	3.25	5.58	4.91	4.58
26.	ADT–NB	2.50	2.25	2.75	2.41	4.41
27.	ADT–Nnge	3.59	5.59	2.08	2.34	3.92
28.	ADT–RF	1.66	0.00	1.67	0.25	0.75
29.	ADT–RBF	1.59	0.08	2.00	1.16	0.34
30.	ADT–REP	1.16	1.50	0.17	1.34	5.09
31.	ADT–SVM	5.92	6.09	2.08	5.75	5.34
32.	ADT–VP	3.41	5.17	3.58	7.25	7.50
33.	ADT–ZeroR	4.91	4.09	6.50	6.67	9.50
34.	BG–BN	0.91	2.75	4.50	3.41	5.59
35.	BG–DTab	3.08	2.5	3.66	2.75	5.92
36.	BG–J48	7.66	3.17	7.58	6.50	7.50
37.	BG–RIPER	7.33	5.17	7.16	6.58	9.34
38.	BG–LR	1.84	2.33	1.09	1.92	0.50
39.	BG–LB	1.00	0.25	0.08	0.42	0.59
40.	BG–MLP	1.09	1.17	0.50	1.00	0.33
41.	BG–NB	1.50	0.17	2.33	1.50	0.16

(Continued)

TABLE 11A.1 (*Continued*)

Nemenyi Test Results

Pair No.	Algorithm Pair	Accuracy	Sensitivity	Precision	*F*-Measure	AUC
42.	BG–Nnge	0.41	7.67	3.00	6.25	8.17
43.	BG–RF	5.66	2.08	6.75	4.16	3.50
44.	BG–RBF	2.41	2.00	3.08	2.75	4.59
45.	BG–REP	5.16	3.58	5.25	5.25	9.34
46.	BG–SVM	1.92	8.17	3.00	9.66	9.59
47.	BG–VP	7.41	7.25	8.66	**11.16**	**11.75**
48.	BG–ZeroR	8.91	6.17	**11.58**	10.58	**13.75**
49.	BN–DTab	2.17	0.25	0.84	0.66	0.33
50.	BN–J48	6.75	0.42	3.08	3.09	1.91
51.	BN–RIPER	6.42	2.42	2.66	3.17	3.75
52.	BN–LR	2.75	5.08	5.59	5.33	6.09
53.	BN–LB	1.91	3.00	4.42	3.83	5.00
54.	BN–MLP	2.00	3.92	5.00	4.41	5.92
55.	BN–NB	0.59	2.92	2.17	1.91	5.75
56.	BN–Nnge	0.50	4.92	1.50	2.84	2.58
57.	BN–RF	4.75	0.67	2.25	0.75	2.09
58.	BN–RBF	1.50	0.75	1.42	0.66	1.00
59.	BN–REP	4.25	0.83	0.75	1.84	3.75
60.	BN–SVM	2.83	5.42	1.50	6.25	4.00
61.	BN–VP	6.50	4.50	4.16	7.75	6.16
62.	BN–ZeroR	8.00	3.42	7.08	7.17	8.16
63.	DTab–J48	4.58	0.67	3.92	3.75	1.58
64.	DTab–RIPER	4.25	2.67	3.50	3.83	3.42
65.	DTab–LR	4.92	4.83	4.75	4.67	6.42
66.	DTab–LB	4.08	2.75	3.58	3.17	5.33
67.	DTab–MLP	4.17	3.67	4.16	3.75	6.25
68.	DTab–NB	1.58	2.67	1.33	1.25	6.08
69.	DTab–Nnge	2.67	5.17	0.66	3.50	2.25
70.	DTab–RF	2.58	0.42	3.09	1.41	2.42
71.	DTab–RBF	0.67	0.50	0.58	0.00	1.33
72.	DTab–REP	2.08	1.08	1.59	2.50	3.42
73.	DTab–SVM	5.00	5.67	0.66	6.91	3.67
74.	DTab–VP	4.33	4.75	5.00	8.41	5.83
75.	DTab–ZeroR	5.83	3.67	7.92	7.83	7.83
76.	J48–RIPER	0.33	2.00	0.42	0.08	1.84
77.	J48–LR	9.50	5.50	8.67	8.42	8.00
78.	J48–LB	8.66	3.42	7.50	6.92	6.91
79.	J48–MLP	8.75	4.34	8.08	7.50	7.83
80.	J48–NB	6.16	3.34	5.25	5.00	7.66
81.	J48–Nnge	7.25	4.50	4.58	0.25	0.67
82.	J48–RF	2.00	1.09	0.83	2.34	4.00
83.	J48–RBF	5.25	1.17	4.50	3.75	2.91
84.	J48–REP	2.50	0.41	2.33	1.25	1.84
85.	J48–SVM	9.58	5.00	4.58	3.16	2.09

(Continued)

TABLE 11A.1 (*Continued*)

Nemenyi Test Results

Pair No.	Algorithm Pair	Accuracy	Sensitivity	Precision	*F*-Measure	AUC
86.	J48–VP	0.25	4.08	1.08	4.66	4.25
87.	J48–ZeroR	1.25	3.00	4.00	4.08	6.25
88.	RIPPER–LR	9.17	7.50	8.25	8.50	9.84
89.	RIPPER–LB	8.33	5.42	7.08	7.00	8.75
90.	RIPPER–MLP	8.42	6.34	7.66	7.58	9.67
91.	RIPPER–NB	5.83	5.34	4.83	5.08	9.50
92.	RIPPER–Nnge	6.92	2.50	4.16	0.33	1.17
93.	RIPPER–RF	1.67	3.09	0.41	2.42	5.84
94.	RIPPER–RBF	4.92	3.17	4.08	3.83	4.75
95.	RIPPER–REP	2.17	1.59	1.91	1.33	0.00
96.	RIPPER–SVM	9.25	3.00	4.16	3.08	0.25
97.	RIPPER–VP	0.08	2.08	1.50	4.58	2.41
98.	RIPPER–ZeroR	1.58	1.00	4.42	4.00	4.41
99.	LR–LB	0.84	2.08	1.17	1.50	1.09
100.	LR–MLP	0.75	1.16	0.59	0.92	0.17
101.	LR–NB	3.34	2.16	3.42	3.42	0.34
102.	LR–Nnge	2.25	10.00	4.09	8.17	8.67
103.	LR–RF	7.50	4.41	7.84	6.08	4.00
104.	LR–RBF	4.25	4.33	4.17	4.67	5.09
105.	LR–REP	7.00	5.91	6.34	7.17	9.84
106.	LR–SVM	0.08	10.5	4.09	**11.58**	10.09
107.	LR–VP	9.25	9.58	9.75	**13.08**	**12.25**
108.	LR–ZeroR	**10.75**	8.50	**12.67**	**12.50**	**14.25**
109.	LB–MLP	0.09	0.92	0.58	0.58	0.92
110.	LB–NB	2.50	0.08	2.25	1.92	0.75
111.	LB–Nnge	1.41	7.92	2.92	6.67	7.58
112.	LB–RF	6.66	2.33	6.67	4.58	2.91
113.	LB–RBF	3.41	2.25	3.00	3.17	4.00
114.	LB–REP	6.16	3.83	5.17	5.67	8.75
115.	LB–SVM	0.92	8.42	2.92	10.08	9.00
116.	LB–VP	8.41	7.50	8.58	**11.58**	**11.16**
117.	LB–ZeroR	9.91	6.42	**11.50**	**11.00**	**13.16**
118.	MLP–NB	2.59	1.00	2.83	2.50	0.17
119.	MLP–Nnge	1.50	8.84	3.50	7.25	8.50
120.	MLP–RF	6.75	3.25	7.25	5.16	3.83
121.	MLP–RBF	3.50	3.17	3.58	3.75	4.92
122.	MLP–REP	6.25	4.75	5.75	6.25	9.67
123.	MLP–SVM	0.83	9.34	3.50	10.66	9.92
124.	MLP–VP	8.50	8.42	9.16	**12.16**	**12.08**
125.	MLP–ZeroR	10.00	7.34	**12.08**	**11.58**	**14.08**
126.	NB–Nnge	1.09	7.84	0.67	4.75	8.33
127.	NB–RF	4.16	2.25	4.42	2.66	3.66
128.	NB–RBF	0.91	2.17	0.75	1.25	4.75
129.	NB–REP	3.66	3.75	2.92	3.75	9.50

(Continued)

TABLE 11A.1 (*Continued*)

Nemenyi Test Results

Pair No.	Algorithm Pair	Accuracy	Sensitivity	Precision	*F*-Measure	AUC
130.	NB–SVM	3.42	8.34	0.67	8.16	9.75
131.	NB–VP	5.91	7.42	6.33	9.66	**11.91**
132.	NB–ZeroR	7.41	6.34	9.25	9.08	**13.91**
133.	NNge–RF	5.25	5.59	3.75	2.09	4.67
134.	NNge–RBF	2.00	5.67	0.08	3.50	3.58
135.	NNge–REP	4.75	4.09	2.25	1.00	1.17
136.	NNge–SVM	2.33	0.50	0.00	3.41	1.42
137.	NNge–VP	7.00	0.42	5.66	4.91	3.58
138.	NNge–ZeroR	8.50	1.50	8.58	4.33	5.58
139.	RF–RBF	3.25	0.08	3.67	1.41	1.09
140.	RF–REP	0.50	1.50	1.50	1.09	5.84
141.	RF–SVM	7.58	6.09	3.75	5.50	6.09
142.	RF–VP	1.75	5.17	1.91	7.00	8.25
143.	RF–ZeroR	3.25	4.09	4.83	6.42	10.25
144.	RBF–REP	2.75	1.58	2.17	2.50	4.75
145.	RBF–SVM	4.33	6.17	0.08	6.91	5.00
146.	RBF–VP	5.00	5.25	5.58	8.41	7.16
147.	RBF–ZeroR	6.50	4.17	8.50	7.83	9.16
148.	REP–SVM	7.08	4.59	2.25	4.41	0.25
149.	REP–VP	2.25	3.67	3.41	5.91	2.41
150.	REP–ZeroR	3.75	2.59	6.33	5.33	4.41
151.	SVM–VP	9.33	0.92	5.66	1.50	2.16
152.	SVM–ZeroR	**10.83**	2.00	8.58	0.92	4.16
153.	VP–ZeroR	1.50	1.08	2.92	0.58	2.00

12

Tools for Analyzing Data

There are many statistical packages available to implement the concepts given in previous chapters. These statistical packages or tools can assist researchers and practitioners to perform operations such as summarizing data, preselecting attributes, hypothesis testing, model creation, and validation. There are various statistical tools available in the market such as SAS, R, Matrix Laboratory (MATLAB®*), SPSS, Stata, and Waikato Environment for Knowledge Analysis (WEKA). An overview and comparison of these tools will help in making decision about selection of an appropriate tool in assisting the research process. In this chapter, we provide an overview of five tools, namely, WEKA, Knowledge Extraction based on Evolutionary Learning (KEEL), SPSS, MATLAB, and R, and summarize their characteristics and available statistical procedures.

12.1 WEKA

WEKA tool was developed at the University of Waikato in New Zealand (http://www.cs.waikato.ac.nz/ml/weka/) and is distributed under GNU public license. The tool was developed in Java language and runs on a number of platforms, be it Linux, Macintosh, or Windows. It provides an easy-to-use interface for using a number of different learning techniques. Moreover, it also provides various methods for preprocessing or postprocessing data. Research has seen wide application of WEKA tool for analyzing the results of different techniques on different data sets. WEKA can be used for multiple purposes, be it analyzing the results of a classification method on data or developing models to obtain predictions on new data or comparing the performances of several classification techniques.

12.2 KEEL

KEEL is a software tool which was developed using the Java language. The tool is open source in nature and aids the user for easy assessment of a number of evolutionary and other soft computing techniques. The tool provides a framework for designing a

---

* MATLAB® is a registered trademark of The MathWorks, Inc. For product information, please contact:

The MathWorks, Inc.
3 Apple Hill Drive
Natick, MA 01760-2098 USA
Tel: +1 508 647 7000
Fax: +1 508 647 7001
E-mail: info@mathworks.com
Web: www.mathworks.com

number of experiments for various data mining tasks such as classification, regression, and pattern mining.

KEEL software tool consists of an extensive number of features that can help researchers and students to perform various data mining tasks in an easy and effective manner. The tool provides a convenient and user-friendly interface to conduct and design various experiments. It also incorporates a library of in-built data sets. KEEL specializes in the use of evolutionary algorithms that can be effectively used for model prediction, preprocessing tasks, and certain postprocessing tasks. The tool also incorporates a number of data preprocessing algorithms for various tasks like noisy data filtering, selection of training sets, discretization, and data transformation among others. It also enables effective analysis and comparisons of results with the help of statistical library. The experiments designed using KEEL can be run both in an offline mode on other or same machine at a later time or an online mode. The tool is designed for two specific types of users: a researcher or a student. It facilitates the researcher by easy automation of experiments and effective result analysis using statistical library. It is a useful learning tool for students as a student can have real-time view of an technique's evolving process with visual feedback (Alcala et al. 2011).

12.3 SPSS

SPSS statistics is a software package that is used for statistical analysis. It was acquired by IBM in 2009 and the current versions (2014) are officially named IBM SPSS Statistics. The software name stands for Statistical Package for Social Sciences, which reflects the original market.

SPSS is one of the most powerful tools that can be used for carrying out almost any type of data analysis. This data analysis could be either in the field of social sciences, natural sciences, or in the world of business and management. This tool is widely used for research and interpretation of the data. It performs four major functions: creates and maintains a data set, analyzes data, produces results after analysis, and graphs them. This tutorial focuses on the main functions and utilities that can be used by a researcher for performing various empirical studies.

12.4 MATLAB®

MATLAB is a high-performance interactive software system that integrates computation and visualization for technical computations and graphics. MATLAB was primarily developed by Cleve Moler in the 1970s. The tool is derived from two FORTRON's subroutine, namely, EISPACK and LINPACK. EISPACK is an eigenvalue system and LINPACK is a linear system. The package was rewritten in 1980s in C. This rewritten package had larger functionality and a number of plotting routines. To commercialize MATLAB and further develop it, the MathWorks Inc. was created in 1984.

MATLAB is specially designed for matrix computations to solve linear equations, factor matrices, and compute eigenvalues and eigenvectors. In addition, it has sophisticated graphical features that are extendable. MATLAB also provides a good programming language environment as it offers many facilities like editing and debugging tools, data structures, and supports object-oriented paradigms.

MATLAB provides a number of built-in routines that aid extensive computations. Also, the results are immediately visualized with the help of easy graphical commands. A MATLAB toolbox consists of a collection of specific applications. There are a number of toolboxes for various tasks such as simulation, symbolic computation, signal processing, and many other related tasks in the field of science and engineering. These factors make MATLAB an excellent tool and it is used at most universities and industries worldwide for teaching and research. However, MATLAB has some weaknesses as it is designed for scientific computing, commands are specific for its usage and is not suitable for other applications like a general purpose programming language C or C++. It is an interpreted language and is therefore slower than compiled language. Mathematica, Scilab, and GNU Octave are some of the competitors of MATLAB.

12.5 R

R was developed by Ross Ihaka and Robert Gentleman at the University of Auckland, New Zealand. It is freely available under the GNU General Public License and can be used with various operating systems.

R is a well-developed, simple, and effective programming language extensively used by the statisticians for statistical computing and data analysis. In addition to this, R includes facilities for data calculation and manipulation, various operators for working with arrays (or matrices), tools and graphical facilities for data analysis, input and output facilities, and so on.

12.6 Comparison of Tools

Table 12.1 summarizes the comparison of WEKA, KEEL, SPSS, MATLAB, and R. Further the table lists the operating system that supports the tool, tool licenses, its interfaces, whether the tool is menu driven or syntax driven, whether the tool is open source in nature, the ease of graphical user interface, its help availability, link of the tool, and its specialty.

Table 12.2 summarizes the comparison of different tools on the basis of number of parameters such as correlation test capability, normality test capability, whether the tool analyzes and provides various descriptive statistics, feature selection techniques used by the tool, various regression, machine learning, and evolutionary algorithms supported by the tool, various cross-validation methods, and the capability to generate receiver operating characteristic (ROC) curves.

TABLE 12.1

Comparison of Tools

Tools	Operating System	License	Interface	Open Source	Lang.	Grap.	Availability of Help	Link	Speciality
WEKA	Mac/Windows	GNU GPL	Syntax/Menu	Yes	Java	Excellent	Good	www.cs.waikato.ac.nz/~ml/weka	Used for machine learning techniques
KEEL	Mac/Windows	GNU GPL	Menu	Yes	Java	Moderate	Moderate	www.keel.es	Used for evolutionary algorithms
SPSS	Mac/Windows	Proprietary	Syntax/Menu	No	Java	Very good	Good	www.ibm.com/software/analytics/spss/	Used for multivariate analysis and statistical testing
MATLAB	Mac/Windows	Proprietary	Syntax/Menu	No	C++/Java	Good	Very good	http://in.mathworks.com/products/matlab/	Best for developing new mathematical techniques, used for image and signal processing
R	Mac/Windows	GNU GPL	Syntax	Yes	Fortron/C	NA	Average	www.r-project.org	Extensive library support

TABLE 12.2

Parameter Comparison

Parameters	WEKA	MATLAB	KEEL	SPSS	R
Correlation	Y	Y	N	Y	Y
Normality tests	N	Y	N	Y	Y
Descriptive Statistics					
Minimum	Y	Y	Y	Y	Y
Maximum	Y	Y	Y	Y	Y
Variance	N	Y	Y	Y	Y
Standard deviation	Y	Y	N	Y	Y
Skewness	N	Y	N	Y	Y
Kurtosis	N	Y	N	Y	Y
Mean	Y	Y	Y	Y	Y
Median	N	Y	N	Y	Y
Mode	N	Y	N	Y	Y
Quartiles	N	Y	N	Y	Y
Feature Selection					
Correlation-based feature selection	Y	N	N	N	Y
Principal component analysis	Y	Y	Y	Y	Y
Statistical and post hoc Tests					
t-Test	Y	Y	Y	Y	Y
Chi-squared test	N	Y	N	Y	Y
f-Test	N	Y	Y	N	Y
ANOVA	N	Y	N	Y	Y
Wilcoxon signed	N	Y	Y	Y	Y
Mann–Whitney	N	Y	Y	Y	Y
Friedman	N	Y	Y	Y	Y
Kruskal–Wallis	N	Y	N	Y	Y
Nemenyi	N	N	Y	N	Y
Bonferroni–Dunn	N	Y	Y	Y	Y
Regression					
Binary logistic regression	Y	Y	Y	Y	Y
Linear regression	Y	Y	Y	Y	Y
Ordinal regression	N	Y	N	Y	Y
Multinominal logistic regression	N	Y	N	Y	Y
Linear discriminant analysis	N	Y	Y	Y	Y
Machine Learning Techniques					
Classification and regression trees	Y	Y	Y	Y	Y
NNge	Y	N	N	N	N
Boosting	Y	Y	Y	N	Y
Radial basis function	Y	Y	Y	Y	Y
Multilayer perceptron	Y	Y	Y	Y	Y
Support vector machine	Y	Y	Y	N	Y
Naïve Bayes	Y	N	Y	N	Y
Bayesian networks	Y	Y	N	Y	N

(Continued)

TABLE 12.2 (*Continued*)

Parameter Comparison

Parameters	WEKA	MATLAB	KEEL	SPSS	R
J48	Y	N	N	N	Y
Alternating decision trees	Y	N	N	N	Y
Voted perceptron	Y	N	N	N	N
Fuzzy logic	Y	Y	Y	N	Y
Convolutional neural network	N	Y	N	N	Y
Probabilistic neural network	N	Y	N	N	Y
Random forest	Y	N	N	N	Y
C4.5	Y	N	Y	N	Y
Chi-squared automatic interaction detection	N	N	N	Y	Y
K-Nearest neighbor	Y	Y	Y	Y	Y
Bagging	Y	Y	Y	N	Y
Evolutionary Algorithms					
Genetic algorithm	Y	Y	Y	N	Y
Genetic programming	Y	Y	N	N	Y
Ant colony optimization	N	Y	Y	N	Y
Ant miner	N	N	Y	N	N
Multi-objective particle swarm optimization	Y	Y	N	N	Y
Artificial immune system	Y	N	N	N	N
Particle swarm optimization linear discriminant analysis	N	N	Y	N	N
Constricted particle swarm optimization	N	N	Y	N	N
Hierarchical decision rules	N	N	Y	N	N
Decision trees with genetic algorithms	N	N	Y	N	N
Neural net evolutionary programming	N	N	Y	N	N
Genetic algorithms with neural networks	N	N	Y	N	N
Genetic fuzzy system logitboost	N	N	Y	N	N
Cross-validation	Y	Y	Y	Y	Y
ROC curve	Y	Y	N	Y	Y

Further Readings

The basic use of WEKA tool is described in:

M. Hall, E. Frank, G. Holmes, B. Pfahringer, P. Reutemann, and I. H. Witten, "The WEKA data mining software: An update," *ACM SIGKDD Explorations Newsletter*, vol. 11, no. 1, pp. 10–18, 2009.

I. H. Witten, and E. Frank, *Data Mining: Practical Machine Learning Tools and Techniques*, Morgan Kaufmann, Boston, MA, 2005.

The illustrations on KEEL tool are presented in:

J. Alcalá, A. Fernández, J. Luengo, J. Derrac, S. García, L. Sánchez, and F. Herrera, "Keel data-mining software tool: Data set repository, integration of algorithms and experimental analysis framework," *Journal of Multiple-Valued Logic and Soft Computing,* vol. 17, no. 11, pp. 255–287, 2010.

J. Alcala-Fdez, L. Sanchez, S. Garcia, M. J. D. Jesus, S. Ventura, J. M. Garrell, J. Otero et al., "KEEL: A software tool to assess evolutionary algorithms for data mining problems," *Soft Computing,* vol. 13, no. 3, pp. 307–318, 2009.

J. Alcalá-Fdez, F. Herrera, S. García, M. J. del Jesus, L. Sánchez, E. Bernadó-Mansilla, A. Peregrín, and S. Ventura, "Introduction to the Experimental Design in the Data Mining Tool KEEL," *Intelligent Soft Computation and Evolving Data Mining: Integrating Advanced Technologies,* vol. 1, pp. 1–25, 2010.

J. Derrac, J. Luengo, J. Alcalá-Fdez, A. Fernández, and S. Garcia, "Using KEEL software as educational tool: A case of study teaching data mining," *7th International Conference on IEEE Next Generation Web Services Practices,* pp. 464–469, Hospender, Spain, 2011.

The classic applications and working of SPSS tool are described in:

S. J. Coakes, and L. Steed, *SPSS: Analysis without Anguish Using SPSS Version 14.0 for Windows,* John Wiley & Sons, Chichester, 2009.

D. George, *SPSS for Windows Step by Step: A Simple Study Guide and Reference, 17.0 Update, 10/e,* Pearson Education, New Delhi, India, 2003.

S. B. Green, N. J. Salkind, and T. M. Jones, *Using SPSS for Windows: Analyzing and Understanding Data,* Prentice Hall, Upper Saddle River, NJ, 1996.

S. Landau, and B. Everitt, *A Handbook of Statistical Analyses Using SPSS,* Chapman & Hall, Boca Raton, FL, vol. 1, 2004.

M. P. Marchant, N. M. Smith, and K. H. Stirling, *SPSS as a Library Research Tool,* School of Library and Information Sciences, Brigham Young University, Provo, UT, 1977.

M. J. Norušis, *SPSS Advanced Statistics: Student Guide,* SPSS, Chicago, IL, 1990.

M. J. Norusis, *SPSS 15.0 Guide to Data Analysis,* Prentice Hall, Englewood Cliffs, NJ, 2007.

J. Pallant, *SPSS Survival Manual,* McGraw-Hill, Maidenhead, 2013.

S. Sarantakos, *A Toolkit for Quantitative Data Analysis: Using SPSS,* Palgrave Macmillan, New York, 2007.

An introduction to the use of commands and interface on MATLAB is provided in:

D. M. Etter, and D. C. Kuncicky, *Introduction to MATLAB,* Prentice Hall, Upper Saddle River, NJ, 2011.

L.N. Trefethen, *Spectral Methods in MATLAB,* SIAM, Philadelphia, PA, vol. 10, 2000.

MATLAB demos and tutorials are present at the following links:

http://math.ucsd.edu/~bdriver/21d -s99/matlab-primer.html.
http://www.mathworks.com/products/demos/.

The basics of R tools are mentioned in:

J. M. Crawley, *Statistics: An Introduction Using R,* John Wiley & Sons, England, 2014.

Appendix: Statistical Tables

This appendix contains statistical tables that are required for examples in Chapter 6. We replicated only a part of the statistical tables used in Chapter 6. To find detailed tables, readers can refer to any statistical book such as Anderson et al. (2002). The various statistical tables included in this appendix are as follows:

- t-Test
- Chi-square test
- Wilcoxon–Mann–Whitney test
- Area under the normal distribution
- F-Test table at 0.05 significance level
- Critical values for two-tailed Nemenyi test at 0.05 significance level
- Critical values for two-tailed Bonferroni test at 0.05 significance level

TABLE A.1
t-Test Table

Level of significance for one-tailed test					
	0.10	0.05	0.02	0.01	0.005
Level of significance for two-tailed test					
Df	0.20	0.10	0.05	0.02	0.01
1	3.078	6.314	12.706	31.821	63.657
2	1.886	2.920	4.303	6.965	9.925
3	1.638	2.353	3.182	4.541	5.841
4	1.533	2.132	2.776	3.747	4.604
5	1.476	2.015	2.571	3.365	4.032
.	.	.	.	.	.
.	.	.	.	.	.
.	.	.	.	.	.
14	1.345	1.761	2.145	2.624	2.977
15	1.341	1.753	2.131	2.602	2.947
.	.	.	.	.	.
.	.	.	.	.	.
.	.	.	.	.	.
20	1.325	1.725	2.086	2.528	2.845
21	1.323	1.721	2.080	2.518	2.831
22	1.321	1.717	2.014	2.508	2.819
.	.	.	.	.	.
.	.	.	.	.	.
.	.	.	.	.	.
120	1.289	1.658	1.980	2.358	2.617
∞	1.282	1.645	1.960	2.326	2.576

TABLE A.2

Chi-Square Table

Df	0.99	0.95	0.50	0.10	0.05	0.02	0.01
1	0.000	0.000	0.455	2.706	3.841	5.412	6.635
2	0.020	0.103	1.386	4.605	5.991	7.824	9.210
3	0.115	0.35	2.366	6.251	7.815	9.837	11.341
4	0.297	0.711	3.357	7.779	9.488	11.668	13.277
5	0.554	0.114	4.351	9.236	11.070	13.388	15.081
6	0.872	1.635	5.348	10.645	12.592	15.033	16.812
7	1.239	2.167	6.346	12.014	14.067	16.622	18.475
.	.	.	.	.	.	.	.
.	.	.	.	.	.	.	.
.	.	.	.	.	.	.	.
17	6.408	8.672	16.338	24.769	27.587	30.995	33.409
18	7.015	9.390	17.338	25.989	28.869	32.346	34.805
.	.	.	.	.	.	.	.
.	.	.	.	.	.	.	.
.	.	.	.	.	.	.	.
29	14.256	17.708	28.336	39.087	42.557	46.693	49.588
30	14.953	18.493	29.336	40.256	43.773	47.962	50.892

TABLE A.3

Wilcoxon–Mann–Whitney Table for $N_2 = 5$

$n_2 = 5$

N_1	1	2	3	4	5
0	0.167	0.047	0.018	0.008	0.004
1	0.333	0.095	0.036	0.016	0.008
2	0.500	0.190	0.071	0.032	0.016
3	0.667	0.286	0.125	0.056	0.028
4		0.429	0.196	0.095	0.048
5		0.571	0.286	0.143	0.075
6			0.393	0.206	0.111
7			0.500	0.278	0.155
8			0.607	0.365	0.210
9				0.452	0.274
10				0.548	0.345
11					0.421
12					0.500
13					0.579

TABLE A.4

Area Under the Normal Distribution

Z	0.00	0.01	0.02	0.03	0.04	0.05	0.06	0.07	0.08	0.09
−3.9	0.00005	0.00005	0.00004	0.00004	0.00004	0.00004	0.00004	0.00004	0.00003	0.00003
−3.8	0.00007	0.00007	0.00007	0.00006	0.00006	0.00006	0.00006	0.00005	0.00005	0.00005
−3.7	0.00011	0.00010	0.00010	0.00010	0.00009	0.00009	0.00008	0.00008	0.00008	0.00008
−3.6	0.00016	0.00015	0.00015	0.00014	0.00014	0.00013	0.00013	0.00012	0.00012	0.00011
−3.5	0.00023	.00022	0.00022	0.00021	0.00020	0.00019	0.00019	0.00018	0.00017	0.00017
−3.4	0.00034	0.00032	0.00031	0.00030	0.00029	0.00028	0.00027	0.00026	0.00025	0.00024
−3.3	0.00048	0.00047	0.00045	0.00043	0.00042	0.00040	0.00039	0.00038	0.00036	0.00035
−3.2	0.00069	0.00066	0.00064	0.00062	0.00060	0.00058	0.00056	0.00054	0.00052	0.00050
−3.1	0.00097	0.00094	0.00090	0.00087	0.00084	0.00082	0.00079	0.00076	0.00074	0.00071
−3.0	0.00135	0.00131	0.00126	0.00122	0.00118	0.00114	0.00111	0.00107	0.00104	0.00100
−2.9	0.00187	0.00181	0.00175	0.00169	0.00164	0.00159	0.00154	0.00149	0.00144	0.00139
−2.8	0.00256	0.00248	0.00240	0.00233	0.00226	0.00219	0.00212	0.00205	0.00199	0.00193
−2.7	0.00347	0.00336	0.00326	0.00317	0.00307	0.00298	0.00289	0.00280	0.00272	0.00264
−2.6	0.00466	0.00453	0.00440	0.00427	0.00415	0.00402	0.00391	0.00379	0.00368	0.00357
−2.5	0.00621	0.00604	0.00587	0.00570	0.00554	0.00539	0.00523	0.00508	0.00494	0.00480
−2.4	0.00820	0.00798	0.00776	0.00755	0.00734	0.00714	0.00695	0.00676	0.00657	0.00639
−2.3	0.01072	0.01044	0.01017	0.00990	0.00964	0.00939	0.00914	0.00889	0.00866	0.00842
−2.2	0.01390	0.01355	0.01321	0.01287	0.01255	0.01222	0.01191	0.01160	0.01130	0.01101
−2.1	0.01786	0.01743	0.01700	0.01659	0.01618	0.01578	0.01539	0.01500	0.01463	0.01426
−2.0	0.02275	0.02222	0.02169	0.02118	0.02068	0.02018	0.01970	0.01923	0.01876	0.01831
−1.9	0.02872	0.02807	0.02743	0.02680	0.02619	0.02559	0.02500	0.02442	0.02385	0.02330
−1.8	0.03593	0.03515	0.03438	0.03362	0.03288	0.03216	0.03144	0.03074	0.03005	0.02938
−1.7	0.04457	0.04363	0.04272	0.04182	0.04093	0.04006	0.03920	0.03836	0.03754	0.03673
−1.6	0.05480	0.05370	0.05262	0.05155	0.05050	0.04947	0.04846	0.04746	0.04648	0.04551
−1.5	0.06681	0.06552	0.06426	0.06301	0.06178	0.06057	0.05938	0.05821	0.05705	0.05592
−1.4	0.08076	0.07927	0.07780	0.07636	0.07493	0.07353	0.07215	0.07078	0.06944	0.06811
−1.3	0.09680	0.09510	0.09342	0.09176	0.09012	0.08851	0.08691	0.08534	0.08379	0.08226
−1.2	0.11507	0.11314	0.11123	0.10935	0.10749	0.10565	0.10383	0.10204	0.10027	0.09853
−1.1	0.13567	0.13350	0.13136	0.12924	0.12714	0.12507	0.12302	0.12100	0.11900	0.11702
−1.0	0.15866	0.15625	0.15386	0.15151	0.14917	0.14686	0.14457	0.14231	0.14007	0.13786
−0.9	0.18406	0.18141	0.17879	0.17619	0.17361	0.17106	0.16853	0.16602	0.16354	0.16109
−0.8	0.21186	0.20897	0.20611	0.20327	0.20045	0.19766	0.19489	0.19215	0.18943	0.18673
−0.7	0.24196	0.23885	0.23576	0.23270	0.22965	0.22663	0.22363	0.22065	0.21770	0.21476
−0.6	0.27425	0.27093	0.26763	0.26435	0.26109	0.25785	0.25463	0.25143	0.24825	0.24510
−0.5	0.30854	0.30503	0.30153	0.29806	0.29460	0.29116	0.28774	0.28434	0.28096	0.27760
−0.4	0.34458	0.34090	0.33724	0.33360	0.32997	0.32636	0.32276	0.31918	0.31561	0.31207
−0.3	0.38209	0.37828	0.37448	0.37070	0.36693	0.36317	0.35942	0.35569	0.35197	0.34827
−0.2	0.42074	0.41683	0.41294	0.40905	0.40517	0.40129	0.39743	0.39358	0.38974	0.38591
−0.1	0.46017	0.45620	0.45224	0.44828	0.44433	0.44038	0.43644	0.43251	0.42858	0.42465
−0.0	0.50000	0.49601	0.49202	0.48803	0.48405	0.48006	0.47608	0.47210	0.46812	0.46414
0.0	0.50000	0.50399	0.50798	0.51197	0.51595	0.51994	0.52392	0.52790	0.53188	0.53586

(Continued)

TABLE A.4 (*Continued*)

Area Under the Normal Distribution

Z	0.00	0.01	0.02	0.03	0.04	0.05	0.06	0.07	0.08	0.09
0.1	0.53983	0.54380	0.54776	0.55172	0.55567	0.55962	0.56356	0.56749	0.57142	0.57535
0.2	0.57926	0.58317	0.58706	0.59095	0.59483	0.59871	0.60257	0.60642	0.61026	0.61409
0.3	0.61791	0.62172	0.62552	0.62930	0.63307	0.63683	0.64058	0.64431	0.64803	0.65173
0.4	0.65542	0.65910	0.66276	0.66640	0.67003	0.67364	0.67724	0.68082	0.68439	0.68793
0.5	0.69146	0.69497	0.69847	0.70194	0.70540	0.70884	0.71226	0.71566	0.71904	0.72240
0.6	0.72575	0.72907	0.73237	0.73565	0.73891	0.74215	0.74537	0.74857	0.75175	0.75490
0.7	0.75804	0.76115	0.76424	0.76730	0.77035	0.77337	0.77637	0.77935	0.78230	0.78524
0.8	0.78814	0.79103	0.79389	0.79673	0.79955	0.80234	0.80511	0.80785	0.81057	0.81327
0.9	0.81594	0.81859	0.82121	0.82381	0.82639	0.82894	0.83147	0.83398	0.83646	0.83891
1.0	0.84134	0.84375	0.84614	0.84849	0.85083	0.85314	0.85543	0.85769	0.85993	0.86214
1.1	0.86433	0.86650	0.86864	0.87076	0.87286	0.87493	0.87698	0.87900	0.88100	0.88298
1.2	0.88493	0.88686	0.88877	0.89065	0.89251	0.89435	0.89617	0.89796	0.89973	0.90147
1.3	0.90320	0.90490	0.90658	0.90824	0.90988	0.91149	0.91309	0.91466	0.91621	0.91774
1.4	0.91924	0.92073	0.92220	0.92364	0.92507	0.92647	0.92785	0.92922	0.93056	0.93189
1.5	0.93319	0.93448	0.93574	0.93699	0.93822	0.93943	0.94062	0.94179	0.94295	0.94408
1.6	0.94520	0.94630	0.94738	0.94845	0.94950	0.95053	0.95154	0.95254	0.95352	0.95449
1.7	0.95543	0.95637	0.95728	0.95818	0.95907	0.95994	0.96080	0.96164	0.96246	0.96327
1.8	0.96407	0.96485	0.96562	0.96638	0.96712	0.96784	0.96856	0.96926	0.96995	0.97062
1.9	0.97128	0.97193	0.97257	0.97320	0.97381	0.97441	0.97500	0.97558	0.97615	0.97670
2.0	0.97725	0.97778	0.97831	0.97882	0.97932	0.97982	0.98030	0.98077	0.98124	0.98169
2.1	0.98214	0.98257	0.98300	0.98341	0.98382	0.98422	0.98461	0.98500	0.98537	0.98574
2.2	0.98610	0.98645	0.98679	0.98713	0.98745	0.98778	0.98809	0.98840	0.98870	0.98899
2.3	0.98928	0.98956	0.98983	0.99010	0.99036	0.99061	0.99086	0.99111	0.99134	0.99158
2.4	0.99180	0.99202	0.99224	0.99245	0.99266	0.99286	0.99305	0.99324	0.99343	0.99361
2.5	0.99379	0.99396	0.99413	0.99430	0.99446	0.99461	0.99477	0.99492	0.99506	0.99520
2.6	0.99534	0.99547	0.99560	0.99573	0.99585	0.99598	0.99609	0.99621	0.99632	0.99643
2.7	0.99653	0.99664	0.99674	0.99683	0.99693	0.99702	0.99711	0.99720	0.99728	0.99736
2.8	0.99744	0.99752	0.99760	0.99767	0.99774	0.99781	0.99788	0.99795	0.99801	0.99807
2.9	0.99813	0.99819	0.99825	0.99831	0.99836	0.99841	0.99846	0.99851	0.99856	0.99861
3.0	0.99865	0.99869	0.99874	0.99878	0.99882	0.99886	0.99889	0.99893	0.99896	0.99900
3.1	0.99903	0.99906	0.99910	0.99913	0.99916	0.99918	0.99921	0.99924	0.99926	0.99929
3.2	0.99931	0.99934	0.99936	0.99938	0.99940	0.99942	0.99944	0.99946	0.99948	0.99950
3.3	0.99952	0.99953	0.99955	0.99957	0.99958	0.99960	0.99961	0.99962	0.99964	0.99965
3.4	0.99966	0.99968	0.99969	0.99970	0.99971	0.99972	0.99973	0.99974	0.99975	0.99976
3.5	0.99977	0.99978	0.99978	0.99979	0.99980	0.99981	0.99981	0.99982	0.99983	0.99983
3.6	0.99984	0.99985	0.99985	0.99986	0.99986	0.99987	0.99987	0.99988	0.99988	0.99989
3.7	0.99989	0.99990	0.99990	0.99990	0.99991	0.99991	0.99992	0.99992	0.99992	0.99992
3.8	0.99993	0.99993	0.99993	0.99994	0.99994	0.99994	0.99994	0.99995	0.99995	0.99995
3.9	0.99995	0.99995	0.99996	0.99996	0.99996	0.99996	0.99996	0.99996	0.99997	0.99997

TABLE A.5

F-Test Table at 0.05 Significance Level

$\frac{v_1}{v_2}$	1	2	3	4	5	6	7	8	9
1	161.44	199.50	215.70	224.58	230.16	233.98	236.76	238.88	240.54
2	18.51	19.00	19.16	19.24	19.29	19.33	19.35	19.37	19.38
3	.	9.55	9.27	9.11	.	.	.	8.84	.
4	.	6.94	6.59	6.38	.	.	.	6.04	.
5	.	5.78	5.40	5.19	.	.	.	4.81	.
6	.	5.14	4.75	4.53	.	.	.	4.14	.
7	.	4.73	4.34	4.12	.	.	.	3.73	.
8	.	4.46	4.06	3.83	.	.	.	3.44	.
9	.	4.26	3.86	3.63	.	.	.	3.23	.

TABLE A.6

Critical Values for Two-Tailed Nemenyi Test at 0.05 Significance Level

Number of Subjects	2	3	4	5	...	...	9	10
$q_{0.10}$	1.645	2.052	2.291	2.459	.	.	2.855	2.920
$q_{0.05}$	1.960	2.344	2.569	2.728	.	.	3.102	3.164
$q_{0.01}$	2.576	2.913	3.113	3.255	.	.	3.590	3.646

TABLE A.7

Critical Values for Two-Tailed Bonferroni Test at 0.05 Significance Level

Number of Subjects	2	3	4	5	...	...	9	10
$q_{0.10}$	1.645	1.960	2.128	2.241	.	.	2.498	2.539
$q_{0.05}$	1.960	2.241	2.394	2.498	.	.	2.724	2.773

TABLE A.8

Data Set Example

WMC	DIT	NOC	CBO	RFC	LCOM	LOC	Fault
28	1	0	32	82	374	926	1
6	1	2	3	7	3	36	0
4	2	0	5	6	4	21	0
4	1	0	9	4	6	4	0
1	1	0	8	1	0	1	0

(Continued)

TABLE A.8 (*Continued*)

Data Set Example

WMC	DIT	NOC	CBO	RFC	LCOM	LOC	Fault
23	2	0	150	67	235	653	1
5	1	0	8	26	0	127	0
25	4	0	18	73	200	666	1
7	1	0	2	13	0	86	0
21	1	0	7	22	154	141	0
2	2	0	2	4	1	11	0
8	1	0	10	21	28	130	0
32	4	0	16	81	406	504	0
13	1	2	4	44	70	208	0
19	1	0	8	42	99	329	1
2	1	0	6	21	1	144	0
1	1	0	4	1	0	1	0
37	6	0	26	120	570	1,123	1
8	1	5	6	22	6	145	0
5	2	2	6	14	2	60	0
7	1	0	10	63	21	1,034	0
2	1	0	15	34	0	326	0
5	2	0	10	44	0	305	1
4	4	2	5	8	6	20	0
5	1	0	6	17	10	112	0
8	1	0	3	35	20	303	1
8	1	0	2	13	14	69	0
47	4	0	15	108	865	896	0
22	2	0	16	59	0	354	0
10	1	0	17	62	11	491	1
2	2	0	3	3	0	14	0
5	1	4	7	8	8	34	0
2	5	3	6	5	1	12	0
11	1	0	4	21	13	68	0
59	1	0	31	148	0	895	0
25	4	0	10	49	224	304	0
6	2	0	6	36	0	165	0
5	1	0	1	22	10	103	0
3	1	0	13	22	0	201	0
57	2	1	56	242	1504	2,550	1
5	1	0	2	6	2	36	1
13	1	0	14	49	24	298	0
2	3	0	3	5	0	15	0
29	2	0	21	104	236	733	1
12	2	0	6	31	20	360	0
3	1	0	16	3	3	3	0
38	4	4	21	104	613	839	1
19	1	0	8	19	171	19	0
2	1	0	9	2	1	4	0

(Continued)

TABLE A.8 (*Continued*)

Data Set Example

WMC	DIT	NOC	CBO	RFC	LCOM	LOC	Fault
8	2	0	9	61	28	544	1
13	6	0	25	69	78	420	0
3	2	0	5	5	3	18	0
2	2	0	1	4	1	10	0
4	1	0	7	17	0	58	0

References

A. Abran and J. Moore, "Guide to the software engineering body of knowledge," In *IEEE Computer Society*, Piscataway, NJ, 2004.

ACM, Computing Machinery, "ACM code of ethics and professional conduct," 2015, http://www.acm.org/constitution/code.html.

W. Afzal, "Metrics in software test planning and test design processes," PhD Dissertation, School of Engineering, Blekinge Institute of Technology, Karlskrona, Sweden, 2007.

W. Afzal, "Using faults-slip-through metric as a predictor of fault proneness," In *Proceedings of the 17th Asia Pacific Software Engineering Conference*, pp. 414–422, 2010.

W. Afzal, and R. Torkar, "Lessons from applying experimentation in software engineering prediction systems," In *Proceedings of the 2nd International. WS on Software Productivity Analysis and Cost Estimation, co-located with APSEC*, vol. 8, 2008.

W. Afzal, R. Torkar, and R. Feldt, "A systematic literature review of search-based software testing for non-functional system properties," *Information and Software Technology*, vol. 51, no. 6, 957–976, 2009.

K. K. Aggarwal, Y. Singh, A. Kaur, and R. Malhotra, "Empirical analysis for investigating the effect of object-oriented metrics on fault proneness: A replicated study," *Software Process: Improvement and Practice*, vol. 16, no. 1, pp. 39–62, 2009.

K. K. Aggarwal, Y. Singh, A. Kaur, and R. Malhotra, "Empirical study of object-oriented metrics," *Journal of Object Technology*, vol. 5, no. 8, pp. 149–173, 2006a.

K. K. Aggarwal, Y. Singh, A. Kaur, and R. Malhotra, "Investigating the effect of coupling metrics on fault proneness in object-oriented systems," *Software Quality Professional*, vol. 8, no. 4, pp. 4–16, 2006b.

K. K. Aggarwal, Y. Singh, A. Kaur, and R. Malhotra, "Investigating the effect of design metrics on fault proneness in object-oriented systems," *Journal of Object Technology*, vol. 6, no. 10, pp. 127–141, 2007.

K. K. Aggarwal, Y. Singh, A. Kaur, and R. Malhotra, "Software reuse metrics for object-oriented systems," In *Proceedings of the 3rd ACIS International Conference on Software Engineering Research, Management & Applications*, Central Michigan University, Mount Pleasant, MI, pp. 48–55, 2005.

J. Al Dallal, "Fault prediction and the discriminative powers of connectivity-based object-oriented class cohesion metrics," *Information and Software Technology*, vol. 54, no. 4, pp. 396–416, 2012a.

J. Al Dallal, "The impact of accounting for special methods in the measurement of object-oriented class cohesion on refactoring and fault prediction activities," *Journal of Systems and Software*, vol. 85, no. 5, pp. 1042–1057, 2012b.

J. Al Dallal, "Improving the applicability of object-oriented class cohesion metrics," *Information and Software Technology*, vol. 53, no. 9, pp. 914–928, 2011.

J. Alcalá, A. Fernández, J. Luengo, J. Derrac, S. García, L. Sánchez, and F. Herrera, "Keel data-mining software tool: Data set repository, integration of algorithms and experimental analysis framework," *Journal of Multiple-Valued Logic and Soft Computing*, vol. 17, no. 11, pp. 255–287, 2011.

M. Alshayeb, and W. Li, "An empirical investigation of object-oriented metrics in two different iterative processes," *IEEE Transactions on Software Engineering*, vol. 29, no. 11, pp. 1043–1049, 2003.

V. Ambriola, L. Bendix, and P. Ciancarini, "The evolution of configuration management and version control," *Software Engineering Journal*, vol. 5, no. 6, pp. 303–310, 1990.

M. D. Ambros, M. Lanza, and R. Robbes, "Evaluating defect prediction approaches: A benchmark and an extensive comparison," *Empirical Software Engineering*, vol. 17, no. 4–5, pp. 531–577, 2012.

M. D. Ambros, M. Lanza, and R. Robbes, "An extensive comparison of bug prediction approaches," In *7th IEEE Working Conference on Mining Software Repositories*, Cape Town, South Africa, pp. 31–41, 2010.

M. D. Ambros, M. Lanza, and R. Robbes, "On the relationship between change coupling and software defects," In *16th Working Conference on Reverse Engineering*, Lille, France, pp. 135–144, 2009.

D. Anderson, D. Sweeney, T. Williams, J. Camm, and J. Cochran, *Statistics for Business & Economics*, Cengage Learning, Mason, OH, 2002.

J. Anvik, L. Hiew, and G. C. Murphy, "Who should fix this bug?," In *28th International Conference on Software Engineering*, Shanghai, China, pp. 361–370, 2006.

A. Arcuri, and G. Fraser, "Parameter tuning or default values? An empirical investigation in search-based software engineering," *Empirical Software Engineering*, vol. 18, no. 3, pp. 594–623, 2013.

E. Arisholm, and L. C. Briand, "Predicting fault-prone components in a java legacy system," In *Proceedings of the 2006 ACM/IEEE International Symposium on Empirical Software Engineering*, New York: ACM pp. 8–17, 2006.

E. Arisholm, L. C. Briand, and A. Foyen, "Dynamic coupling measures for object-oriented software," *IEEE Transactions on Software Engineering*, vol. 30, no. 8, pp. 491–506, 2004.

E. Arisholm, L. C. Briand, and E. B. Johanessen, "A systematic and comprehensive investigation of methods to build and evaluate fault prediction models," *Journal of Systems and Software*, vol. 83, no. 1, pp. 2–17, 2010.

D. Ary, L. C. Jacobs, and A. Razavieh, "Introduction to Research in Education," New York: Holt Rinehart & Winston, vol. 1, pp. 9–72, 1972.

D. Azar, and J. Vybihal, "An ant colony optimization algorithm to improve software quality prediction models: Case of class stability," *Information and Software Technology*, vol. 53, no. 4, pp. 388–393, 2011.

E. R. Babbie, *Survey Research Methods*, Wadsworth, OH: Belmont, 1990.

A. W. Babich, *Software Configuration Management: Coordination for Team Productivity*, Addison-Wesley, Boston, MA, 1986.

T. Ball, J. M. Kim, A. A. Porter, and H. P. Siy, "If your version control system could talk," In *ICSE Workshop on Process Modelling and Empirical Studies of Software Engineering*, vol. 11, Boston, MA, 1997.

R. K. Bandi, V. K. Vaishnavi, and D. E. Turk, "Predicting maintenance performance using object-oriented design complexity metrics," *IEEE Transactions on Software Engineering*, vol. 29, no. 1, pp. 77–87, 2003.

J. Bansiya, and C. Davis, "A hierarchical model for object-oriented design quality assessment," *IEEE Transactions on Software Engineering*, vol. 28, no. 1, pp. 4–17, 2002.

G. M. Barnes, and B. R. Swim, "Inheriting software metrics," *Journal of Object Oriented Programming*, vol. 6, no. 7, 27–34, 1993.

V. Barnett, and T. Lewis, *Outliers in Statistical Data*, New York: Wiley, 1994.

M. O. Barros, and A. C. D. Neto, "Threats to validity in search-based software engineering empirical studies," Technical Report TR 0006/2011, UNIRIO—Universidade Federal do Estado do, Rio de Janeiro, Brazil, 2011.

V. R. Basili, L. C. Briand, and W. L. Melo, "A validation of object-oriented design metrics as quality indicators," *IEEE Transactions on Software Engineering*, vol. 22, no. 10, pp. 751–761, 1996.

V. R. Basili, and D. M. Weiss, "A methodology for collecting valid software engineering data," *IEEE Transactions on Software Engineering*, vol. 6, pp. 728–738, 1984.

V. R. Basili and R. Reiter, "Evaluating automable measures of software models," In *IEEE Workshop on Quantitative Software Models*, New York, pp. 107–116, 1979.

V. R. Basili, R. W. Selby, and D. H. Hutchens, "Experimentation in software engineering," *IEEE Transactions on Software Engineering*, vol. 12, no. 7, pp. 733–743, 1986.

U. Becker-Kornstaedt, "Descriptive software process modeling how to deal with sensitive process information," *Empirical Software Engineering*, vol. 6, no. 4, pp. 353–367, 2001.

J. Bell, *Doing Your Research Project: A Guide for First-Time Researchers in Education, Health and Social Science*, Maidenhead: Open University Press, 2005.

D. A. Belsley, E. Kuh, and R. Welsch, *Regression Diagnostics: Identifying Influential Data and Sources of Collinearity*, New York: Wiley, 1980.

R. Bender, "Quantitative risk assessment in epidemiological studies investigating threshold effects," *Biometrical Journal*, vol. 41, no. 3, pp. 305–319, 1999.

S. Benlarbi, and W. L. Melo, "Polymorphism measures for early risk prediction," In *Proceedings of the 21st International Conference on Software Engineering*, Los Angeles, CA, pp. 335–344, 1999.

S. Benlarbi, K. El Emam, N. Goel, and S. Rai, "Thresholds for object-oriented measures," In *Proceedings of 11th International Symposium on Software Reliability Engineering*, San Jose, CA, pp. 24–38, 2000.

E. H. Bersoff, V. D. Henderson, and S. G. Siegel, "Software configuration management," *ACM SIGSOFT Software Engineering Notes*, vol. 3, no. 5, pp. 9–17, 1978.

N. Bevan, "Measuring usability as quality of use," *Software Quality Journal*, vol. 4, no. 2, pp. 115–150, 1995.

J. Bieman, and B. Kang, "Cohesion and reuse in an object oriented system," In *Proceedings of the ACM Symposium Software Reusability*, Seattle, WA: ACM, pp. 259–262, 1995.

J. Bieman, G. Straw, H. Wang, P. W. Munger, and R. T. Alexander, "Design patterns and change proneness: An examination of five evolving systems," In *Proceedings of the 9th International Software Metrics Symposium*, Sydney, Australia, pp. 40–49, 2003.

A. Binkley, and S. Schach, "Validation of the coupling dependency metric as a risk predictor," In *Proceedings of the International Conference on Software Engineering*, Kyoto, Japan, pp. 452–455, 1998.

A. Birk, T. Dingsøyr, and T. Stålhane, "Postmortem: Never leave a project without it," *IEEE Software*, vol. 19, pp. 43–45, 2002.

L. D. Bloomberg, and M. Volpe, *Completing Your Qualitative Dissertation: A Roadmap from Beginning to End*, London: Sage Publications, 2012.

G. Booch, *Object-Oriented Analysis and Design with Applications*, 2nd edition, Benjamin/Cummings, San Francisco, CA, 1994.

M. Bramer, *Principles of Data Mining*, Springer, London, 2007.

P. Brereton, B. Kitchenham, D. Budgen, M. Turner, and M. Khalid, "Lessons from applying the systematic literature review process within the software engineering domain," *Journal of Systems and Software*, vol. 80, no. 4, pp. 571–583, 2007.

L. C. Briand, J. W. Daly, and J. K. Wust, "A unified framework for cohesion measurement in object-oriented systems," *Empirical Software Engineering*, vol. 3, no. 1, pp. 65–117, 1998.

L. C. Briand, J. W. Daly, and J. K. Wust, "A unified framework for coupling measurement in object-oriented systems," *IEEE Transactions on Software Engineering*, vol. 25, no. 1, pp. 91–121, 1999a.

L. C. Briand, S Morasca, V. R. Basili, "Defining and validating measures for object-based high-level design," *IEEE Transactions on Software Engineering*, vol. 25, no. 5, pp. 722–743, 1999b.

L. C. Briand, P. Devanbu, and W. Melo, "An investigation into coupling measures for C++," In *Proceedings of the ICSE*, Boston, MA: ACM, pp. 412–421, 1997.

L. C. Briand, and J. W. Wüst. "Empirical studies of quality models in object-oriented systems," *Advances in Computers*, vol. 56, pp. 97–166, 2002.

L. C. Briand, J. W. Wüst, J. W. Daly, and D. V. Porter, "Exploring the relationships between design measures and software quality in object-oriented systems," *Journal of Systems and Software*, vol. 51, no. 3, pp. 245–273, 2000.

L. C. Briand, J. W. Wust, and H. Lounis, "Replicated case studies for investigating quality factors in object-oriented designs," *Empirical Software Engineering*, vol. 6, no. 1, pp. 11–58, 2001.

L. Breiman, "Bagging predictors," *Machine Learning*, vol. 24, no. 2, pp. 123–140, 1996.

L. Breiman, "Random forests," *Machine Learning*, vol. 45, no. 1, pp. 5–32, 2001.

L. Breiman, J. Friedman, C. J. Stone, and R. A. Olshen, *Classification and Regression Tree*, CRC Press, Boca Raton, FL, 1984.

F. B. Abreu, and W. Melo, "Evaluating the impact of object-oriented design on software quality," In *Proceedings of the 3rd International Symposium on Software Metrics*, Berlin, Germany, pp. 90–99, 1996.

V. R. Caldiera, G. Caldiera, and H. D. Rombach, "The goal question metric approach," *Encyclopedia of Software Engineering*, vol. 2, pp. 528–532, 1994.

D. T. Campbell, and J. C. Stanley, "Experimental and quasi-experimental designs for research," Boston, MA: Houghton Miffin Company, 1963.

G. Canfora and L. Cerulo, "How software repositories can help in resolving a new change request," In *Proceedings of Workshop on Empirical Studies in Reverse Engineering*, Paolo Tonella, Italy, 2005.

G. Canfora, A. De Lucia, M. Di Penta, R. Oliveto, A. Panichella, and S. Panichella, "Multi-objective cross-project defect prediction," In *Proceedings of the IEEE 6th international Conference on Software Testing, Verification and Validation*, Luxembourg City, Luxembourg, pp. 252–261, 2013.

M. Cartwright, and M. Shepperd, "An empirical investigation of an object-oriented software system," *IEEE Transactions on Software Engineering*, vol. 26, no. 8, pp. 786–796, 2000.

C. Catal, "Software fault prediction: A literature review and current trends," *Expert Systems with Applications*, vol. 38, no. 4, pp. 4626–4636, 2011.

C. Catal, and B. Diri, "Investigating the effect of dataset size, metrics sets, and feature selection techniques on software fault prediction problem," *Information Sciences*, vol. 179, no. 8, pp. 1040–1058, 2009.

P. Cederqvist, *Version Management with CVS*, 1992, http://www.cvshome.org/docs/manual/.

M. A. Chaumum, H. Kabaili, R. K. Keller, and F. Lustman, "A change impact model for changeability assessment in object oriented software systems," In *Proceedings of the 3rd European Conference on Software Maintenance and Reengineering*, Amsterdam, The Netherlands, pp. 130–138, 1999.

N. Chen, S. C. Hoi, and X. Xiao, "Software process evaluation: A machine learning framework with application to defect management process," *Empirical Software Engineering*, vol. 19, no. 6, pp. 1531–1564, 2013.

J. K. Chhabra, and V. Gupta, "A survey of dynamic software metrics," *Journal of Computer Science and Technology*, vol. 25, no. 5, pp. 1016–1029, 2010.

S. Chidamber, and C. Kemerer, "A metrics suite for object-oriented design," *IEEE Transactions on Software Engineering*, vol. 20, no. 6, pp. 476–493, 1994.

S. Chidamber, and C. Kemerer, "Towards a metrics suite for object oriented design," In *Proceedings of the Conference on Object-Oriented Programming: Systems, Languages and Applications*, Phoenix, AZ, vol. 26, no. 11, pp. 197–211, 1991.

S. Chidamber, D. Darcy, and C. Kemerer, "Managerial use of metrics for object-oriented software: An exploratory analysis," *IEEE Transactions on Software Engineering*, vol. 24, no. 8, pp. 629–639, 1998.

W. Cohen, "Fast effective rule induction," In *Proceedings of the 12th International Conference on Machine Learning*, Tahoe City, CA, pp. 115–123, 1995.

D. Coleman, B. Lowther, and P. Oman, "The application of software maintainability models in industrial software systems," *Journal of System and Software*, vol. 29, no. 1, pp. 3–16, 1995.

S. K. Conte, H. E. Dunsmore, and V. Y. Shen, *Software Engineering Metrics and Models*, Menlo Park, CA: Benjamin Cummings Publishing, 1986.

T. D. Cook, and D. T. Campbell, *Quasi-Experimentation—Design and Analysis for Field Settings*," Boston, MA: Houghton Miffin, 1979.

J. C. Coppick, and T. J. Cheatham, "Software metrics for object-oriented systems," In *Proceedings of the ACM Annual Computer Science Conference*, Kansas City, MO, pp. 317–322, 1992.

C. Cortes, and V. Vapnik, "Support-vector networks," *Machine Learning*, vol. 20, no. 3, pp. 273–297, 1995.

J. W. Creswell, *Research Design: Qualitative and Quantitative Approaches*, London: SAGE, 1994.

N. Cristianini, and J. Shawe-Taylor, *An Introduction to Support Vector Machines and Other Kernel-Based Learning Methods*, Cambridge: Cambridge University Press, 2000.

A. B. De Carvalho, A. Pozo, and S. R. Vergilio, "A symbolic fault prediction model based on multiobjective particle swarm optimization," *Journal of Systems and Software*, vol. 83, no. 5, pp. 868–882, 2010.

F. De Comité, R. Gilleron, and M. Tommasi, "Learning multi-label alternating decision trees from texts and data," In *Machine Learning and Data Mining in Pattern Recognition*, Berlin, Germany: Springer pp. 35–49, 2003.

K. Dejager, T. Verbraken, and B. Baesens, "Toward comprehensible software fault prediction models using Bayesian network classifiers," *IEEE Transactions on Software Engineering*, vol. 39, no. 2, pp. 237–257, 2013.

T. DeMarco, *Controlling Software Projects: Management, Measurement & Estimation*, Upper Saddle River, NJ: Prentice Hall, 1982.

J. Demšar, "Statistical comparisons of classifiers over multiple data sets," *The Journal of Machine Learning Research*, vol. 7, pp. 1–30, 2006.

L. Di Geronimo, F. Ferrucci, A. Murolo, and F. Sarro, "A parallel genetic algorithm based on hadoop mapreduce for the automatic generation of junit test suites," In *Proceedings of 5th International Conference on Software Testing, Verification and Validation*, Montreal, Canada, pp. 785–793, 2012.

S. Di Martino, F. Ferrucci, C. Gravino, and F. Sarro, "A genetic algorithm to configure support vector machines for predicting fault prone components," In *Proceedings of the 12th International Conference on Product-Focused Software Process Improvement*, Limerick, Ireland, pp. 247–261, 2011.

K. Dickersin, "The existence of publication bias and risk factors for its occurrence," *Journal of the American Medical Association*, vol. 263, no. 10, pp. 1385–1395, 1990.

W. Ding, P. Liang, A. Tang, and H. V. Vilet, "Knowledge-based approaches in software documentation: A systematic literature review," *Information and Software Technology*, vol. 56, no. 6, pp. 545–567, 2014.

E. Duman, "Comparison of decision tree algorithms in identifying bank customers who are likely to buy credit cards," In *Proceedings of the 7th International Baltic Conference on Databases and Information Systems*, Kaunas, Lithuania, July 3–6, 2006.

R. P. Duran, M. A. Eisenhart, F. D. Erickson, C. A. Grant, J. L. Green, L. V. Hedges, F. J. Levine, P. A. Moss, J. W. Pellegrino, and B. L. Schneider, "Standards for reporting on empirical social science research in AERA publications american educational research association," *Educational Researcher*, vol. 35, no. 6, pp. 33–40, 2006.

B. Eftekhar, K. Mohammad, H. Ardebili, M. Ghodsi, and E. Ketabchi, "Comparision of artificial neural network and logistic regression models for prediction of mortality in head trauma based on initial clinical data," *BMC Medical Informatics and Decision Making*, vol. 5, no. 3, pp. 1–8 2005.

K. El Emam, "Ethics and open source," *Empirical Software Engineering*, vol. 6, no. 4, pp. 291–292, 2001.

K. El Emam, S. Benlarbi, N. Goel, and S. N. Rai, "The confounding effect of class size on the validity of object-oriented metrics," *IEEE Transactions on Software Engineering*, vol. 27, no. 7, pp. 630–650, 2001a.

K. El Emam, S. Benlarbi, N. Goel, and S. Rai, "The Optimal Class Size for Object-Oriented Software: A Replicated Case Study," Technical Report ERB-1074, National Research Council of Canada, Canada, 2000a.

K. El Emam, S. Benlarbi, N. Goel, and S. Rai, "A validation of object- oriented metrics," Technical Report ERB-1063, National Research Council of Canada, Canada, 1999.

K. El Emam, N. Goel, and S. Rai, "Thresholds for object oriented measures," In *Proceedings of the 11th International Symposium on Software Reliability Engineering*, San Jose, CA, pp. 24–38, 2000b.

K. El Emam, W. Melo, and J. C. Machado, "The prediction of faulty classes using object-oriented design metrics," *Journal of Systems and Software*, vol. 56, no. 1, pp. 63–75, 2001b.

A. E. Eiben and J. E. Smith, *Introduction to Evolutionary Computing*, Natural Computing Series, New York: Springer-Verlag, 2003.

M. O. Elish, and M. A. Al-Khiaty, "A suite for quantifying historical changes to predict future change-prone classes in object-oriented software," *Journal of Software: Evolution and Process*, vol. 25, no. 5, pp. 407–437, 2013.

M. O. Elish, A. Al-Yafei, and M. Al-Mulhem, "Empirical comparison of three metrics suites for fault prediction in packages of object-oriented systems: A case study of Eclipse," *Advances in Engineering Software*, vol. 42, no. 10, pp. 852–859, 2011.

K. Erni, and C. Lewerentz, "Applying design-metrics to object-oriented frameworks," In *Proceedings of the 3rd International in Software Metrics Symposium*, New York: IEEE, pp. 64–74, 1996.

L. H. Etzkorn, J. Bansiya, and C. Davis, "Design and code complexity metrics for OO classes," *Journal of Object-Oriented Programming*, vol. 12, no. 1, pp. 35–40, 1999.

N. Fenton, and M. Neil, "A critique of software defect prediction models," *IEEE Transactions on Software Engineering*, vol. 25, no. 3, pp. 1–15, 1999.

N. E. Fenton, and S. L. Pfleeger, *Software Metrics—A Rigorous & Practical Approach*, International Thomson Computer Press, 1996.

M. Fowler, *Refactoring: Improving the Design of Existing Code*, New Delhi, India: Pearson Education, 1999.

V. French, "Establishing software metric thresholds," In *Proceedings of the 9th International Workshop on Software Measurement*, Quebec, Canada, 1999.

Y. Freund, and L. Mason, "The alternating decision tree algorithm," In *Proceedings of 16th International Conference on Machine Learning*, Bled, Slovenia, pp. 124–133, 1999.

Y. Freund, and R. E. Schapire, "Experiments with a new boosting algorithm," *Proceedings of the 13th International Conference on Machine Learning*, San Francisco, CA, vol. 96, pp. 148–156, 1996.

Y. Freund, and R. E. Schapire, "Large margin classification using the perceptron algorithm," *Machine Learning*, vol. 37, no. 3, pp. 277–296, 1999.

J. Friedman, T. Hastie, and R. Tibshirani, "Additive logistic regression: A statistical view of boosting," *The Annals of Statistics*, vol. 28, no. 2, pp. 337–407, 2000.

M. Friedman, "A comparison of alternative tests of significance for the problem of m rankings," *The Annals of Mathematical Statistics*, vol. 11, no. 1, pp. 86–92, 1940.

H. Gall, K. Hajek, and M. Jazayeri, "Detection of logical coupling based on product release history," In *Proceedings of the 14th International Conference on Software Maintenance*, Bethesda, MD, pp. 190–198, 1998.

H. Gall, M. Jazayeri, and J. Krajewski, "CVS release history data for detecting logical couplings," In *Proceedings of IEEE 6th International Workshop on Software Evolution*, Helsinki, Finland, pp. 13–23, 2003.

H. Gall, M. Jazayeri, R. R. Klosch, and G. Trausmuth, "Software evolution observations based on product release history," In *Proceedings of the International Conference on Software Maintenance*, pp. 160–166, Bari, Italy, 1997.

S. García, A. D. Benítez, F. Herrera, and A. Fernández, "Statistical comparisons by means of non-parametric tests: A case study on genetic based machine learning," *Algorithms*, vol. 13, pp. 95–104, 2007.

D. Glassberg, K. El-Emam, W. Melo, and N. Madhavji, *Validating Object-Oriented Design Metrics on a Commercial Java Application*, Technical Report, NRC-ERB-1080, National Research Council of Canada, Canada, 2000.

I. Gondra, "Applying machine learning to software fault-proneness prediction," *Journal of Systems and Software*, vol. 81, no. 2, pp. 186–195, 2008.

P. Goodman, *Practical Implementation of Software Metrics*, London: McGraw-Hill, 1993.

C. Grosan, and A. Abraham, "Hybrid evolutionary algorithms: Methodologies, architectures and reviews, studies in computational intelligence," In *Hybrid Evolutionary Algorithms*, Berlin, Germany: Springer, pp. 1–17, 2007.

T. Gyimothy, R. Ferenc, and I. Siket, "Empirical validation of object-oriented metrics on open source software for fault prediction," *IEEE Transactions on Software Engineering*, vol. 31, no. 10, pp. 897–910, 2005.

J. Hair, R. Anderson, R. Tatham, and W. Black, *Multivariate Data Analysis*, Upper Saddle River, NJ: Pearson, 2006.

M. Hall, "Benchmarking attribute selection techniques for discrete class data mining," *IEEE Transactions on Knowledge and Data Engineering*, vol. 15, no. 3, pp. 1–16, 2003.

M. A. Hall, "Correlation-based feature selection for discrete and numeric class machine learning," In *Proceedings of the 7th International Conference on Machine Learning*, pp. 359–366, 2000.

A. R. Han, S. Jeon, D. H. Bae, and J. Hong, "Measuring behavioural dependency for improving change-proneness prediction in UML-based design models," *Journal of Systems and Software*, vol. 83, no. 2, pp. 222–234, 2010.

J. Han, and M. Kamber, *Data Mining: Concepts and Techniques*, San Francisco, CA: Morgan Kaufmann, 2001.

J. A. Hanley, and B. J. McNeil, "The meaning and use of the area under a receiver operating characteristic ROC curve," *Radiology*, vol. 143, no. 1, pp. 29–36, 1982.

M. Harman, E. K. Burke, J. A. Clark, and Xin Yao, "Dynamic adaptive search based software engineering," In *Proceedings of IEEE International Symposium on Empirical Software Engineering and Measurement*, Lund, Sweden, pp. 1–8, 2012a.

M. Harman, and J. A. Clark, "Metrics are fitness functions too," In *Proceedings of 10th IEEE International Symposium on Software Metrics*, Chicago, IL, 2004.

M. Harman, Y. Jia, and Y. Zhang, "App store mining and analysis: MSR for app stores," In *Proceedings of 9th IEEE Working Conference on Mining Software Repositories*, Zurich, Switzerland, pp. 108–111, June 2012b.

M. Harman, and B. F. Jones, "Search-based software engineering," *Information and Software Technology*, vol. 43, no. 14, pp. 833–839, 2001.

M. Harman, S. A. Mansouri, and Y. Zhang, "Search-based software engineering: Trends, techniques and applications," *ACM Computing Survey*, vol 45, no. 1, pp. 1–64, 2012c.

M. Harman, P. McMinn, J. Teixeira de Souza, and S. Yoo, "Search based software engineering: Techniques, taxonomy and tutorial," In *Empirical Software Engineering and Verification*, Lecture Notes in Computer Science, Berlin, Germany: Springer-Verlag, pp. 1–59, 2012d.

D. L. Harnett, and J. L. Murphy, *Introductory Statistical Analysis*, Don Mills, Ontario, Canada: Addison-Wesley, 1980.

R. Harrison, S. Counsell, and R. Nithi, "Experimental assessment of the effect of inheritance on the maintainability of object oriented systems," *Journal of Systems and Software*, vol. 52, pp. 173–179, 1999.

A. Hassan, "The road ahead for mining software repositories," In *Frontiers of Software Maintenance*, pp. 48–57, Beijing, People's Republic of China, 2008.

E. Hassan, and R. C. Holt, "Predicting change propagation in software systems," In *Proceedings of the 20th International Conference on Software Maintenance*, Chicago, IL, pp. 284–293, 2004a.

E. Hassan, and R. C. Holt, "Using development history sticky notes to understand software architecture," In *Proceedings of the 12th International Workshop on Program Comprehension*, Bari, Italy, pp. 183–192, 2004b.

O. Hauge, C. Ayala, and R. Conradi, "Adoption of open source software in software-intensive organizations—a systematic literature review," *Information and Software Technology*, vol. 52, no. 11, pp. 1133–1154, 2010.

S. Haykin, and R. Lippmann, "Neural networks, A comprehensive foundation," *International Journal of Neural Systems*, vol. 5, no. 4, pp. 363–364, 1994.

S. Haykin, *Neural Network: A Comprehensive Foundation*, Prentice Hall, New Delhi, India, vol. 2, 1998.

H. He, and E. A. Garcia, "Learning from imbalanced data," *IEEE Transactions on Knowledge and Data Engineering*, vol. 21, no. 9, pp. 1263–1284, 2009.

Z. He, F. Shu, Y. Yang, M. Li, and Q. Wang, "An investigation on the feasibility of cross-project defect prediction," *Automated Software Engineering*, vol. 19, no. 2, pp. 167–199, 2012.

B. Henderson-Sellers, *Object-Oriented Metrics: Measures of Complexity*, Prentice Hall, NJ, 1996.

B. Henderson-Sellers, "Some metrics for object-oriented software engineering," In *Proceedings of the 1st IEEE International Conference on New Technology and Mobile Security*, Beirut, Lebanon, 2007.

S. Herbold, "Training data selection for cross-project defect prediction," In *Proceedings of the 9th International Conference on Predictive Models in Software Engineering*, Baltimore, MD, 2013.

M. Hitz, and B. Montazeri, "Measuring coupling and cohesion in object-oriented systems," In *Proceedings of the International Symposium on Applied Corporate Computing*, Monterrey, Mexico, 1995.

R. P. Hooda, *Statistics for Business and Economics*, New Delhi, India: Macmillan, 2003.

W. G. Hopkins, "A new view of statistics," *Sport Science*, 2003. http://www.sportsci.org/resource/stats/.

D. W. Hosmer, and S. Lemeshow, *Applied Logistic Regression*, New York: Wiley, 1989.

S. K. Huang, and K. M. Liu, "Mining version histories to verify the learning process of legitimate peripheral participants," In *Proceedings of the 2nd International Workshop on Mining Software Repositories*, St. Louis, MO, pp. 84–78, 2005.

IEEE, *IEEE Guide to Software Configuration Management*, IEEE/ANSI Standard 1042–1987, IEEE, 1987.

IEEE, *IEEE Standard Classification for Software Anomalies*, IEEE Standard 1044–1993, IEEE, 1994.

M. Jorgenson, and M. Shepperd, "A systematic review of software development cost estimation studies," *IEEE Transactions on Software Engineering*, vol. 33, no. 1, 33–53, 2007.

H. H. Kagdi, I. Maletic, and B. Sharif, "Mining software repositories for traceability links," In *Proceedings of 15th IEEE International Conference on Program Comprehension*, pp. 145–154, 2007.

S. Kanmani, V. R. Uthariraj, V. Sankaranarayanan, and P. Thambidurai, "Object-oriented software fault prediction using neural networks," *Information and Software Technology*, vol. 49, no. 5, Alberta, Canada, pp. 483–492, 2007.

T. M. Khoshgaftar F. D. Allen, J. P. Hudepohl, and S. J. Aud, "Application of neural networks to software quality modelling of a very large telecommunications system," *IEEE Transactions on Neural Networks*, vol. 8, no. 4, pp. 902–909, 1997.

T. M. Khoshgoftaar, J. C. Munson and D. L. Lanning, "Dynamic system complexity," In *Proceedings of Software Metrics Symposium*, Baltimore, MD, pp. 129–140, 1993.

B. Kitchenham, L. Pickard, and S. L. Pfleeger, "Case studies for method and tool evaluation," *IEEE Software*, vol. 12, no. 4, pp. 52–62, 1995.

B. A. Kitchenham, "Guidelines for performing systematic literature review in software engineering," Technical report EBSE-2007-001, London, 2007.

B. A. Kitchenham, E. Mendes, and G. H. Travassos, "Cross versus within-company cost estimation studies: A systematic review," *IEEE Transactions on Software Engineering*, vol. 33, no. 5, pp. 316–329, 2007.

A. G. Koru, and H. Liu, "Building effective defect-prediction models in practice," *IEEE Software*, vol. 22, no. 6, pp. 23–29, 2005.

A. G. Koru, and H. Liu, "Identifying and characterizing change-prone classes in two large-scale open-source products," *Journal of Systems and Software*, vol. 80, no. 1, pp. 63–73, 2007.

A. G. Koru, and J. Tian, "Comparing high-change modules and modules with the highest measurement values in two large-scale open-source products," *IEEE Transactions on Software Engineering*, vol. 31, no. 8, pp. 625–642, 2005.

C. V. Koten, and A. R. Gray, "An application of Bayesian network for predicting object-oriented software maintainability," *Information and Software Technology*, vol. 48, no. 1, pp. 59–67, 2006.

C. R. Kothari, *Research Methodology. Methods and Techniques*, New Delhi, India: New Age International Limited, 2004.

S. Kpodjedo, F. Ricca, P. Galnier, Y. G. Gueheneuc, and G. Antoniol, "Design evolution metrics for defect prediction in object-oriented systems," *Empirical Software Engineering*, vol. 16, no. 1, pp. 141–175, 2011.

A. Lake, and C. Cook, "Use of factor analysis to develop OOP software complexity metrics," In *Proceedings of the 6th Annual Oregon Workshop on Software Metrics*, Silver Falls, OR, 1994.

Y. Lee, B. Liang, S. Wu, and F. Wang, "Measuring the coupling and cohesion of an object-oriented program based on information flow," In *Proceedings of the International Conference on Software Quality*, Maribor, Slovenia, 1995.

S. Lessmann, B. Baesans, C. Mues, and S. Pietsch, "Benchmarking classification models for software defect prediction: A proposed framework and novel finding," *IEEE Transactions on Software Engineering*, vol. 34. no. 4, pp. 485–496, 2008.

T. C. Lethbridge, S. E. Sim, and J. Singer, "Studying software engineers: Data collection techniques for software field studies," *Empirical Software Engineering*, vol. 10, pp. 311–341, 2005.

W. Li, and S. Henry, "Object-oriented metrics that predict maintainability," *Journal of Systems and Software*, vol. 23, no. 2, pp. 111–122, 1993.

W. Li, and R. Shatnawi, "An empirical study of the bad smells and class error probability in the post-release object-oriented system evolution," *Journal of Systems and Software*, vol. 80, no. 7, pp. 1120–1128, 2007.

M. Li, H. Zhang, R. Wu, and Z. H. Zhou, "Sample-based software defect prediction with active and semi-supervised learning," *Automated Software Engineering*, vol. 19, no. 2, pp. 201–230, 2012.

P. Liang, and J. Li, "A change-oriented conceptual framework of software configuration management," In *International Conference on Service Systems and Service Management*, Chendu, People's Republic of China, pp. 1–4, 2007.

Y. Liu, T. M. Khoshgoftaar, and N. Seliya, "Evolutionary optimization of software quality modeling with multiple repositories," *IEEE Transactions on Software Engineering*, vol. 36, no. 6, pp. 852–864, 2010.

V. B. Livshits, and T. Zimmermann, "DynaMine: Finding common error patterns by mining software revision histories," In *Proceedings of the 10th European Software Engineering Conference held jointly with 13th ACM SIGSOFT International Symposium on Foundations of Software Engineering,* Lisbon, Portugal, pp. 296–305, 2005.

M. Lorenz, and J. Kidd, "Object-oriented software metrics," Prentice Hall, NJ, 1994.

H. Lu, Y. Zhou, B. Xu, H. Leung, and L. Chen, "The ability of object-oriented metrics to predict change-proneness: A meta-analysis," *Empirical Software Engineering Journal,* vol. 17, no. 3, pp. 200–242, 2012.

Y. Ma, G. Luo, X. Zeng, and A. Chen, "Transfer learning for cross-company software defect prediction," *Information and Software Technology,* vol. 54, no. 3, pp. 248–256, 2012.

R. Malhotra, "Empirical validation of object-oriented metrics for predicting quality attributes," PhD Dissertation, New Delhi, India: Guru Gobind Singh Indraprastha University, 2009.

R. Malhotra, "A systematic review of machine learning techniques for software fault prediction," *Applied Software Computing,* vol. 27, pp. 504–518, 2015.

R. Malhotra, and A. J. Bansal, "Investigation on feasibility of machine learning algorithms for predicting software change using open source software," *International Journal of Reliability, Quality and Safety Engineering,* World Scientific, Singapore, 2014a.

R. Malhotra, and A. Jain, "Fault Prediction Using statistical and machine learning methods for improving software quality," *Journal of Information Processing System,* vol. 8, pp. 241–262, 2012.

R. Malhotra, and A. Jain, "Software Effort Prediction using Statistical and Machine Learning Methods," *International Journal of Advanced Computer Science and Applications,* vol. 2, no. 1, pp. 145–152, 2011.

R. Malhotra, and M. Khanna, "The ability of search-based algorithms to predict change-prone classes," *Software Quality Professional,* vol. 17, no. 1, pp. 17–31, 2014b.

R. Malhotra, and M. Khanna, "Investigation of relationship between object-oriented metrics and change proneness," *International Journal of Machine Learning and Cybernetics,* vol. 4, no. 4, pp. 273–286, 2013.

R. Malhotra, and M. Khanna, "Software engineering predictive modeling using search-based techniques: Systematic review and future directions," In *1st North American Search-Based Software Engineering Symposium,* Dearborn, MI, 2015.

R. Malhotra, and Y. Singh, "A defect prediction model for open source software," In *Proceedings of the World Congress on Engineering,* London, Vol II, 2012.

R. Malhotra, and Y. Singh, "On the applicability of machine learning techniques for object-oriented software fault prediction," *Software Engineering: An International Journal,* vol. 1, pp. 24–37, 2011.

R. Malhotra, Y. Singh, and A. Kaur, "Empirical validation of object-oriented metrics for predicting fault proneness at different severity levels using support vector machines," *International Journal of System Assurance Engineering Management,* vol. 1, pp. 269–281, 2010.

R. Malhotra, "Search based techniques for software fault prediction: Current trends and future directions," In *Proceedings of 7th International Workshop on Search-Based Software Testing,* Hyderabad, India, pp. 35–36, 2014a.

R. Malhotra, "Comparative Analysis of statistical and machine learning methods for predicting faulty modules," *Applied Soft Computing,* vol. 21, pp. 286–297, 2014b.

R. Malhotra and A. Bansal, "Fault prediction considering threshold effects of object oriented metrics," *Expert Systems,* vol. 32, no. 2, pp. 203–219, 2015.

R. Malhotra and, M. Khanna, "A new metric for predicting software change using gene expression programming", In *Proceedings of 5th International Workshop on Emerging Trends in Software Metrics,* Hyderabad, India, pp. 8–14, 2014a.

R. Malhotra, N. Pritam, K. Nagpal, and P. Upmanyu, "Defect collection and reporting system for git based open source software," In *Proceedings of International Conference on Data Mining and Intelligent Computing,* Delhi, India, pp. 1–7, 2014.

R. Malhotra and R. Raje, "An empirical comparison of machine learning techniques for software defect prediction," In *Proceedings of 8th International Conference on Bio-Inspired Information and Communication Technologies,* Boston, MA, pp. 320–327, 2014.

A. Marcus, D. Poshyvankyk, and R. Ferenc, "Using the conceptual cohesion of classes for fault prediction in object-oriented systems," *IEEE Transactions on Software Engineering*, vol. 34, no. 2, pp. 287–300, 2008.

F. Marini, R. Bucci, A. L. Magri, and A. D. Magri, "Artificial neural networks in chemometrics: History, examples and perspectives," *Microchemical Journal*, vol. 88, no. 2, pp. 178–185, 2008.

R. C. Martin, *Agile Software Development: Principles, Patters, and Practices*. Prentice Hall, NJ, 2002.

G. Mausa, T. G. Grbac, and B. D. Basic, "Multivariate logistic regression prediction of fault proneness in software modules," In *Proceedings of the IEEE 35th International Convention on MIPRO*, Adriatic Coast, Craotia, pp. 698–703, 2012.

T. J. McCabe, "A complexity measure," *IEEE Transactions on Software Engineering*, vol. SE-2, no. 4, pp. 308–320, 1976.

Metrics Data Program, 2006, http://sarpresults.ivv.nasa.gov/ViewResearch/107.jsp.

T. Menzies, A. Butcher, D. Cok, A. Marcus, L. Layman, F. Shell, B. Turhan, and T. Zimmermann, "Local versus global lessons for defect prediction and effort estimation," *IEEE Transactions on Software Engineering*, vol. 39, no. 6, pp. 822–834, 2013.

T. Menzies, J. Greenwald, and A. Frank, "Data mining static code attributes to learn defect predictors," *IEEE Transactions on Software Engineering*, vol. 33, no. 1, pp. 2–13, 2007.

T. Menzies, and A. Marcus, "Automated severity assessment of software fault reports," *IEEE International Conference on Software Maintenance*, 2008.

T. Menzies, Z. Milton, B. Turhan, B. Cukic, Y. Jiang, and A. Bener, "Defect prediction from static code features: Current reults, limitations, new approaches," *Automated Software Engineering*, vol. 17, no. 4, pp. 375–407, 2010.

B. Meyer, H. Gall, M. Harman, and G. Succi, "Empirical answers to fundamental software engineering problems (panel)," In *ESEC/SIGSOFT FSE*, Saint Petersburg, Russia, pp. 14–18, 2013.

L. S. Meyers, G. C. Gamst, and A. J. Guarino. *Applied Multivariate Research: Design and Interpretation*, Thousand Oaks, CA: Sage, 2013.

J. Michura, and M. A. M. Capretz, "Metrics suite for class complexity," In *Proceedings of the International Conference on Information Technology: Coding and Computing*, Las Vegas, CL, pp. 404–409, 2005.

A. T. Misirh, A. B. Bener, and B. Turhan, "An industrial case study of classifier ensembles for locating software defects," *Software Quality Journal*, vol. 19, no. 3, pp. 515–536, 2011.

A. Mitchell, and J. F. Power, "An empirical investigation into the dimensions of run-time coupling in Java programs," In *Proceedings of the 3rd Conference on the Principles and Practice of Programming in Java*, Las Vegas, NV, pp. 9–14, 2004.

A. Mitchell, and J. F. Power, "Run-time cohesion metrics for the analysis of Java programs," Technical Report, Series No. NUIM-CS-TR-2003–08, Kildare, Ireland: National University of Ireland, Maynooth, Co., 2003.

W. N. H. W. Mohamed, M. N. M. Salleh, and A. H. Omar, "A comparative study of reduced error pruning method in decision tree algorithms," In *Proceedings of the IEEE International Conference on Control System, Computing and Engineering*, IEEE, Penang, Malaysia, pp. 392–397, 2012.

R. Moser, W. Pedrycz, and G. Succi, "A Comparative analysis of the efficiency of change metrics and static code attributes for defect prediction," In *Proceedings of International Conference on Software Engineering*, Leipzig, Germany, pp. 181–190, 2008.

T. R. G. Nair, and R. Selvarani, "Defect proneness estimation and feedback approach for software design quality improvement," *Information and Software Technology*, vol. 54, no. 3, pp. 274–285, 2012.

J. Nam, S. J. Pan, and S. Kim, "Transfer defect learning," In *Proceedings of the International Conference on Software Engineering*, San Francisco, CA, pp. 382–391, 2013.

NASA, Metrics data Repository, 2004, www.mdp.ivv.nasa.gov.

P. Naur and B. Randell (eds.), *Software Engineering: Report of a Conference Sponsored by the NATO Science Committee*, Garmisch, Germany. Brussels, Belgium: Scientific Affairs Division, NATO, 1969.

A. A. Neto, and T. Conte, "A conceptual model to address threats to validity in controlled experiments," In *Proceedings of the 17th International Conference on Evaluation and Assessment in Software Engineering*, Porto de Galinhas, Brazil, pp. 82–85, 2013.

M. Ohira, N. Ohsugi, T. Ohoka, and K. I. Matsumoto, "Accelerating cross-project knowledge collaboration using collaborative filtering and social networks," In *Proceedings of the 2nd International Workshop on Mining Software Repositories*, New York, pp. 111–115, 2005.

A. Okutan, and O. T. Yildiz, "Software defect prediction using bayesian networks," *Empirical Software Engineering*, vol. 19, no. 1, pp. 154–181, 2014.

H. Olague, L. Etzkorn, S. Gholston, and S. Quattlebaum, "Empirical validation of three software metric suites to predict the fault-proneness of object-oriented classes developed using highly iterative or agile software development processes," *IEEE Transactions on Software Engineering*, vol. 33, no. 10, pp. 402–419, 2007.

H. M. Olague, L. H. Etzkorn, S. L. Messimer, and H. S. Delugach, "An empirical validation of object-oriented class complexity metrics and their ability to predict error-prone classes in highly iterative, or agile, software: A case study," *Journal of Software Maintenance and Evolution: Research and Practice*, vol. 20, no. 3, pp. 171–197, 2008.

A. D. Oral, and A. B. Bener, "Defect prediction for embedded software," In *Proceedings of the IEEE 22nd International Symposium on Computer and Information Science*, Ankara, Turkey, pp: 1–6, 2007.

G. J. Pai, and J. Bechta Dugan, "Empirical analysis of software fault content and fault proneness using Bayesian methods," *IEEE Transactions on Software Engineering*, vol. 33, no. 10, pp. 675–686, 2007.

L. Pelayo, and S. Dick, "Evaluating stratification alternatives to improve software defect prediction," *IEEE Transactions on Software Reliability*, vol. 61, no. 2, pp. 516–525, 2012.

F. Peters, T. Menzies, and A. Marcus, "Better cross company defect prediction," In *Proceedings of the 10th IEEE Working Conference on Mining Software Repositories*, San Francisco, CA, pp. 409–418, 2013.

S. L. Pfleeger, "Experimental design and analysis in software engineering," *Annals of Software Engineering*, vol. 1, no. 1, pp. 219–253, 1995.

M. F. Porter, "An algorithm for suffix stripping," *Program*, vol. 14, no. 3, pp. 130–137, 1980.

A. Porter, and R. Selly, "Empirically guided software development using metric-based classification trees," *IEEE Software*, vol. 7, no. 2, pp. 46–54, 1990.

R. S. Pressman, *Software Engineering: A Practitioner's Approach*, New York: Palgrave Macmillan, 2005.

PROMISE, 2007, http://promisedata.org/repository/.

J. R. Quinlan, *C4.5 Programs for Machine Learning*, Morgan Kaufmann, San Francisco, CA, 1993.

D. Radjenović, M. Hericko, R. Torkar, and A. Zivkovic, "Software fault prediction metrics: A systematic literature review," *Information and Software Technology*, vol. 55, no. 8, pp. 1397–1418, 2013.

F. Rahman, D. Posnett, and P. Devanbu, "Recalling the imprecision of cross-project defect prediction," In *Proceedings of the ACM SIGSOFT 20th International Symposium on the Foundations of Software Engineering*, Cary, NC, p. 61, 2012.

J. Ratzinger, M. Fischer, and H. Gall, "Improving evolvability through refactoring," In *Proceedings of the 2nd International Workshop on Mining Software Repositories*, St. Louis, MO, pp. 69–73, 2005.

M. Riaz, E. Mendes, and E. Tempero, "A systematic review of software maintainability prediction and metrics," In *Proceedings of the 3rd International Symposium on Empirical Software Engineering and Measurement*, Lake Buena Vista, FL, pp. 367–377, 2009.

P. C. Rigby, and A. E. Hassan, "What can OSS mailing lists tell us? a preliminary psychometric text analysis of the apache developer mailing list," In *Proceedings of the 4th International Workshop on Mining Software Repositories*, Minneapolis, MN, pp. 23, 2007.

D. Rodriguez, R. Ruiz, J. C. Riquelme, and J. S. Agular-Ruiz, "Searching for rules to detect defective modules: A subgroup discovery approach," *Information Sciences*, vol. 191, pp. 14–30, 2012.

F. Rosenblatt, "The perceptron: A probabilistic model for information storage and organization in the brain," *Psychological Review*, vol. 56, no. 6, pp. 386–408, 1958.

M. Sassano, "An experimental comparison of the voted perceptron and support vector machines in Japanese analysis tasks," In *Proceedings of the 3rd International Conference on Natural Language Processing,* Hyderabad, India, pp. 829–834, 2008.

H. J. Seltman, *Experimental Design and Analysis,* 2012, http://www. stat. cmu. edu/, hseltman/309/ Book/Book. Pdf.

R. Shatnawi, "A quantitative investigation of the acceptable risk levels of object-oriented metrics in open-source systems," *IEEE Transactions on Software Engineering,* vol. 36, no. 2, 2010.

R. Shatnawi, "The validation and threshold values of object-oriented metrics," Dissertation, Huntsville, AL: Department of Computer Science, University of Alabama, 2006.

R. Shatnawi, and W. Li, "The effectiveness of software metrics in identifying error-prone classes in post-release software evolution process," *Journal of Systems and Software,* vol. 81, no. 11, 1868–1882, 2008.

R. Shatnawi, W. Li, J. Swain, and T. Newman, "Finding software metrics threshold values using ROC curves," *Journal of Software Maintenance and Evolution: Research and Practice,* vol. 22, no. 1, pp. 1–16, 2010.

P. H. Sherrod, "Predictive Modeling Software," 2003, http://www.dtreg.com.

J. Singer, and N. Vinson, "Why and how research ethics matters to you. Yes, you!," *Empirical Software Engineering,* vol. 6, no. 4, pp. 287–290, 2001.

Y. Singh, *Software Testing,* New York: Cambridge University Press, 2011.

Y. Singh, A. Kaur, and R. Malhotra, "A comparative study of models for predicting fault proneness in object-oriented systems," *International Journal of Computer Applications in Technology,* vol. 49, no. 1, pp. 22–41, 2014.

Y. Singh, A. Kaur, and R. Malhotra, "Empirical validation of object-oriented metrics for predicting fault proneness models," *Software Quality Journal,* vol. 18, no. 1, pp. 3–35, 2010.

Y. Singh, A. Kaur, and R. Malhotra, "Software fault proneness prediction using support vector machines," *Procedings of the World Congress on Engineering,* London, pp. 240–245, 2009b.

Y. Singh, and R. Malhotra, *Object-Oriented Software Engineering,* New Delhi, India: PHI Learning, 2012.

J. Snoek, H. Larochelle, and R. P. Adams, "Practical Bayesian optimization of machine learning algorithms," In *Proceedings of Advances in Neural Information Processing Systems,* Nevada, pp. 2951–2959, 2012.

J. Spolsky, *Painless Bug Tracking,* 2000, http://www.joelonsoftware.com/articles/fog0000000029 .html.

G. E. Stark, R. C. Durst, and T. M. Pelnik, "An evaluation of software testing metrics for NASA's mission control center," *Software Quality Journal,* vol. 1, no. 2, pp. 115–132, 1992.

K. J. Stol, M. A. Babar, B. Russo, and B. Fitzgerald, "The use of empirical methods in open source software research: Facts, trends and future directions," In *Proceedings of the ICSE Workshop on Emerging Trends in Free/Libre/Open Source Software Research and Development,* Washington, DC: IEEE Computer Society, pp. 19–24, 2009.

M. Stone, "Cross-validatory choice and assessment of statistical predictions," *Journal of the Royal Statistical Society. Series B (Methodological),* vol. 36, pp. 111–147, 1974.

R. Subramanyam, and M. S. Krishnan, "Empirical analysis of CK metrics for object-oriented design complexity: Implications for software defects," *IEEE Transactions on Software Engineering,* vol. 29, no. 4, pp. 297–310, 2003.

M. H. Tang, M. H. Kao, and M. H. Chen, "An empirical study on object-oriented metrics," In *Proceedings of the 6th International Software Metrics Symposium,* Boca Raton, FL, pp. 242–249, 1999.

D. Tegarden, S. Sheetz, and D. Monarchi, "A software complexity model of object-oriented systems," *Decision Support Systems,* vol. 13, no. 3–4, pp. 241–262, 1995.

D. Tegarden, S. Sheetz, and D. Monarchi, "The effectiveness of traditional software metrics for object-oriented systems," In *Proceedings of the 25th Hawaii International Conference on Systems Sciences,* Kauai, HI, pp. 359–368, 1992.

S. W. Thomas, A. E. Hassan, and D. Blostein, "Mining unstructured software repositories," In *Evolving Software Systems,* Berlin, Germany: Springer-Verlag, 2014.

M. M. T. Thwin, and T. Quah, "Application of neural networks for software quality prediction using object-oriented metrics," *Journal of Systems and Software*, vol. 76, no. 2, pp. 147–156, 2005.

A. Tosun, B. Turhan, and A. Bener, "Validation of network measures as indicators of defective modules in software systems," In *Proceedings of the 5th International Conference on Predictor Models in Software Engineering*, Vancouver, Canada, 2009.

B. Turhan, and A. Bener, "Analysis of naïve bayes assumptions on software fault data: An empirical study," *Data and Knowledge Engineering*, vol. 68, no. 2, pp. 278–290, 2009.

B. Turhan, G. Kocak, and A. Bener, "Software defect prediction using call graph based ranking (CGBR) framework," In *Proceedings of the 34th EUROMICRO Conference on Software Engineering and Advanced Applications*, Parma, Italy, pp. 191–198, 2008.

B. Turhan, T. Menzies, A. Bener, and J. Di Stefano, "On the relative value of cross-company and within-company data for defect prediction," *Empirical Software Engineering*, vol. 14, no. 5, pp. 540–578, 2009.

K. Ulm, "A statistical method for assessing a threshold in epidemiological studies," *Statistics in Medicine*, vol. 10, no. 3, pp. 341–349, 1991.

Y. Uzun, and G. Tezel, "Rule learning with machine learning algorithms and artificial neural networks," *Journal of Selcuk University Natural and Applied Science*, vol. 1, no. 2, pp. 54, 2012.

N. G. Vinson, and J. Singer, "A practical guide to ethical research involving humans," In *Guide to Advanced Empirical Software Engineering*, London: Springer, pp. 229–256, 2008.

S. Wasserman, *Social Network Analysis: Methods and Applications*, Cambridge: Cambridge University Press, 1994.

J. Wen, S. Li, Z. Lin, Y. Hu, and C. Huang, "Systematic literature review of machine learning based software effort estimation models," *Information and Software Technology*, vol. 54, no. 1, pp. 41–59, 2012.

E. J. Weyuker, "Evaluating software complexity measures," *IEEE Transactions on Software Engineering*, vol. 14, no. 9, pp. 1357–1365, 1998.

D. R. White, "Cloud computing and SBSE," In G. Ruhe and Y. Zhang (eds.), *Search Based Software Engineering*, Berlin, Germany: Springer, pp.16–18, 2013.

F. Wilcoxon, "Individual comparisons by ranking methods," *Biometrics*, vol. 1, no. 6, pp. 80–83, 1945.

I. H. Witten, and E. Frank, *Data Mining: Practical Machine Learning Tools and Techniques*, Morgan Kaufmann, Burlington, MA, 2005.

C. Wohlin, P. Runeson, M. Host, M. C. Ohlsson, B. Regnell, and A. Wesslen, *Experimentation in Software Engineering*, Berlin, Germany: Springer-Verlag, 2012.

H. K. Wright, M. Kim, and D. E. Perry, "Validity concerns in software engineering research," In *FSE/SDP Workshop on Future of Software Engineering Research*, Santa Fe, NM, pp. 411–414, 2010.

S. M. Yacoub, H. H. Ammar, and T. Robinson, "Dynamic metrics for object-oriented designs," In *Proceedings of the 5th International Software Metrics Symposium*, Boca Raton, FL, pp. 50–61, 1999.

L. M. Yap, and B. Henderson-Sellers, "Consistency considerations of object-oriented class libraries," Centre for Information Technology Research Report 93/3, University of New South Wales, 1993.

R. K. Yin, *Case Study Research: Design and Methods*, Sage publications, New York, 2002.

P. Yu, X. X. Ma, and J. Lu, "Predicting fault-proneness using OO metrics: An industrial case study," In *CSMR*, Budapest, Hungary, pp. 99–107, 2002.

Z. Yuming, and L. Hareton, "Empirical analysis of object oriented design metrics for predicting high servility faults," *IEEE Transactions on Software Engineering*, vol. 32, no. 10, pp. 771–784, 2006.

L. Zhao, and N. Takagi, "An application of Support vector machines to Chinese character classification problem," In *IEEE International Conference on systems, Man and Cybernetics*, Montreal, Canada, pp. 3604–3608, 2007.

Y. Zhao, and Y. Zhang, "Comparison of decision tree methods for finding active objects," *Advances in Space Research*, vol. 41, no. 12, pp. 1955–1959, 2008.

Y. Zhou, and H. Leung, "Empirical analysis of object-oriented design measures for predicting high and low severity faults," *IEEE Transactions on Software Engineering*, vol. 32, no. 10, pp. 771–789, 2006.

Y. Zhou, H. Leung, and B. Xu, "Examining the potentially confounding effect of class size on the associations between object-oriented metrics and change proneness," *IEEE Transactions on Software Engineering*, vol. 35, no. 5, pp. 607–623, 2009.

Y. Zhou, B. Xu, and L. Hareton, "On the ability of complexity metrics to predict fault-prone classes in object-oriented systems," *Journal of Systems and Software*, vol. 83, no. 4, pp. 660–674, 2010.

T. Zimmermann, N. Nagappan, H. Gall, E. Giger, and B. Murphy, "Cross-project defect prediction: A large scale experiment on data vs domain vs process," In *Proceedings of the 7th Joint Meeting of the European Software Engineering Conference and the ACM SIGSOFT Symposium on the Foundation of Software Engineering*, Amsterdam, the Netherlands, pp. 91–100, 2009.

T. Zimmermann, P. Weibgerber, S. Diehl, and A. Zeller, "Mining version histories to guide software changes," *IEEE Transactions on Software Engineering*, vol. 31, no. 6, pp. 429–445, 2005.

H. Zuse, *Software Complexity: Measures and Methods*, New York: Walter de Cruyter, 1991.

Index

Note: Locators followed by 'f' and 't' denote figures and tables in the text.

A

Absolute effect, measure of. *See* Risk difference
Abstract data types (ADTs), 80
Abstracts, 391
  basic, 391–392
  method, 392
  reporting results, 355t, 356
  result, 392
Academic software, 24, 143, 251–252
Accuracy, 404
  aspects of data analysis methods, 138
  and error rate, 302
  measure
    cross-validation results, 408–409, 409t
    Friedman mean ranks using, 416, 417t
    hypothesis for, 397, 416
  performance values, 246, 246t
  positive and negative, 303
  and precision, 295
Acknowledgment, reporting results, 356t, 359
Across-release validation result, 321, 322t
Adaptive boosting (AB), 394
Alternating decision tree (ADT), 394, 400
Alternative hypothesis, 123, 125–126, 419
Analysis of variance (ANOVA) test, 244
Analysis process, 401
Analysis results, 404
  CFS results, 408
  descriptive statistics, 404–407
  hypothesis testing and evaluation, 416–419
  Nemenyi results, 419–420
  outlier analysis, 408
  tenfold cross-validation results, 408–416
Android OS, 190
  deleted source files, 196–197, 197f
  Mms application, 191f, 194f, 195f, 196f, 197f, 198f, 199f
  newly added source files, 195–196, 196f
Appendix, reporting results, 356t, 359
Area under normal distribution, 437, 439t–440t
Area under ROC curve (AUC), 47, 300, 304, 312
  across-release validation results, 322f, 322t
  measure

Friedman mean ranks using, 419t
  hypothesis for, 398
  performance, 413t
  of models predicted, 309t
  results of five studies, 50t
  tenfold cross-validation results, 318t, 322f
  values, 312, 312t
Attribute reduction methods
  attribute extraction, 220, 222–223
  attribute selection, 220, 221–222
    CFS techniques, 221–222
    univariate analysis, 222
  classification, 220f
  discussion, 223
  procedure, 220f
Attributes in software engineering research, 23
AUC. *See* Area under ROC curve (AUC)
Average inheritance depth (AID), 85
Average lines of code (LOC), 210–211

B

Back propagation algorithm, 283–284
Backward elimination method, 281
Bagging (BG) techniques, 284t, 289t, 309, 392, 402
Bag-of-words approach, 368
Bauer, Fritz, 2
Bayesian learner (BL) technique, 282
Bayesian network (BN), 282, 282f, 394, 402
Bender method, 96
Bimodal distribution, 209
Binary classification, 366
Binary variables
  contingency table for, 46t, 47t
  dependent variable, 118, 137
  effects of interest, 46–48
Bivariate outliers, 214
Blind data, 12
Bonferroni–Dunn test, 261–263, 437, 441t
Boosting technique, 284t, 289t, 402–403
Box length, 215
Box plots, 44, 45
Bug life cycle, 156
Bug prediction data set, 183–184
Bug repositories, 149, 178

Bug tracking systems, 155–156, 379
Bugzilla, 166–169
    blocks, 169
    bug ID, 168
    bug severity, 169
    bug status, 168
    component, 169
    database schema, 167*f*
    depends on, 169
    integrating with VCS, 169–170
    life cycle of bug, 168*f*
    product, 168
    sample bug report, 169
    target milestone, 169

C

Categorical dependent variable, performance
        measures
    accuracy and precision, 295
    confusion matrix, 292–294
    F-measure, *G*-measure, and *G*-mean, 295–296
    guidelines for using, 302–304
    kappa coefficient, 295
    ROC analysis, 298
        AUC, 300, 304, 312
        curve, 298, 300
        cutoff point and co-ordinates of ROC
            curve, 300–301
        sensitivity and specificity, 294–295, 301,
            310–311
Causal validity. *See* Internal validity threats
Centralized VCS (CVCS), 153–154, 154*f*
Central tendency, measures of
    choice, 209–211
        descriptive statistics, 210*t*
        faults severity levels, 209*t*
        graphical skew and symmetrical
            distributions, 210*f*
        maintenance effort, 209*t*
        sample data of LOC, 211*t*
        statistical measures with scale types, 210*t*
    mean, 207–208
    median, 208–209
    mode, 209
Change control board (CCB), 18, 144
Change logs, 188–189
Change prediction models, 391
Change-proneness models, 118, 178–179, 203,
        391, 392, 393
Chidamber and kemerer java metrics (CKJM), 156
Chidamber and Kemerer (CK) metric suite, 172
Chi-squared test, 235–242, 437, 438*t*

CKJM calculator, 173
Class–attribute (CA) interaction, 81–82
Classification process
    example of, 22*f*
    and prediction, 22
    steps in, 23*f*
Class–method (CM) interaction, 81–82
    types of, 82
Class stack, 84*f*
Class template factor (CTF) metric, 78
    defined, 87
    source code for calculating, 88*f*
Class to leaf depth (CLD), 86
Cohesion metrics, 77, 83–85
    faults in software development life
        cycle, 83
    LCOM metric, 83
Computing type of interaction, example
        for, 82*f*
Conclusions, reporting results, 356*t*, 359
Conclusion validity threats, 331–333, 342*t*
    evaluation, 423
    fault prediction system, 337, 339
Conference articles, 354
Confidence interval (CI), 49
Configuration management systems
    accounting, 146
    control
        change cycle, 145*f*
        change request form, 145*f*
        software change notice, 146*f*
    identification, 144
Confounding effect, 333, 334
Confounding variables, 23–24, 333, 339
Construct validity threats, 334–335, 345*t*–346*t*
    evaluation, 423
    fault prediction system, 339–340
Continuous dependent variable, 48, 137
    performance measures
        MARE, 304–305
        MRE, 304
        Pred (A), 305
    in software engineering, 118
Correlation analysis, 218
Correlation-based feature selection (CFS)
        method, 221–222
Coupling between object (CBO), 396
    defined, 80
    hypothesis, 130
    metrics, 80, 80*f*, 219
Coupling metrics, 79–82
    Chidamber and Kemerer defined, 80
    between classes, 81–82

export or import coupling (EC/IC), 81
    relationship, 81
    type of interaction, 82–83
  fan-in and fan-out metrics, 79, 79*f*
  MPC, 80
Critical region, 127
Critical value, 127
Cross-company analysis, 25, 25*f*
Cross-validation technique, 276–277, 306–307, 431
  hold-out, 307
  k-fold, 307
  LOO, 307, 308*f*
  tenfold, 308*f*, 340, 393, 395, 401, 404, 423
CVS
  features
    branching and merging, 157–158
    revision numbers, 157
    version control data, 158
  files stored, 159*f*
    author, 158
    date, 158
    description, 158
    free text, 159
    lines, 159
    locks and accesslist, 158
    RCS file, 158
    revision number, 158
    state, 158
    symbolic names, 158
Cyclomatic complexity (CC), 171

D

Daily rolling append policy, 201
Data
  extraction form, 43, 44*f*, 53, 54*f*
  hiding, 76
  mining, 12
  qualitative, 22–23
  quantitative, 22–23
  set example, 437, 441*t*–443*t*
  synthesis, 53, 55
    methods, 58*t*–60*t*
    tools and techniques, 43
  test, 12
  training, 12
  validation, 12
Data abstraction coupling (DAC) metric, 80
Data analysis
  method, selection, 136–139
    aspects, 138–139
    data set nature, 137–138
    dependent variable type, 137

tools, 429
  comparison, 431, 432*t*
  KEEL, 429–430
  MATLAB, 430–431
  parameter comparison, 433*t*–434*t*
  R, 431
  SPSS, 430
  WEKA, 429
Data-collection procedure, 24, 358
Decision-making processes, 147, 150
Decision table (DTab), 394, 402
Decision tree (DT) technique, 282
  algorithm, 282, 283*f*
  characteristics, 289*t*
Defect collection and reporting system (DCRS), 187–203, 187*f*
  cloning operation, 200*f*–201*f*
  data source and dependencies, 190–191
  defect reports, 191–199
    consolidated change and, 197–198
    defect count and metrics report, 193–194
    deleted source files, 196–197, 197*f*
    details, 191–192
    generation by DCRS, 192*f*
    LOC changes report, 194–195
    newly added source files, 195–196
    statistics report for incorporated metrics, 198–199
  descriptive statistics report records, 199*f*
  features, 199–202
    cloning of Git-based software repositories, 199–201
    self-logging, 201–202
  flowchart, 193*f*
  generation of change logs, 190*f*
  motivation, 188
  potential applications of, 202–203
    change-proneness studies, 203
    defect prediction studies, 202
    statistical comparison, 203
  self-log file, 202*f*
  tool, 393
  working mechanism, 188–190
Defect density metric, 72–73
Defect IDs, 189
Defect prediction models. *See* Defect proneness models
Defect proneness models, 178, 202
Defect removal effectiveness (DRE), 73
Defect tracking system. *See* Bug tracking systems
Degrees of freedom (DOF), 230

Dependent variable, 10, 23–24, 23*f*, 117, 118
  binary, 118
  change proneness, 24
  continuous, 137
  discrete, 137
  examples of, 24
  performance measures, 278
    categorical, 292–304
    continuous, 304–306
  vs. independent variables, 11, 117*t*
Deployment logs. *See* Run-time repositories
Depth of inheritance tree (DIT), 396
  hypothesis, 130, 314
  metric, 85, 118, 173
  values of, 218–219
Descriptive statistics, 404–407
  fault prediction system, 218–219
  for metrics, 219*t*
  used to describe data, 207
Digital portals, 110–111
Directional hypothesis, 125
Directional test. *See* One-tailed test
Discrete-dependent variable, 137
Discussion and interpretation
  reporting results, 356*t*, 359
  of result, 420–423
Dispersion, measures of, 211–212
Distributed VCS (DVCS), 154–155, 155*f*
Dynamic complexity metrics, 90
Dynamic coupling metrics, 89
Dynamic software metrics
  dynamic cohesion metrics, 89–90
  dynamic complexity metrics, 90
  dynamic coupling metrics, 89, 90*t*
  static and, 89*t*

E

Eclipse bug data (EBD), 184
Eclipse metrics plugin, 174
EISPACK, 430
Empirical data collection, 10–11,
        399–400
  fault prediction system, 134–136
  repositories, 132–134
  strategies, 131–132, 132*f*
Empirical elements, 316*f*
Empirical error, 290
Empirical evidence, 33
Empirical methods, 11
Empirical procedures, 391
  abstract, 391
    basic, 391–392

    method, 392
    result, 392
  analysis results, 404
    CFS results, 408
    descriptive statistics, 404–407
    hypothesis testing and evaluation,
        416–419
    Nemenyi results, 419–420
    outlier analysis, 408
    tenfold cross-validation results, 408–416
  discussion and interpretation of result,
        420–423
  experimental design
    analysis process, 401
    empirical data collection, 399–400
    hypothesis formulation, 397–398
    problem definition, 395
    research questions, 395–396
    statistical tests, 398
    technique selection, 400–401
    variables selection, 396–397
  method, 393
  motivation, 392
  objectives, 392–393
  related work, 394
  research methodology, 401
    performance measures and validation
        method, 404
    techniques, 401–404
  subject selection, 394
  technique selection, 393–394
  validity evaluation, 423–424
Empirical research
  basic elements in, 17–18
  ethics of, 13–16
  importance of, 16–17
  in software engineering, 13–14
Empirical software engineering (ESE), 1–2
  ethical principles, 14
  results of, 16
Empirical studies
  benefits of, 2
  characteristics of good, 12–13
  classification, 3
  definition of goals and objectives, 9
  overview of, 2–3
  process, 8–13
    experiment design, 10–11
    phases in, 9*f*
  scope of, 9
  in software engineering, 2, 143
  steps, 3, 3*f*, 5
  types of, 3–8, 5*f*

case study, 5–6, 6*f*
experiment, 4–5, 5*f*
postmortem analysis, 8
survey research, 6–7
systematic review, 7–8, 7*f*
Encapsulation, 76
Ensemble learner (EL) technique, 282–283, 284*f*, 284*t*
Evidence-based medicine (EBM), 33
Evolution phase, prediction during, 19, 19*f*
Experimental designs
analysis process, 401
data analysis method, selection, 136–139
empirical data collection, 131–136, 399–400
hypothesis formulation, 120–131, 397–398
literature review, 109–116, 113*t*–116*t*
problem definition, 395
reporting results, 355*t*, 357–358
research questions, 106–108, 395–396
research variables, 117–118
statistical tests, 398
steps, 103, 104*f*
technique selection, 400–401
terminology used, 118–120
types, 120–121
variables selection, 396–397
Experimental process, 358*f*
Experimental study tests, 4–5
Exploratory surveys, 6
Export object coupling (EOC) metric, 89
Export or import coupling (EC/IC), 81
External attributes, 68–69
External validity threats, 331, 335–337, 347*t*–349*t*
evaluation, 423–424
fault prediction system, 340

F

Fabrication, 361
False positive rate (FPR), 295–296
False positive. *See* Type I error
Falsification, 361
Fan-in and fan-out metrics, 79, 79*f*
Fault coverage metric (FCM), 75
Fault prediction system (FPS)
comparing ML techniques, 315–323
empirical data collection, 134–136
hypothesis formulation, 129–131
hypothesis testing results, 312, 313*t*, 314–315
literature review, 12–116
motivation, 104–105
objective, 104
research questions, 108

research variables, 105
results, 105
studies, 341*t*
study context, 105
validity threats, 338*t*
conclusion, 337, 339
construct, 339–340
external, 340
internal, 339
Fault proneness, 118
Faults category, 105, 112, 135, 136*f*
Feature selection in text mining, 372–375
Filter method, procedure of, 221, 221*f*
FLOSSMetrics, 181–182
FLOSSmole, 180–181
F-measure, 295, 303, 404, 415
Forest plots, 49–50
Forward stepwise method, 281
Friedman's statistical test, 257–259, 318–319, 319*t*, 393, 398, 417*t*, 418*t*, 419*t*
F-test, 242–243, 437, 441*t*
Function template factor (FTF) metric, 78
defined, 87
source code for calculating, 87*f*
Funnel plot, 51, 51*f*

G

Generalization error, 290
Git
blob, 162
branching and merging, 164
clone command, 170
commit record, 162, 164–165
example log file for android, 165*f*
free text, 165
local operations, 163–164
object, 162
repositories, 165–166
revision numbers, 163
storage structure, 163*f*
tag, 163
tree, 162
version control data, 164
Global dictionary, 367
G-mean, 296
G-measure, 295, 303
Goal/Question/Metric (GQM) method, 26–27
elements, 26
framework of, 27*f*
phases of, 27, 27*f*
Graphmania tool, 179

H

Hash–object pair, 162
Helix data set, 186
Histogram analysis, 213
Historical repositories, 148–149
Hold-out validation, 307
Hyperplane, 285
Hypothesis, 121, 314–315
    alternative, 123, 125–126
    directional, 125
    formulation, 10, 120–129, 397–398
        experiment design types, 120–121
        form, 122–124
        FPS, 129–131
        purpose and importance in empirical
            research, 121–122
    generation, 121, 122*f*
    nondirectional, 125
    null, 123, 125
    sets, 397–398
    testing, 12, 124*f*
        choosing test of significance, 126
        computing test statistic and associated
            p-value, 126
        defining significance level, 127
        deriving conclusions, 127–129
        and evaluation, 416–419
        stating null and alternative hypothesis,
            124–126
        steps, 224–225
    verification test, 126

I

Imbalanced data, 291
Import object coupling (IOC) metric, 89
Independent variable, 10, 23–24, 23*f*, 117, 118, 119
    dependent and, 11, 117*t*
    examples of, 24
    OO metrics, 130
    software metrics, 24
Infogain measure, 373
Information flow-based cohesion (ICH)
        metric, 85
Information flow-based coupling (ICP)
        metric, 82
Information flow-based inheritance coupling
        (IH-ICP) metrics, 82
Information flow metrics, 79
Inheritance hierarchy, 85*f*
Inheritance metrics, 85–86, 86*f*

Integrated development environment (IDE), 173
Interactions, type of, 81–82
    CA interaction, 81–82
    CM interaction, 82
    example for computing, 82*f*
    MM interaction, 82
Internal attribute, 68–69
Internal validity threats, 331, 333–334,
        343*t*–344*t*
    evaluation, 423
    fault prediction system, 339
International Software Benchmarking
        Standards Group (ISBSG), 184
Interquartile range (IQR), 211
Introduction, reporting results, 355*t*, 357

J

J48 decision tree, 394, 402
Java programming language
Java programming language, application
    code analyzer, 175
    DCRS system, 189
    Eclipse metrics plugin, 174
    phpUnderControl and PHPUnit, 175
    qualitas corpus (QC), 182
    SonarQube, 174
Journal papers, 354

K

Kappa coefficient, 295
KEEL, 429–430
Kernel function, 282*f*, 285–286, 286*f*
K-fold cross-validation, 307
    tenfold, 308*f*
Kruskal–Wallis test, 254–256

L

Lack of cohesion in methods (LCOM)
    hypothesis, 130
    metrics, 83, 219, 397
        example of, 83*f*
        variations of, 89
Learning algorithm/technique
    EL technique, 282–283, 284*f*, 284*t*
    guidelines for selecting, 291
    ML techniques, 36, 391
        analysis of non-ML and, 42
        Bayesian Learner (BL), 282
        benefits, 281

categories, 281
characteristics, 289*f*
classification, 281, 282*f*
DT algorithm, 282, 283*f*, 284*t*
ensemble learner (EL), 282–283, 284*f*, 284*t*
goal, 281
in literature, 36
neural network (NN), 283–285
rule-based learning (RBL), 286
search-based technique (SBT), 287–288
for SFP, 37, 39
support vector machine (SVM), 285–286
properties, 291, 292*f*
Leave-one-out (LOO) cross-validation, 307, 308*f*
Lines of code (LOC), 11, 210
changes report, 194–195
maintenance level, 236–237
metric, 68
added and deleted for classes, 69*t*
definition of, 71
evolution, 91
function of find-maximum, 71, 72*f*
in software development, 71
number of, 150
records of changes report, 196*f*
LINPACK, 430
Literature review, experimental design, 109–110
benefit, 110
FPS, 112–116
goal of conducting, 110
guidelines for writing, 111–112
key questions, 109*f*
purpose, 110
steps, 110–111
Literature survey
empirical evidence from, 37
systematic reviews and, 34*t*
Local dictionary approach, 368
Local VCS, 153, 153*f*
Logistic regression (LR) technique, 224, 280, 309, 392, 400
LogitBoost (LB), 394, 403
Loose class cohesion (LCC), 77
LR–ZeroR technique, 420

M

Machine learning (ML) techniques, 36, 391
analysis of non-ML and, 42
Bayesian learner (BL), 282
benefits, 281
categories, 281

characteristics, 289*f*
classification, 281, 282*f*
DT algorithm, 282, 283*f*, 284*t*
ensemble learner (EL), 282–283, 284*f*, 284*t*
goal, 281
in literature, 36
neural network (NN), 283–285
rule-based learning (RBL), 286
search-based technique (SBT), 287–288
for SFP, 37, 39
support vector machine (SVM), 285–286
Mahalanobis Jackknife distance, 216
Maintainability process, 76
Maintenance effort, 118
Masters and doctoral students, guidelines for, 359–361
MATLAB, 430–431
Mean, 207–208
difference, 48
Friedman ranks
using AUC measure, 418–419, 419*t*
using *F*-measure, 418, 418*t*
using precision measure, 418, 418*t*
z-score, 216
Mean absolute relative error (MARE), 304–305
Mean relative error (MRE), 304
Measurements
defined, 22
measures, metrics and, 20–22
scales, 70*t*
Median, 208–209
Message passing coupling (MPC) metric, 80, 81*f*
Method–method (MM) interaction, 82
Metric data, 69
correlation analysis, 218
data distributions, 212–213
descriptive statistics of FPS, 218–219
histogram analysis, 213
measures
of central tendency, 207–211
of dispersion, 211–212
outlier analysis, 213–218
Mining software repositories (MSR), 147
Mining unstructured data, 365–366
characteristics, 367–368
classifications, 366
defined, 366
Min–max normalization, 283
ML. *See* Machine learning (ML) techniques

Mms application
  change record from change log file, 191*f*
  example of report records
    defect and change, 198*f*
    defect count, 195*f*
    defect details, 194*f*
    deleted files, 197*f*
    descriptive statistics, 199*f*
    LOC changes, 196*f*
    newly added files, 197*f*
Mode, 209
Model development
  attribute reduction, 277
  construction using learning algorithms/
    technique, 277
  data partition, 276–277, 277*f*
  hypothesis testing, 278
  process, 275–276
  result interpretation, 278
  software quality prediction system,
    278–279, 279*f*
  validation, 277–278
Model error rates, 290
Model prediction, 275, 276*f*
  comparison test, 309–310, 309*f*
  concerns
    imbalanced data, 291
    multicollinearity analysis, 290–291
    parameters tuning, 292
    problems, 290, 291*t*
    selecting learning technique, guidelines,
      291, 292*f*
Mozilla Firefox, 365
Multicollinearity analysis, 290–291
Multilayer perceptron (MLP), 380–381,
  392, 403
Multiple classifications, 366
Multivariate LR formula, 280
Multivariate outliers, 214, 215

N

Naïve Bayes (NB) technique, 289*t*, 394, 403
NDepend, 174
Nemenyi test, 259–261
  critical values for two-tailed, 437, 441*t*
  results, 393, 419–420
Neural network (NN) technique, 40, 283–285
  architecture, 285*f*
  characteristics, 289*t*
Nominal scale, 70
Nondirectional hypothesis, 125

Nondirectional test. *See* Two-tailed test
Noninheritance information flow-based
    coupling (NIH-ICP) metrics, 82
Nonmetric data, 69–70
Non-nested generalized exemplars (NNge),
  394, 403
Nonparametric data analysis method, 139
Nonparametric tests, 26, 26*t*
Normal curve, 212
Normal distribution, area under, 437,
  439*t*–440*t*
Null hypothesis, 123, 125, 397, 419
Number of ancestors (NoA) metric, 85
Number of attributes (NOA) metric, 88
Number of children (NOC) metric, 85, 130,
  218–219, 396
Number of descendants (NOD) metric, 85
Number of methods added (NMA)
  metric, 85
Number of methods inherited (NMI) metric,
  85, 86
Number of methods (NOM) metric, 88
Number of methods overridden (NMO)
  metric, 85
Number of parents (NOP) metric, 85

O

Object-oriented (OO) languages, 76
Object-oriented (OO) metrics, 36, 218, 356, 357,
  391, 394
  cohesion, 83–85
  coupling, 79–82
  inheritance, 85–86
  mere properties for, 92–93
  paradigm, features of, 76
  polymorphism metrics, 76, 79*t*
  reuse, 86–88
  size, 88
  suites, 76–88
    Briand, 78*t*
    Chidamber and Kemerer, 77*t*
    Lee, 78*t*
    Li and Henry, 77*t*
    Lorenz and Kidd, 78*t*
  used in study, 316*t*–317*t*
Odds ratio (OR), 47
Ohloh, 185
One-tailed test, 226
One-way ANOVA, 244–247
  problem on, 245
  treatments, 244

Open source software (OSS), 24–25, 133
  application
    CKJM calculator, 173
    Eclipse metrics plugin, 174
    phpUnderControl, 175
    Pylint, 175
    SonarQube, 174
  use of, 143
Outlier analysis, 213–218, 408
  box plots, 214–216, 216f, 217f
  z-score, 216–218, 217t–218t
Overfitting, 290
Oversampling method, 291

P

Parametric data analysis method, 139
Parametric tests, 26, 26t
P.C. method. *See* Principal component (P.C.)
    method
Performance analyzer selection, 277
Performance measure, 18
PhpUnderControl and PHPUnit tool, 175
Plagiarism, 361, 362
Plots
  box, 44, 45
  forest, 49–50
  funnel, 51, 51f
Polymorphism metrics, 76, 79t
Precision, 303, 311, 404
  accuracy and, 295
  measure
    Friedman mean ranks using, 418t
    hypothesis for, 398, 417
    validation results using, 410, 411t
Pred (A), 305
Prediction error, 280
Prediction models, 22–24
PRedictOr Models In Software Engineering
    (PROMISE), 182
Predictor variable, 23, 119
Preprocessing techniques, 368–372, 369
Primary studies, 34
  distribution according to metrics
      used, 55f
  identification of, 41
  results in SR, 44
  selection of, 53
  statistical measures, 44
Principal component (P.C.) method, 222–223,
    291, 372
Problem definition, experimental design, 395

Process metrics, 68
  to measure testing, 75
  metric and nonmetric data, 69
Product metrics, 68
  defect density, 72–73
  defect rate, 72
  metric and nonmetric data, 69
Project and issue tracking system (PITS), 379
Proprietary software, 24–25
Publication bias, 51
Publication venue, 354
Published works, 353–354
p-value, 126–127
Pylint tool, 175

Q

Qualitas corpus (QC), 182
Qualitative data, 22–23
Qualitative research, 4, 4t
Quality, 74
  assessment questions, 42, 42t, 43t
  attributes in literature, 66
  scores for quality assessment questions, 43t
Quality metrics for OO design (QMOOD), 172,
    194
Quantitative analysis, 358
Quantitative data, 22–23
Quantitative research, 4–5, 4t
Quantity, 74
Quartiles
  example, 212f
  lower, 211
  upper, 211

R

R (programming language), 431
Radial basis function (RBF) network, 282f, 285,
    286f, 394
Random forest (RF) technique, 40, 282f, 284t,
    394, 403
  characteristics, 289t
  results of, 55t
Recall (Rec), measure, 294, 303, 311
Receiver operating characteristic analysis. *See*
    ROC analysis
Reference management system, 52
References, reporting results, 356t, 359
Region of rejection, 127
Related work, reporting results, 357
Relative risk (RR), 46–47

Removal of stop words, 369, 370
Repeated incremental pruning to produce error
        reduction (RIPPER), 394, 402
Report results, 354
Report structure, 355–356
    abstract, 355t, 356
    acknowledgment, 356t, 359
    appendix, 356t, 359
    conclusions, 356t, 359
    discussion and interpretation, 356t, 359
    experimental design, 355t, 357–358
    introduction, 355t, 357
    references, 356t, 359
    related work, 357
    research methods, 355t, 358
    research results, 355t, 358
    threats to validity, 356t, 359
REP tree (REP), 394, 403
Research
    ethics and misconduct, 361–362
    hypothesis, 359
    methodology, 401
        performance measures and validation
            method, 404
        techniques, 401–404
    plagiarism, 362
    problem, 106
    reports. *see* Results, reporting and
        presenting
    types, 107
    variables, experiment design
        FPS, 118
        independent and dependent, 117
        multivariate analysis, 119
        selection, 118
        software engineering, 118
Research question (RQ), experimental design,
        395–396
    characteristics, 107–108
    context, 106f
    formation, 106–107
    to hypothesis transition, 123–124
Response for a class (RFC) metric, 80, 81f, 219, 396
Response variable, 24
Results
    of five studies, 50t
    interpretation
        conclusions from hypothesis testing, 312
        FPS, hypothesis testing results for, 312,
            313t, 314–315
        issues to be addressed, 310, 311f
        performance measure analysis, 310–312
        qualitative and quantitative result, 312

    methods for presenting, 46–51
    of NN technique, 56t
    of RF technique, 55t
Results, reporting and presenting, 353–354
    dissemination, 354
    structure, 355–356
        abstract, 355t, 356
        acknowledgment, 356t, 359
        appendix, 356t, 359
        conclusions, 356t, 359
        discussion and interpretation, 356t, 359
        experimental design, 355t, 357–358
        introduction, 355t, 357
        references, 356t, 359
        related work, 357
        research methods, 355t, 358
        research results, 355t, 358
        threats to validity, 356t, 359
Reuse metrics, 76, 86–88
    FTF and CTF, 78
Review protocol
    development of, 39, 40f
    document sections, 45
    evaluation of, 45
    search process, 52f
    steps involved in, 40f
Revision control system (RCS), 153
RFC hypothesis, 130
Risk difference, 47
Risk ratio. *See* Relative risk (RR)
ROC analysis, 47, 178, 298, 377
    AUC, 300, 304, 312
    curve, 298, 300
    cutoff point and co-ordinates of ROC curve,
        300–301
    method, 96
Rule-based learning (RBL) technique,
        286, 289t
Runtime call-weighted LCOM (RWLCOM), 89
Run-time repositories, 149
Runtime simple LCOM (RLCOM), 89

S

Sarbanes-Oxley Act (2002), 149
Search-based technique (SBT), 11,
        287–288, 289t
Search process, 52f
Secondary studies, 34
Self-logging, 201–202
Sensitivity, 294, 301, 310–311, 404
Significance test, 126
Significance value, 127

Significant metrics, constant (α) and coefficient
(β) of, 97*t*
Simply logging. *See* Self-logging
Size metrics, 88
Software
    constructs, 66
    defect reports, automated severity
        assessment, 379–380
        data source, 380
        experimental design, 380–382
        result analysis, 382–385
        threats to validity, 387
    development
        never-ending cycle, 151
        risk of estimation, 180
    evolution, 18–19, 151
        cycle, 19*f*
        research avenues in, 19*f*
    evolution metrics
        code churn based, 91
        other related, 92
        related to refactoring and bug-fixes, 91
    historical analysis
        change impact, 177
        change prediction, 178–179
        change propagation, 177
        change smells and refactoring, 180
        defect proneness models, 178
        dependencies in system, 176–177
        effort estimation, 180
        mining textual descriptions, 179
        social network, 179
        user and team dynamics
            understanding, 178
    library, 365
    maintenance, 391
    measurement steps, 67*f*
    planning team, 1
    project artifact, 144, 157
    project management team, 144
    quality, 18–19
        attribute, 20, 20*f*, 21*t*
        metrics, 72–76
        prediction models, 392
        prediction system, 278–279, 279*f*, 287
Software-artifact infrastructure repository
    (SIR), 184–185
Software development life cycle, 1, 66,
    143–144
    empirical models for cost and effort
        estimation, 71
    product metrics, 68
    software quality measuring, 72

Software engineering
    categories of, 68
    conferences on, 52*t*
    defined, 2
    empirical research in, 13
    empirical studies, 2, 143
    features, 23
    IEEE Computer Society defined, 2
    journals, 52*t*, 111
    research variables, 118
    steps in, 3*f*
    systematic reviews in, 58, 58*t*–60*t*
    text mining in, 378
        change logs to predict effort, 378
        fault reports to predict faults
            severity, 378
        SRS document to classify requirements
            into NFR, 378–379
Software fault prediction (SFP), 36, 356
    ML techniques for, 37, 39
    RQ in SRML case study for, 44
Software metrics, 22, 65–100, 104, 118
    application areas of, 66–67
    calculation tools
        CKJM calculator, 173
        code analyzer, 175
        Eclipse metrics plugin, 174
        NDepend, 174
        phpUnderControl and PHPUnit, 175
        Pylint, 175
        SonarQube, 174
        source monitor, 173–174
        understand, 173
        Vil, 174
    categories, 55, 69*f*
    characteristics of, 67
    defined, 66
    dynamic, 89–92
    as independent variables, 24
    industrial relevance of, 100
    to measure customer satisfaction, 65
    measurement basics, 67–71
    measuring size, 71–72
    object-oriented (OO), 76–88
    practical relevance, 93–100
    product and process, 68–71
    selection of, 94–95
    in software engineering, 66
    software quality measuring, 72–76
    testing, 66, 75
    uses, 94, 99–100
    validation of, 92–93
    in various domains, 66

Software repositories, 28
    applications of data mine, 177*f*
    benefits, 147
    cloning of Git-based, 199–201
    data analysis procedure, 148*f*
    extracting data from
        Bugzilla, 166–169
        CVS, 157–159
        Git, 162–166
        integrating Bugzilla with VCS, 169–170
        SVN, 159–162
    mining importance, 147
    open research data sets, 181*t*
        bug prediction, 183–184
        EBD, 184
        FLOSSMetrics, 181–182
        FLOSSmole, 180–181
        Helix data set, 186
        ISBSG, 184
        Ohloh, 185
        PROMISE, 182
        qualitas corpus, 182
        SECOLD, 186–187
        SIR, 184–185
        Sourcerer project, 182–183
        SRDA, 185–186
        Tukutuku, 186
        UDD, 183
    overview, 143
    types, 28
        historical repositories, 148–149
        run-time repositories, 149
        source code repositories, 150
    used, 148*f*
Software requirements specification (SRS), 378
SonarQube tool, 174
Source code, 170. *See also* Static source code
        analysis
Source code ECO system Linked Data
        (SECOLD), 186
Source code repositories, 150
Source control repositories, 149. *See also* Version
        control systems (VCS)
SourceForge research data archive (SRDA),
        185–186
Source LOC (SLOC), importance of, 239–240
Source monitor, 173–174
Sourcerer project, 182–183
Specialization index (SIX), 86
Specificity, measure, 294, 301, 310–311
SPSS statistics, 430
SRML review, quality assessment questions, 43*t*
Stack class, 84, 84*f*

Standard deviation (SD), 44
Standardized mean difference, 48
Static metrics, 89
Static source code analysis
    levels, 171–172
        class level, 171
        file level, 171
        method level, 171
        system level, 171
    metrics, 172–173
    software metrics calculation tools
        CKJM calculator, 173
        code analyzer, 175
        Eclipse metrics plugin, 174
        NDepend, 174
        phpUnderControl and PHPUnit, 175
        Pylint, 175
        SonarQube, 174
        source monitor, 173–174
        understand, 173
        Vil, 174
    tools, 176*t*
Statistical analysis, 358
Statistical measures, 44
Statistical multiple regression technique
    coefficients and variable selection, 280–281
    multivariate analysis, 280
Statistical significance of metrics, 97*t*
Statistical tests/testing, 398
    ANOVA test, 244–247
    Bonferroni–Dunn test, 261–263
    categories, 225–226, 226*f*
    chi-squared test, 235–242
    Friedman test, 257–259
    F-test, 242–243
    interpreting significance results, 229
    Kruskal–Wallis test, 254–256
    Nemenyi test, 259–261
    one-tailed and two-tailed tests, 226–228
    overview, 225
    summary, 227*t*
    t-test, 229–235
    type I and type II errors, 228–229
    univariate analysis, 263
    Wilcoxon–Mann–Whitney test (*U*-test), 250–254
    Wilcoxon signed-ranks test, 247–250
Stemming algorithm, 369, 371–372
Study quality assessment, 53
Supervised learning, 281
Support vector machine (SVM) model,
        285–286, 403
    characteristics, 289*t*
    purpose, 285

Support vector using sequential minimal
    optimization (SVM-SMO), 394
Survey or systematic reviews, 354
Survey research, 6–7
SVN, 159–162
    features
        branching and merging, 160
        revision numbers, 160
        version control data, 160–161
    files stored
        actions, 161
        added, 162
        author, 161
        Bugzilla ID (optional), 161–162
        log file from Apache Tomcat project, 161*f*
        message, 162
        modified, 162
        revision numbers, 161
Systematic review of machine learning (SRML)
    techniques, 37
    case study, 36–37
        data extraction form for, 54*t*
        identification of primary studies in, 40–41
        ML techniques for, 44*f*
        research questions, 39*t*, 44
        software metrics categories in, 55
    inclusion and exclusion criteria in, 41
    quality assessment questions for, 43*t*
    RQ with motivation for, 39*t*
Systematic reviews (SRs), 7–8
    aim of, 7
    characteristics of, 34
    checklist for evaluating existing, 38
    conducting review, 51–56
    defined, 33
    EBM defined, 33
    format and contents of, 56*t*–57*t*
    identification of necessity, 37
    importance of, 35
    inclusion and exclusion criteria for, 41
    independent studies in, 49
    literature survey *vs.*, 33–34, 34*t*
    overview of, 33
    phases of, 7*f*
    planning review, 37–45
    presenting results methods, 46–51
    process, 36*f*
    quality assessment questions for, 42*t*
    quality of, 37
    report format of, 56*t*–57*t*
    reporting results of, 56–57
    reporting review, 56–57
    results of primary studies in, 44

secondary studies, 34
    in software engineering, 58, 58*t*–60*t*
    stages of, 35–36
    structure of, 38
    tools and techniques, 46–49

T

Task effectiveness, 74
Technical reports, 353
Technique selection, 400–401
Tenfold cross-validation, 308*f*, 318*t*, 395, 408–416
Term frequency (TF), 375
Term frequency inverse document frequency
    (TFIDF), 375, 376
Test data, 12
Test focus (TF) metric, 75
Testing effort, 118
Testing metrics, 75
Text classifier, predicting and validating, 377
Text documents, representation of, 368
Text mining, 365
    characteristics, 367–368
    importance, 367
    in software engineering, 378
        change logs to predict effort, 378
        fault reports to predict faults severity, 378
        SRS document to classify requirements
            into NFR, 378–379
    steps in, 368, 368*f*
        feature selection, 372–375
        preprocessing techniques, 368–372
        representation of text documents, 368
        text classifier, predicting and
            validating, 377
        vector space model, 375–377
Thousands of lines of code (KLOC), 72
Threats, validity, 332*f*
    conclusion, 331–333
        evaluations, 423
        in FPS, 337, 339
    construct, 334–335
        evaluations, 423
        in FPS, 339–340
    essential, 337
    external, 335–337
        evaluations, 423
        in FPS, 340
    internal, 333–334
        evaluations, 423
        in FPS, 339
    reporting results, 356*t*, 359
    and their countermeasures, 341–350

Threshold
    computing, 95–98
    values
        computation of, 97
        logistic regression method, 98*t*
        ROC curve method, 99*t*
        roc curve to calculate, 98
Thesis, formats, 360
Tight class cohesion (TCC), 77
    and LCC, 84
Tokenization, 368, 370
Tortoise-SVN, 160
Training data, 12, 275, 277, 306
Treatment effect, 244
True positive rate (TPR). *See* Sensitivity
T-test, 128–129, 129*f*, 229–235, 358
    one sample, 229–232
    paired, 233–235
    table, 437, 437*t*
    two sample, 232–233
Tukutuku, 186
Two-tailed test, 226–227, 437, 441
Type I error, 228–229
Type II error, 228–229

U

Ultimate Debian Database (UDD), 183
Underfitting, 290
Undersampling method, 291
Understanding systems
    system characteristics, 150–151
        application/domain, 151
        company, 150
        number of LOC, 150
        number of source files, 150
        platform, 150
        programming language, 150
        versions and editions, 150–151
    system evolution, 151
Unethical research, examples of, 13*t*
Unified model language (UML) tool, 6
Univariate outliers, 214
University software, 24–25
Unknown variables, 24
Unstructured data, 365–366
    characteristics, 367–368
    classifications, 366
    defined, 366
Unsupervised learning, 281
Usability metrics, 74

V

Validation data, 12
Validity threats, 332*f*
    conclusion, 331–333
        evaluations, 423
        in FPS, 337, 339
    construct, 334–335
        evaluations, 423
        in FPS, 339–340
    essential, 337
    external, 335–337
        evaluations, 423
        in FPS, 340
    internal, 333–334
        evaluations, 423
        in FPS, 339
Value of an acceptable risk level (VARL), 96
Variables selection, 396–397
Variance inflation factor (VIF), 291
Varimax rotation, 223
Vector space model, 375–377
Version control systems (VCS)
    classification
        CVCS, 153–154, 154*f*
        DVCS, 154–155, 155*f*
        local VCS, 153, 153*f*
    functionalities of, 152
    general terms for, 152
Vil tool, 174
Voted perceptron (VP), 394, 403

W

Weighted methods per class (WMC) metric, 88,
    219, 396
WEKA, 429
Wilcoxon–Mann–Whitney test (*U*-test), 250–254,
    437, 438*t*
Wilcoxon matched pairs test. *See* Wilcoxon
    signed-ranks test
Wilcoxon signed-ranks test, 247–250
Wilcoxon test results, 319, 320*t*, 321*f*
Within-company analysis, 25, 25*f*
WMC hypothesis, 130
Wrapper methods, procedure of,
    221, 221*f*

Z

ZeroR techniques, 394, 404